6.0
5.0
4.0
frequency (Hz x 10^14)
solar spectrum
potassium
calcium
strontium
barium
mercury
450
500
550
600
650
700
750
wavelength (nm)

Nat Phil 5

By the same author

Physics is Fun Books 1–4

Nat Phil Workbooks 3, 4 and 5

Physics Workbooks 3, 4 and 5

In preparation

Nat Phil O (Text for years 3 and 4)

Nat Phil 5

TEXT

Jim Jardine

Heinemann Educational Books
London and Edinburgh

Heinemann Educational Books Ltd
LONDON EDINBURGH MELBOURNE AUCKLAND TORONTO
HONG KONG SINGAPORE KUALA LUMPUR
IBADAN NAIROBI JOHANNESBURG NEW DELHI

ISBN 0 435 68220 2

First published 1973

Published by Heinemann Educational Books Ltd
48 Charles Street, London W1X 8AH

Filmset by Keyspools Ltd, Golborne, Lancs
Printed in Great Britain by BAS Printers Limited,
Wallop, Hampshire

Preface

NATURAL PHILOSOPHY

Over two thousand years ago, Aristotle and other Greek thinkers tried to solve the riddles of nature by pondering, speculating, and discussing, rather than by experimenting and measuring. These *philosophers* (from Greek words meaning 'lovers of wisdom') believed that it was possible to understand nature only by hard thinking and careful reasoning. For them there was no need to test each new theory by experiment. Their concern was to find or invent *causes* for natural phenomena and an experiment, they argued, could not produce a cause.

Galileo Galilei (1564–1642) may not have been the first scientist to appreciate the importance of experimentation but his ingenious and convincing experiments on motion showed clearly the weakness in Aristotelian physics and established *experimental science* once and for all. In the years that followed, explanations which could not be tested by experiment—for example, that the planets are kept going by angels pushing them—lost popularity. Carefully controlled repeatable experiments involving accurate measurements gradually became the criterion of validity and the hallmark of *science*.

Why then resurrect the ancient label—natural philosophy? Partly because physics is a human activity in which the *history* of ideas ought to play a part and partly because, in addition to our knowing the rules, some sense of *understanding* the naturally occurring events around us should be encouraged.

A change of name will not alter the nature of the subject but it may emphasize that physics is not just a 'collection of facts' nor is it a form of 'applied mathematics'. A study of physics should help us to make some sense of the physical world. And it has failed miserably when it leads Arthur Koestler to cry in despair:

> 'Each of the "ultimate" and "irreducible" primary qualities of the world of physics proved in its turn to be an illusion . . . Compared to the modern physicist's picture of the world the Ptolemaic universe of epicycles and crystal spheres was a model of sanity. The chair on which I sit seems a hard fact, but I know that I sit on a nearly perfect vacuum. . . .
>
> These waves, then, on which I sit, coming out of nothing, travelling through a non-medium in multi-dimensional non-space, are the ultimate answer modern physics has to offer to man's question after the nature of reality.'

Making Sense

If one of our aims is to make sense of the physical world, perhaps we ought to retain something of the natural philosopher's approach and marry it to the rigours of experimental testing. Physics is not, after all, merely a matter of knowing or even of 'discovering' the rules that govern the behaviour of things: it must look beyond the rules. We are not satisfied with the bare facts—*what* happens—we want to know *why* it happens. And we must learn to differentiate between 'Why?' questions which can be answered usefully, and those which cannot. Aristotle asked, 'Why does a body keep moving?' and Galileo, 'Why does it stop?' We must try to see if both questions are equally helpful and not be too shocked if the only answer we can find to some 'Why?' questions is, 'Because it is so!!' Mystery will always remain. J. Robert Oppenheimer has said,

> 'Both the man of science and the man of art live always at the edge of mystery, surrounded by it; both . . . have had to do with the harmonization of what is new with what is familiar, with the balance between novelty and synthesis, with the struggle to make partial order in total chaos.'

By inventing concepts, constructing theories, and devising analogies and conceptual models, scientists have tried to discover ordered patterns within the chaos.

There is, however, no simple recipe for the creative process by which a scientist devises a new hypothesis or theory. Writing on *Scientific Method* J. J. Davies suggests four stages: preparation, incubation, illumination, and verification.

Preparation can be a lengthy process. Phenomena must be studied, data collected, experiments devised and conducted, measurements taken. An important part of the preparation is the precise formulation of the problem. For this, experience, intuition, and inspiration are needed.

During a period of relaxation the problem may not consciously be considered except perhaps at odd times. Illumination can be quite sudden—a flash of inspiration—or it can result from a series of insights.

Finally an idea emerges which, when translated into measureable quantities, is tested during the stage of verification.

In practice, however, a scientific discovery may be quite different! Professor David Horrobin says in his book *Science is God,*

> 'Every schoolboy knows that science progresses by means of a logical, ordered sequence of events called the scientific method. Every schoolboy is wrong. Science is a thoroughly disordered and illogical activity. The making of a great scientific discovery is as personal and idiosyncratic as the writing of a great poem.'

Physics, then, is essentially a way of looking at nature which has evolved and is evolving as a result of human activity. If we can see further than Aristotle, it is because we stand, in Newton's words, 'on the shoulders of giants'.

Making Use

Science is sometimes unfortunately divided into two areas: pure and applied. For many, making sense of the physical world—attempting to interpret phenomena in terms of the minimum number of basic assumptions—is a challenging intellectual pursuit. This search for understanding is pure science.

The main concern of others is to use their knowledge in a practical way; for example, to build an amplifier, repair a sewing machine or devise some spectacular lighting for a party. This is applied science.

Both are equally important and both demand equal ability and dedication. In general the two aspects are complementary; they are not alternatives. Each draws from and contributes to the other.

In this book we will try to make sense of at least some of the things that happen in the physical world. We will study some of the simple concepts, models, and analogies used in physics and we wil take a look at a few of the practical applications c physics in industry, medicine, communication and in the home and in recreation.

CONTENT

In the Scottish physics syllabus there are thr cycles which normally correspond to the work dor in (1) years one and two, (2) years three and fou and (3) year 5. This book covers the *additional* wor needed for the third (Higher) cycle. Where it h been thought appropriate, some revision work h been included in the text, and the summaries an problems at the end of the chapters contai materials from the second (O Grade) cycle. It i however, impossible to include all the O Gra material, and in particular the following topics a not discussed other than incidentally in some the problems: sound and water waves, hea electrostatics, electromagnetism, electronics, ar radioactivity.

In the historical notes and the applications various physical principles, the book quite inte tionally goes beyond the bare examination syllabu In this sense it is a background book rather than straight textbook. It is expected that teachers w prescribe the sections which they consider appr priate to particular classes.

The book is intended to be used in conjuncti with a practical laboratory course such as th found in *Nat Phil 5 Workbook*. Details of suitab experiments are given also in the *Nuffield Guid to Experiments* and *Physics is Fun*.

Answers to all the questions, except those fro S.C.E. Higher papers, will be found at the end the book.

Units, abbreviations, and symbols recommend by the Royal Society and the Scottish Certificate Education Examination Board have been used. T solidus has been used throughout the text and t negative index in the questions at the end of ea chapter.

The following prefixes have the meanings in cated.

Prefix	*Symbol*	*Multiple*	*Prefix*	*Symbol*	*Multi*
atto	a	10^{-18}	kilo	k	10^3
femto	f	10^{-15}	mega	M	10^6
pico	p	10^{-12}	giga	G	10^9
nano	n	10^{-9}	tera	T	10^1
micro	μ	10^{-6}			
milli	m	10^{-3}			
centi	c	10^{-2}			

THANKS

It would be impossible for me adequately to express my thanks for all the help I have received in the production of this book. Many friends and colleagues have read the manuscript and made helpful comments and criticisms.

I am particularly grateful to Geoffrey Salter and George Hartfield for once again producing a series of delightful illustrations, to the publishers for their patience and good humour, and to Dr Davidson of the Department of Natural Philosophy, Aberdeen University, for so carefully reading and editing the entire manuscript. To them all and to many others who helped I express my most sincere thanks.

I am indebted also to the Scottish Certificate of Education Examination Board for permission to reprint questions from Higher Papers.

1973 J.J.

ACKNOWLEDGEMENTS

The author and publishers are grateful to the following organizations and individuals who have kindly provided photographs for this book.

A.E.R.E. (Harwell), 5–8, 19–27
Advance Components, 5–15
J. Allan Cash, 3–17, 4–25
American Telephone and Telegraph Co., 13–30
Asahi Optical Co. Ltd., 15–35, 15–36
B.B.C., 5–18, 18–23
C. Bille, 5–7
Lord Blackett, F.R.S., 3–14
British Aircraft Corporation Ltd., 4–23
British Sub-Aqua Club, 4–42
C.E.R.N., 19–61
Cavendish Laboratory, University of Cambridge, 19–7, 19–9, 19–24, 19–33, 19–34
De Beers Consolidated Mines Ltd., 14–5
Department of Mechanical Engineering, Loughborough University of Technology, 18–29
Department of The Environment, Hydraulics Research Station, (Crown Copyright), 16–2
Doubleday and Co. Ltd., 19–31
Professor H. E. Edgerton, p. 2 (golf), 5–28
E.M.I. Ltd., 18–31
Features International, 4–7
Ferranti Ltd., 9–19
G.P.O., 13–1
Glaverbel S.A., 2–22
Griffin and George Ltd., 16–1, 17–4
Professor Henry Hill, 18–34
Hitachi Ltd., 18–32
A. H. Hunts (Capacitors) Ltd., 10–15
Hughes International, 18–18
Italian Tourist Office, p. 2 (skiing)
Jackson Brothers (London) Ltd., 10–31
Jodrell Bank, University of Manchester, 13–14
Dr C. Jönsson, 16–9
Kodansha Ltd., 1–22, 1–29, 2–28, 5–26, 12–3, 12–4, 12–5
E. D. Lacey, p. 2 (water skiing)
Landis and Gyr Ltd., 9–20
Laser Associates Ltd., 18–25
Laser Technique S.A., 18–28
Lawrence Radiation Laboratory, Berkeley, p. 143, 19–62
J. J. Lloyd Instruments Ltd., 8–21, 8–25
Longines S.A., 12–26
Robert Madden, 14–18 and front cover
Mansell Collection, 14–9, 14–11, 19–4
Marconi Co. Ltd., 13–11, 13–21
Metropolitan Police, p. 2 (judo)
Ministry of Defence (Royal Navy), 19–48
N.A.S.A., p. 2 (astronaut working in space), 1–19, 2–17, 2–18, 4–15, 12–16, 13–15, 18–24
National Accelerator Laboratory, Batavia, 19–36, 19–58
National Physical Laboratory, New Delhi, 14–6
National Physical Laboratory, Teddington, 18–30
New Scientist, 14–21
North-Eastern Fire Brigade, Aberdeen, 2–19
Novosti Press Agency (A.P.N.), 19–49
Omega, 2–35
P.S.S.C., 16–10, 16–11
Professor L. G. Paleg, 18–26
Planair, 3–1
Plessey, 4–34
Paul Popper Ltd., 4–8, 5–5
Professor J. G. Powles, 5–14
Queens Ice Skating Rink, 4–1
Radio Times Hulton Picture Library, 12–1, 12–2, 19–2, 19–5, 19–6, 19–8, 19–10, 19–22
Rank Precision Industries Ltd., 14–25
Raytheon Co. Ltd., 13–17
Road Research Laboratory, 2–23
Royal Institution of Great Britain, 6–8
R. P. Scherer Ltd., 4–33
The Science Museum, 6–9, 16–3, 17–1, 19–15
Scientific Apparatus, 19–53
The Scotsman Publications Ltd., 4–6, 4–32
South of Scotland Electricity Board, 9–1
Sport and General, 4–14
Professor F. R. Stannard, 19–57
Dr K. A. Stetson, p. 105
I. D. Taylor, 19–29
Teltron Ltd., 6–5, 6–6, 6–7
Raymond Thatcher, p. 2 (racing car)
Trinity College Library, Cambridge, 2–1
U.K.A.E.A., 10–27, 19–32, 19–45, 19–46, 19–47, 19–54 and back cover
U.S.I.S., pp 1 and 53, 4–2, 4–9, 13–13, 13–19, 18–22, 19–42, 19–43, 19–56, 19–60
U.S. Navy, 19–50
Ralph Wyckoff, 5–6
Yashica Co. Ltd., 15–34
Yerkes Observatory, University of Chicago, 14–12

Contents

UNIT ONE

Matter in Motion

The Kinetic Model

15
MATRA F1

1 Motion

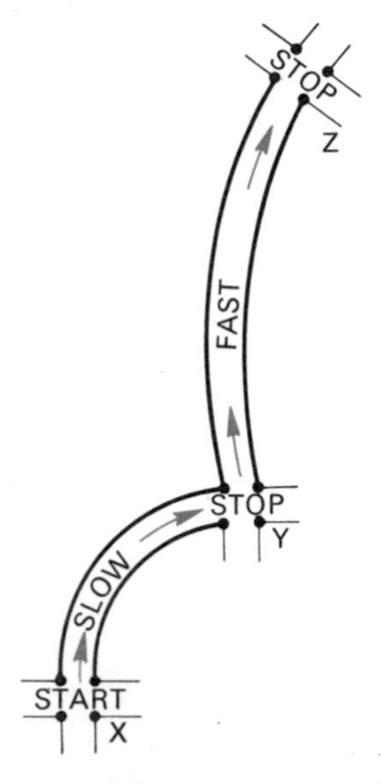

Fig. 1–1

Life is motion! Every minute of every day we are involved with things that move. Our most exciting experiences are concerned with them and our interest often flags the minute they stop. Many types of motion such as those illustrated here are extremely complex and difficult to analyse. In this course we will concentrate on simpler movements, which include motion in a straight line. First, however, let us glance at some of the devices we will be using; they include graphs, vectors, and equations.

To illustrate our use of graphs and vectors, consider two simple situations. A car is standing at traffic lights X (Fig. 1–1). When the lights change to green, the car moves off slowly along the path indicated. It is soon stopped by a second set of lights Y. When they change to green the car turns left and moves more quickly to the next corner Z, when it again stops. In the second situation (Fig. 1–2), a batsman runs at a steady speed eastwards from one end of a cricket pitch to the other. He then turns and comes two-thirds of the way back before stopping when he is stumped.

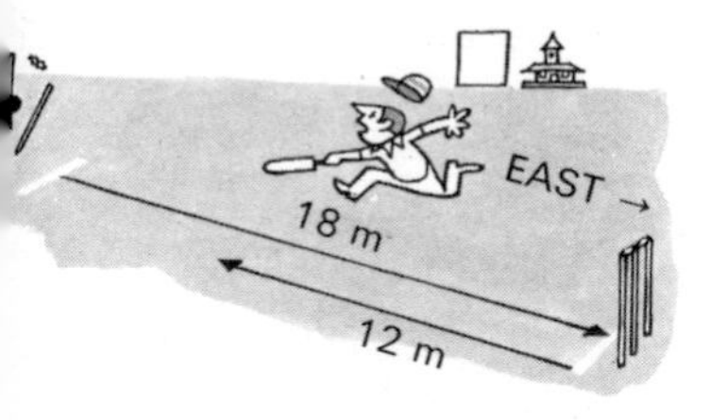

Fig. 1–2

DISTANCE AND DISPLACEMENT

Distance

By *distance* we shall mean the total length of the path travelled. It is represented by d or s (for space) and is measured in units of length, for example, metres. It is a scalar quantity.

Distance = total path length

The distance the car has travelled in a given time is represented on Fig. 1–3. The distance the batsman moves in a given time is represented on Fig. 1–4.

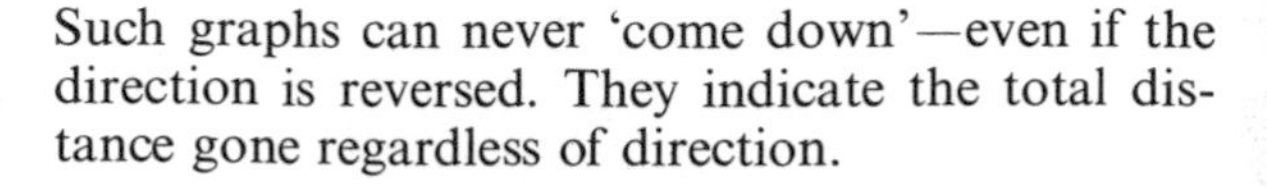

Such graphs can never 'come down'—even if the direction is reversed. They indicate the total distance gone regardless of direction.

Displacement

By *displacement* we shall mean the change in position. If a body moves from A to B, we can represent its displacement by a straight line drawn from A to B. The direction can be shown by putting an arrow head at B. For example, if you moved from one end of a 15-metre laboratory to the other, your displacement could be 15 metres west, yet you might have walked 50 metres back and forth between benches. In addition to a unit of length a direction must always be stated, e.g. metres west. Displacement is a *vector quantity* and is represented by $\boldsymbol{d}$ or $\boldsymbol{s}$.

Displacement = distance in a stated direction from a point

In general we cannot draw graphs of displacement/time but we can use such graphs in the special case of straight line motion. In this case the graphs merely indicate the *sense* of the displacement. For example, displacement to the east of the starting point could be represented by the positive direction of the graph and displacement to the west of the starting point represented by the negative direction. We could not use a displacement/time graph to represent the complex motion of the car in the above example but the relatively simple displacement of the batsman could be represented as shown in Fig. 1–5. Notice that in the second part of the graph the displacement of the batsman is still to the east, even although he is running westwards.

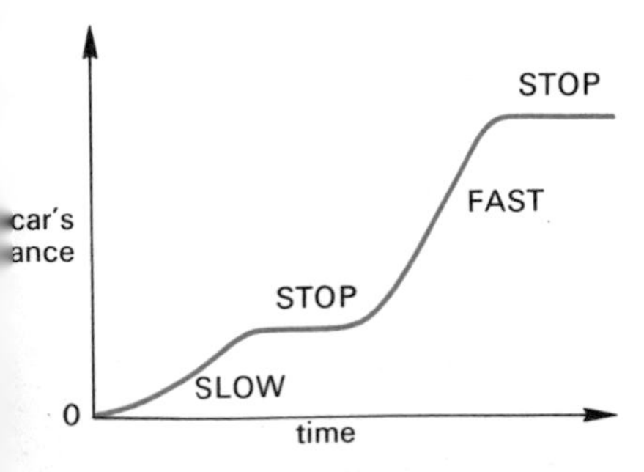

Fig. 1–3

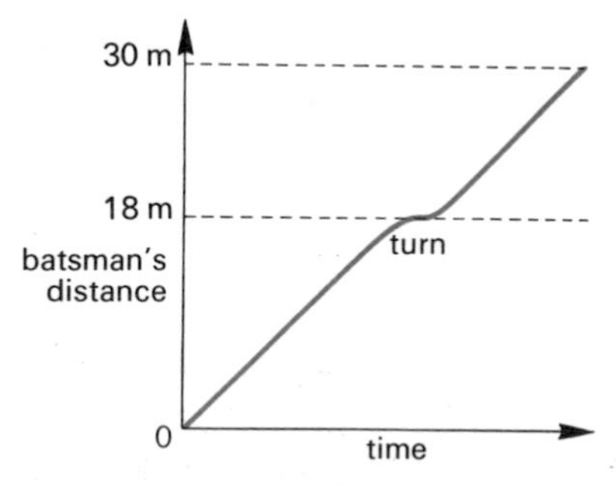

Fig. 1–4

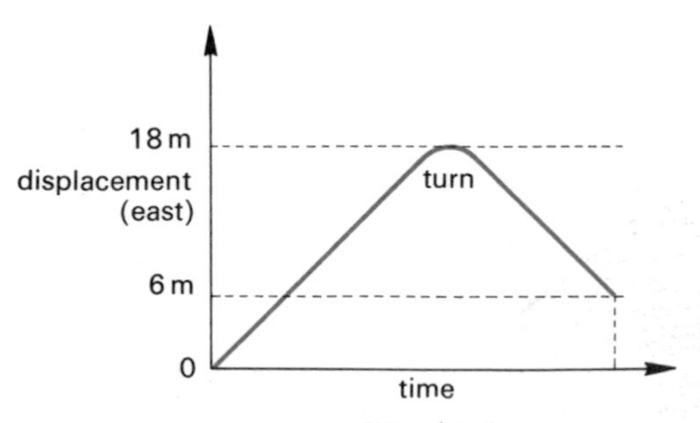

Fig. 1–5

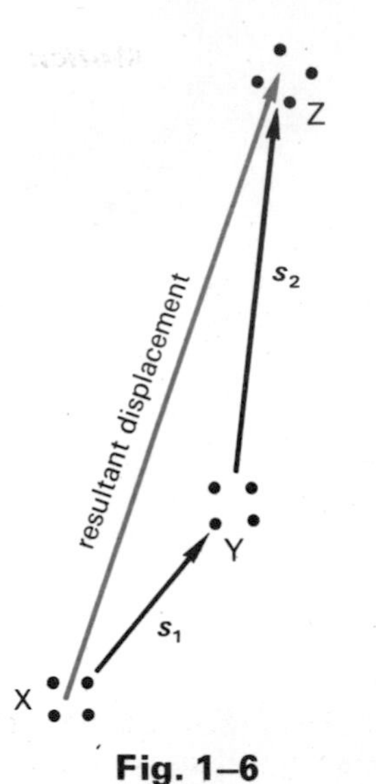

Fig. 1–6

Vectors

Displacements can be represented by vectors and the 'tip to tail' rule used to find the resultant displacement. For example the displacements of the car between traffic lights could be represented as shown in Fig. 1–6. The two displacements s_1 and s_2 are then added vectorially to give the resultant displacement indicated. Vectors showing the final displacement of the batsman can also be used in the same way. His final displacement is 6 metres east (Fig. 1–7).

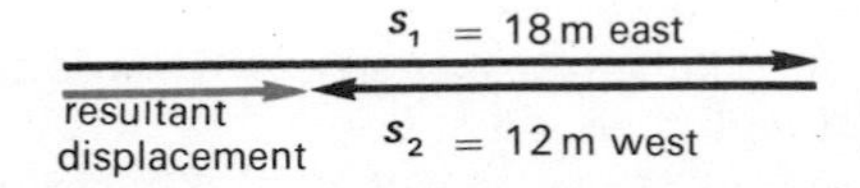

Fig. 1–7

SPEED AND VELOCITY

Speed

Speed refers to motion in which direction is not taken into account.

$$\text{Speed} = \text{distance travelled per unit time}$$

It is measured in metres per second (written m/s or m s^{-1}) and is a scalar quantity. If a car is travelling at a steady speed and it covers 300 metres in 10 seconds, it is easy to see that its speed is 30 m/s, which is about 66 miles per hour. The distance travelled is represented by the area under the graph (Fig. 1–8). If, however, the speed is continually changing, it is not easy to see what a reading on the speedometer of, say, 35 miles per hour means at a particular *instant*. Clearly we cannot use the relationship speed = distance/time, since, if the time interval is zero, the distance travelled must also be zero!

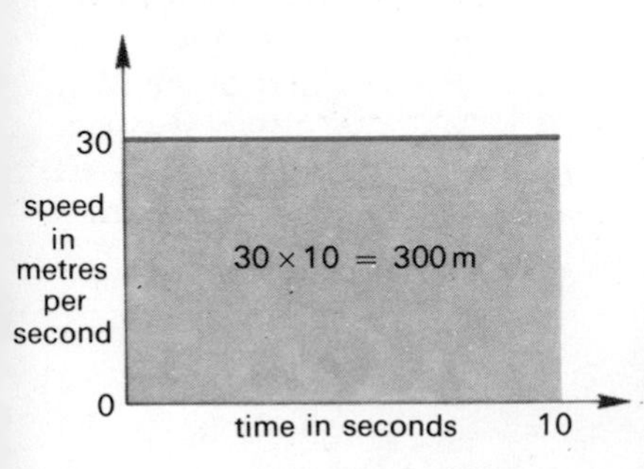

Fig. 1–8

We could, of course, think of the average speed over a finite time interval, i.e. total distance/total time. Returning to the car in the above example, we might find that the distance travelled between the first two sets of traffic lights was 100 metres and the time interval 10 seconds. The average speed would then be 100/10 = 10 m/s. Suppose, however, that we wanted to find the speed during one particular second—say the fifth second—of its motion. We would have to know the distance travelled during that second. Even then we would be finding the *average* speed during that second, i.e. $\Delta s/\Delta t$ (Fig. 1–9).

To find the *instantaneous speed*, say after 4.5 seconds of travel, we have to imagine Δs and Δt shrinking and shrinking but never disappearing

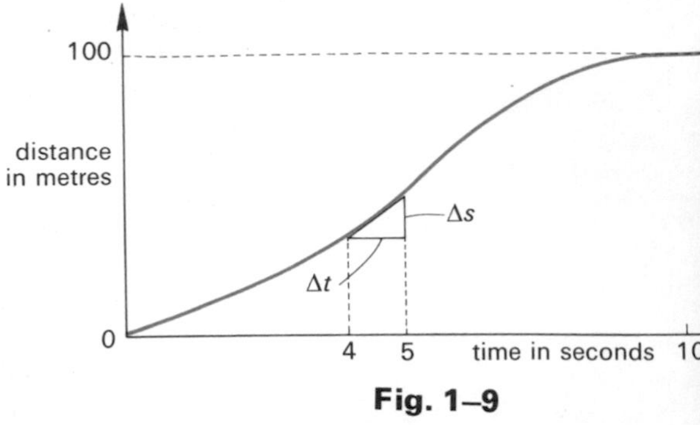

Fig. 1–9

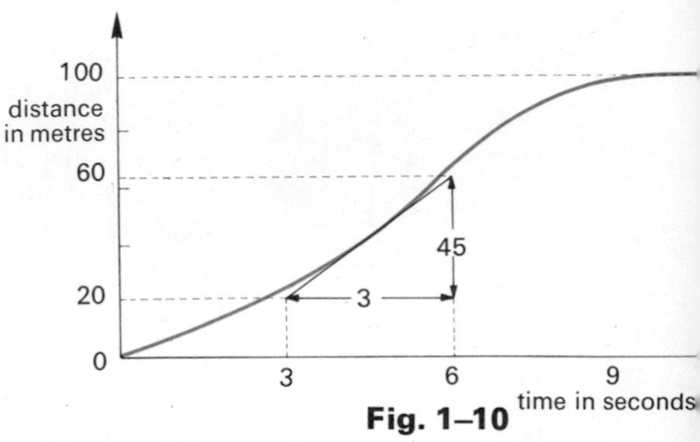

Fig. 1–10

completely. They can be represented on the graph by the gradient of a tangent to the curve at the point in question. The tangent at 4.5 seconds is shown in Fig. 1–10. The gradient of this tangent, $\Delta s/\Delta t$, is 45/3 = 15 m/s. At that instant the speedometer would be reading the equivalent of 15 m/s.

A speed/time graph of the car's journey from X to Z would look something like Fig. 1–11. The speed/time graph of the batsman is shown in Fig 1–12. Notice that the graphs tell us nothing whatever about the direction of motion.

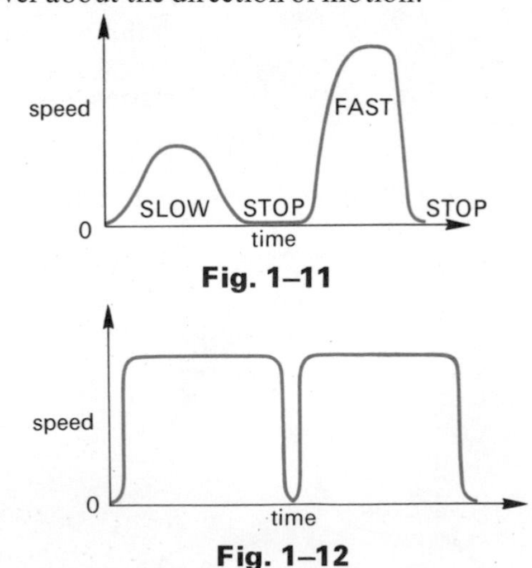

Fig. 1–11

Fig. 1–12

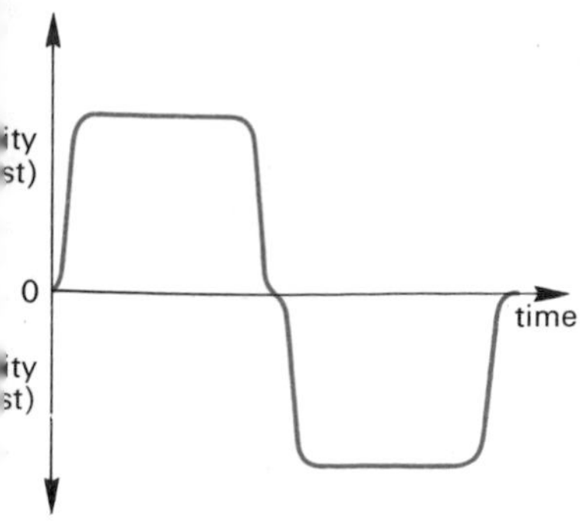

Fig. 1–13

Velocity

Velocity refers to motion in which direction is taken into account. It is measured in speed units with a direction added, e.g. m/s east or m s^{-1} east. It is a vector quantity.

$$\text{Average velocity} = \frac{\text{displacement}}{\text{time interval in which displacement occurs}}$$

$$v = \frac{\Delta s}{\Delta t}$$

The instantaneous velocity would be the limiting value of this expression as $\Delta t \to 0$.

As with displacement, a graph of velocity against time can be used only for the special case of straight-line motion, in which case the sign (up or down) indicates the sense. A graph of the batsman's velocity eastwards could look like Fig. 1–13. The curve below the axis indicates a velocity westwards.

Relative Velocity

Have you ever tried to walk along the corridor of a train so that you stay beside someone standing on the station platform? You are at rest relative to your friend on the platform but moving with respect to the train. The train is, of course, moving with respect to the platform at exactly the same speed. We can represent the train's velocity relative to the platform thus

and your velocity relative to the train thus

Combining them gives us this

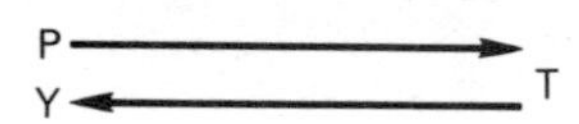

The resultant is zero. You are therefore at rest relative to the platform. Once the train speeds up, so that it is going faster than you can walk, the situation alters like this

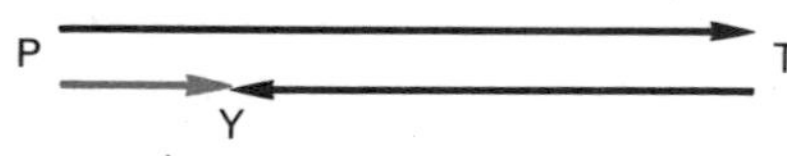

The vector $\overrightarrow{PY}$ then represents your velocity with respect to the platform.

Fig. 1–14

If you walk across the deck of a moving ship (Fig. 1–14), you have a velocity relative to the ship, say 1 m/s south, and the ship has a velocity relative to the water, say 6 m/s east. As the two velocities are simultaneous, they can be combined vectorially to find your velocity relative to the water. The ship's velocity relative to the water may be represented as follows

Your velocity relative to the ship may be represented thus

Combining the two vectors gives us this

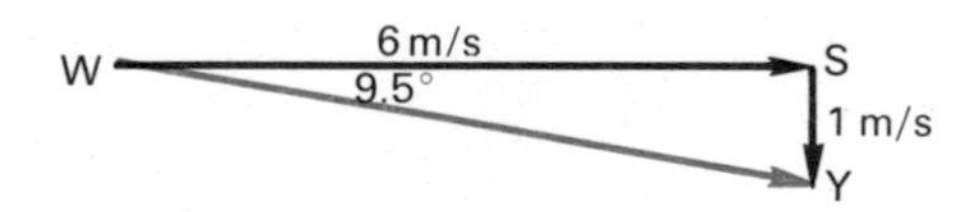

Your velocity relative to the water is thus seen to be

$$\sqrt{6^2 + 1^2} = \sqrt{37} = 6.1 \text{ m/s} \quad 9.5° \text{ south of east.}$$

ACCELERATION

Pressing your foot on the accelerator pedal causes a car to speed up. In everyday language to accelerate simply means to increase speed. In physics the word is more precisely defined.

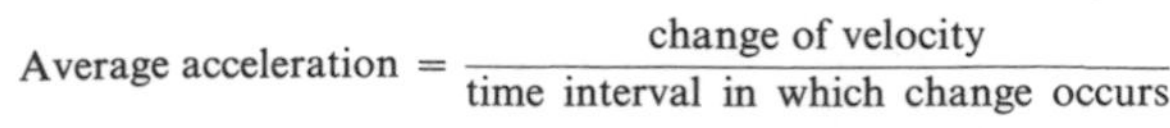

$$\text{Average acceleration} = \frac{\text{change of velocity}}{\text{time interval in which change occurs}}$$

$$a = \frac{\Delta v}{\Delta t}$$

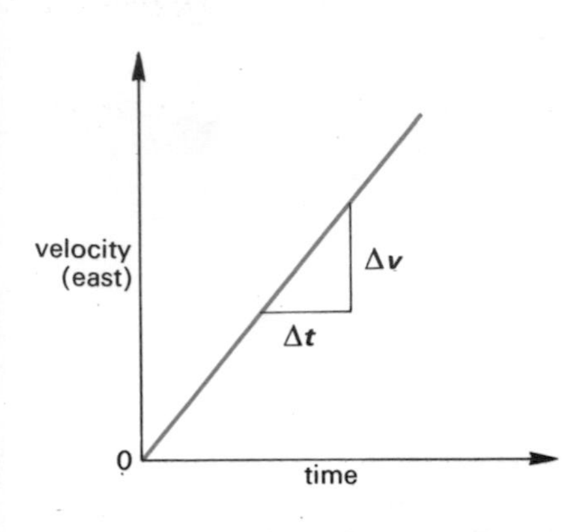

Fig. 1–15

The instantaneous acceleration is the limiting value of this expression as $\Delta t \to 0$.

In this course we will be dealing mainly with *constant* acceleration. Then instantaneous acceleration has the same value. Acceleration is a vector quantity and is measured in (metres per second) per second, written m/s^2 or $m\,s^{-2}$ plus a stated direction. Unlike velocity and displacement, which have scalar counterparts, we have no scalar term which means 'increase in speed per unit time'. On a velocity/time graph, acceleration would be represented by the gradient $\Delta v/\Delta t$ (Fig. 1–15).

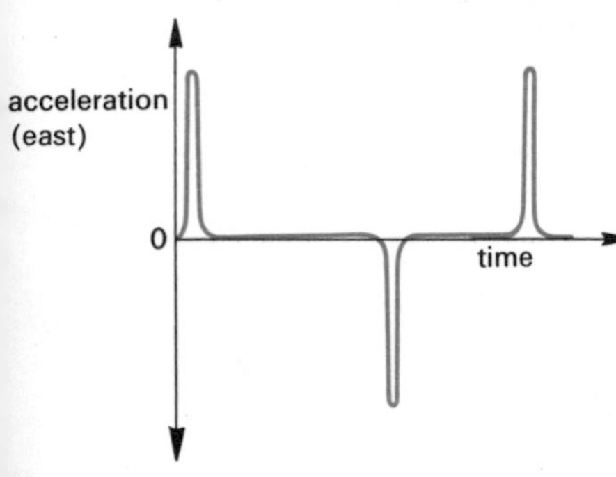

Fig. 1–16

A graph of the batsman's acceleration would indicate acceleration eastwards or deceleration westwards as positive, and deceleration eastwards and acceleration westwards as negative (Fig. 1–16). Constant velocity and rest mean, of course, zero acceleration.

Change in Velocity

Imagine a football rolling along at velocity v_1 east when it is kicked so that its velocity increases to v_2 east. We can represent the velocities by the vectors shown in Fig. 1–17. The red line shows the

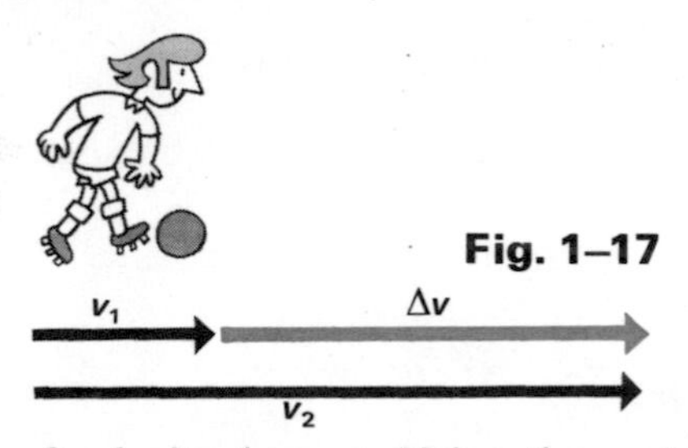

Fig. 1–17

change of velocity Δv east. Using the vector rule, we can say that

$$v_1 + \Delta v = v_2$$

Suppose, however, that the kick caused the ball to change direction. Then the new velocity of the ball could be found by using the 'tip to tail' vector rule. Again we have

$$v_1 + \Delta v = v_2$$

The change of velocity is represented by the red vector in Fig. 1–18.

$$\Delta v = v_2 - v_1$$

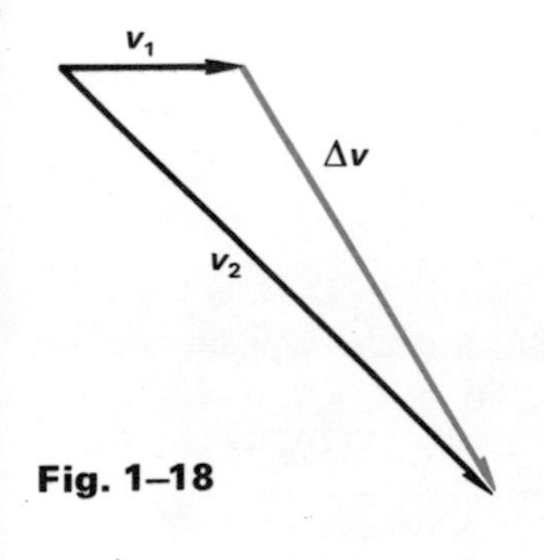

Fig. 1–18

EARTH SATELLITES

The difference between scalar and vector quantitie can be illustrated dramatically by considering th motion of an Earth satellite (Fig. 1–19). Its *distanc* from the centre of the Earth may be the same at and B but its *displacement* from the centre is di ferent because it lies in a different direction (Fi 1–20). Its *speed* at A may be exactly the same as i

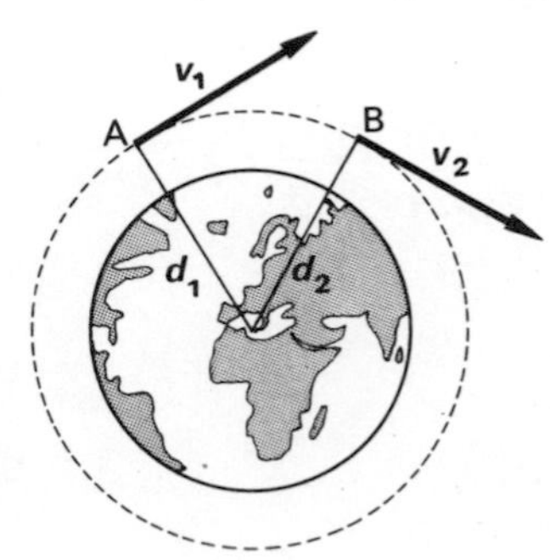

Fig. 1–20

speed at B but its *velocity* at B is different from th at A because the direction is different. Finally, an perhaps most perplexing, if the satellite's speeds A and B are the same, there has been no increa in speed yet the satellite has been accelerated!

If v_1 and v_2 represent the original and the fin velocities, then the *difference*, i.e. the change velocity, will be represented by Δv (Fig. 1–21).

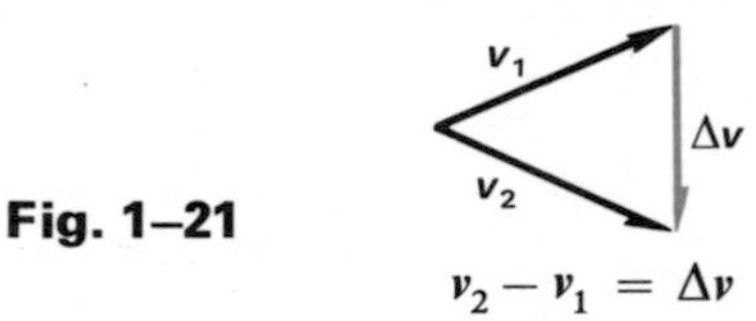

Fig. 1–21

$$v_2 - v_1 = \Delta v$$

There has therefore been a change of velocity the time taken for the satellite to move from A B. If this time interval is Δt, the average acceler tion is $\Delta v/\Delta t$. This acceleration is directed towar the centre of the Earth and is, of course, caused the gravitational pull on the satellite. The satelli is accelerating towards the Earth yet going faster and getting no closer to the Earth! The stro photograph in Fig. 1–22 shows by analogy the effe you would expect if gravity could be switched c suddenly. *What do you think is shown in this phot graph? (1.1)*

BACK TO THE STRAIGHT

As we have spent so much time bamboozling y with vectors you may be relieved to know th much of the work that follows deals with moti

g. 1–19

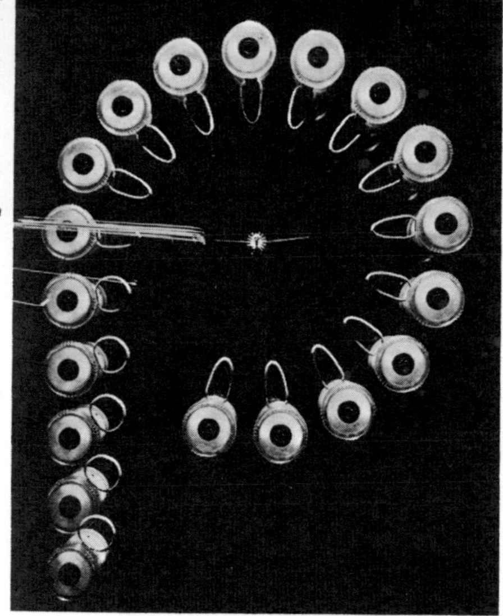

Fig. 1–22

in a straight line. When this is the case, we will be concerned only with the magnitude and sign of displacement, velocity, and acceleration. In most cases s is simply the distance travelled, v is the speed, and a the rate of change of speed. In some physics books, the terms velocity and acceleration are often used without stating a direction. In such cases you are expected to assume that the direction remains unchanged.

Equations of Motion for Constant Acceleration in a Straight Line

Suppose a truck is accelerating down a track with a uniform acceleration a. As it passes point A we start a stop watch and find the time t taken to reach point B (Fig. 1–23).

time = 0
A
u
time = t
B
v
displacement s

Fig. 1–23

The time interval $= t-0 = t$.

If the velocity at A is u and at B it is v, the acceleration is easily calculated.

$$a = \frac{v-u}{t}$$

$$\Leftrightarrow v = u+at$$

A graph of this motion would look like Fig. 1–24.

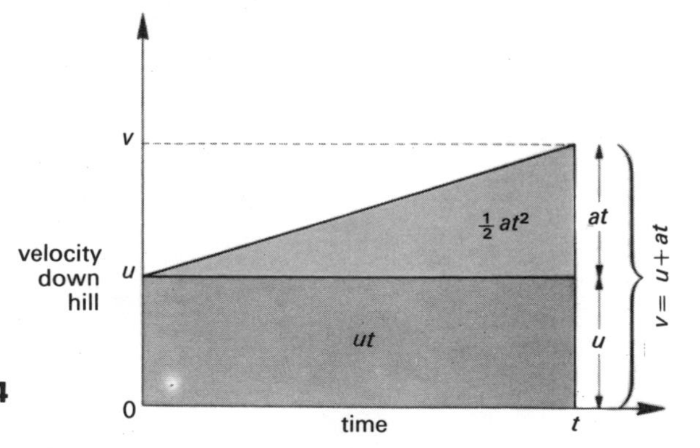

Fig. 1–24

The distance travelled between A and B would be represented by the total shaded area.

$$s = ut+\tfrac{1}{2}at^2$$

Alternatively

$$s = \text{average velocity} \times \text{time}$$

$$= \left\{\frac{u+v}{2}\right\}t$$

Finally another equation, which does not contain t, can be derived as follows.

$$v = u+at$$

$$\Rightarrow v^2 = (u+at)^2$$

$$\Leftrightarrow v^2 = u^2+2uat+a^2t^2$$

$$\Leftrightarrow v^2 = u^2+2a(ut+\tfrac{1}{2}at^2)$$

$$\Rightarrow v^2 = u^2+2as$$

SUMMARY

Vector addition

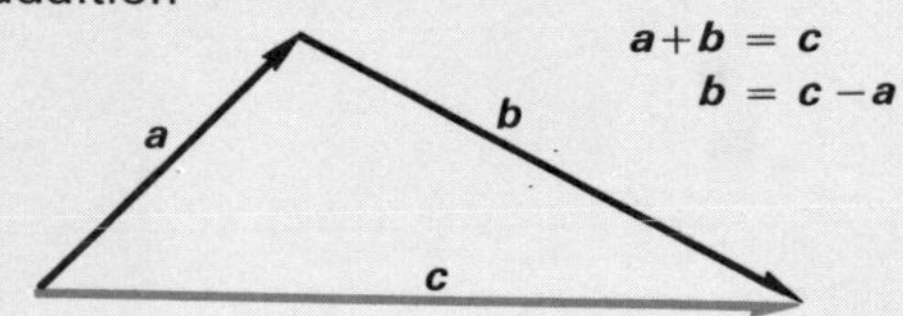

$\boldsymbol{a}+\boldsymbol{b} = \boldsymbol{c}$

$\boldsymbol{b} = \boldsymbol{c}-\boldsymbol{a}$

Relative velocity

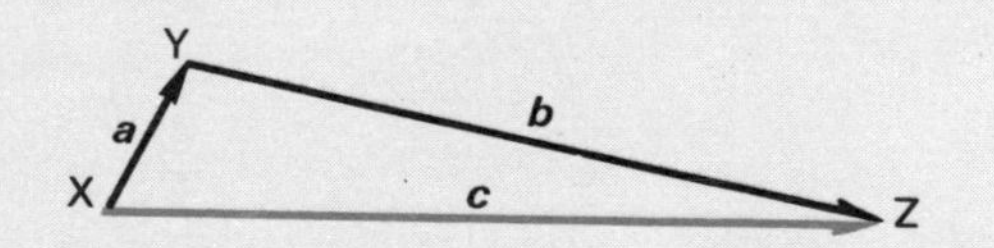

If $\boldsymbol{a}$ represents the velocity of Y with respect to X and $\boldsymbol{b}$ represents the velocity of Z with respect to Y then $\boldsymbol{c}$ represents the velocity of Z with respect to X.

Rectilinear motion

$$v = u+at$$

$$s = ut+\tfrac{1}{2}at^2$$

$$v^2 = u^2+2as$$

$$v = \frac{u+v}{2}$$

Area under a speed/time graph represents distance travelled.

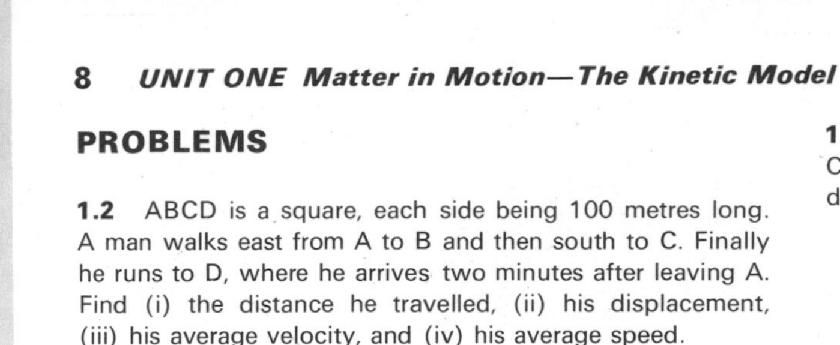

Fig. 1–25

PROBLEMS

1.2 ABCD is a square, each side being 100 metres long. A man walks east from A to B and then south to C. Finally he runs to D, where he arrives two minutes after leaving A. Find (i) the distance he travelled, (ii) his displacement, (iii) his average velocity, and (iv) his average speed.

1.3 A racing car travels 10 km east at 400 km h^{-1} assisted by a tail wind. On his return trip west the driver clocks only 300 km h^{-1}. Find his average speed and his average velocity.

1.4 How long will a satellite take to circle the Earth if it travels at 8×10^3 m s^{-1} in a circular orbit of radius 7.6×10^6 m?

1.5 A solenoid is attached to the top of a board mounted on a trolley (Fig. 1–25).

A ball bearing is held by the solenoid and can be released while the trolley is in motion. Fig. 1–26 shows a strobe photograph of the ball after its release.

Fig. 1–26

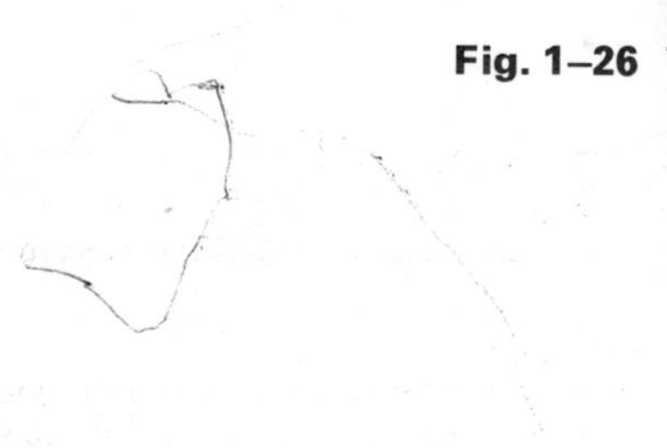

How was the trolley moving? Was the camera fixed to the bench or to the trolley? Taking g, the acceleration due to gravity, as 10 m s^{-2} and the flash frequency as 25 per second, find, approximately, the time taken for the ball to fall (in seconds), the vertical distance through which the ball falls (in metres), and the speed of the trolley (in metres per second).

What can you say about the horizontal motion of the ball before and during its fall?

1.6 A spacecraft is travelling at 2000 m s^{-1} when a rocket motor is fired. If it produces a constant acceleration of 10 m s^{-2} and burns for 300 s, find the final speed of the spacecraft. How far would it travel during this time?

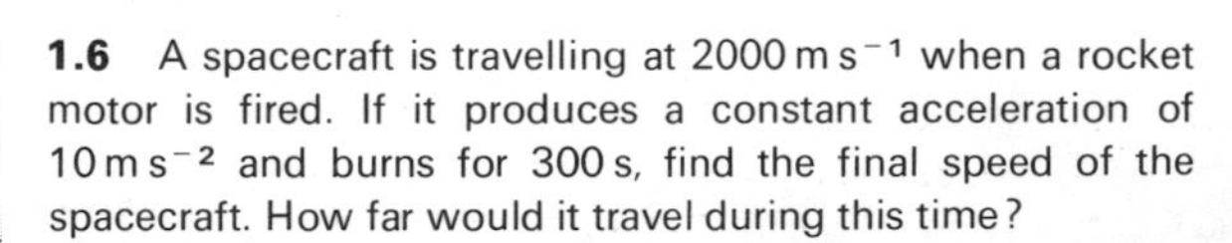

1.7 If the speed of a spacecraft is reduced from 2000 m s^{-1} to 200 m s^{-1} in 30 s, find the distance travelled in that time.

1.8 Fig. 1–27 represents the velocity of a particle moving in a straight line. Draw corresponding displacement/time and acceleration/time graphs.

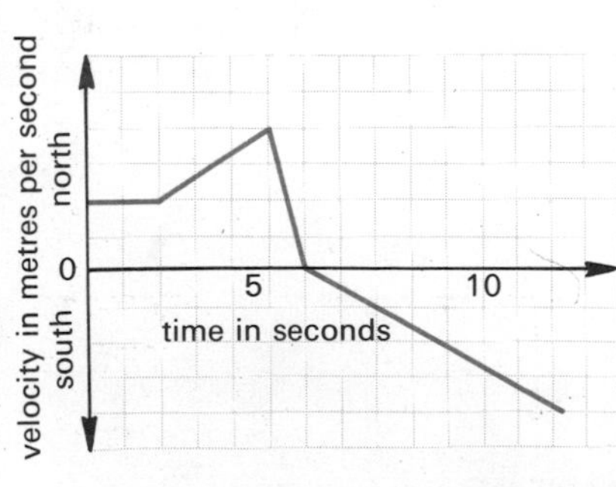

Fig. 1–27

1.9 Discuss the motions represented by graphs A, B, and C in Fig. 1–28. Compare the total distances travelled and the displacements.

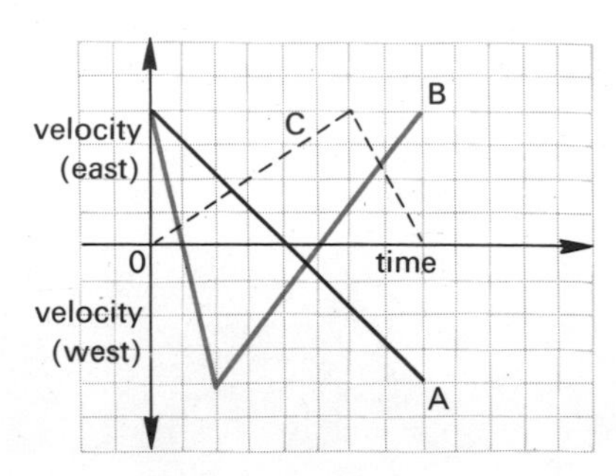

Fig. 1–2

1.10 You are running at your maximum speed of 6 m s to catch a bus which is standing at the bus stop. The b moves off with an acceleration of 1 m s^{-2} when you a 20 metres from it. Find graphically if you can catch the bu

1.11 A car passes you at 10 m s^{-1} and accelerates ur formly for 10 seconds at 2 m s^{-2}. What is its average spe for the 10 seconds? What is its speed at the end of the seconds?

What speed will the car reach if it continues acceleratir uniformly at 2 m s^{-2} for 375 metres after passing you?

1.12 A juggler throws a ball into the air and when it is its highest point he throws up a second ball. Just as t juggler catches the first ball he releases a third ball and on. If he throws one ball each second and each rises to t same height, find how high they go.

1.13 How much information can you get from the stro picture of a bouncing ball shown in Fig. 1–29?

Fig. 1–29

Fig. 1–30

1.14 From the events depicted in Fig. 1–30 draw a vector diagram showing possible velocities of the woman and bird and hence the velocity of the bird relative to the woman!

1.15 A canoe is being paddled upstream at $2.5\,m\,s^{-1}$ relative to the river which is flowing east at $1.5\,m\,s^{-1}$. If an Indian on the canoe shoots an arrow due north at $15\,m\,s^{-1}$, relative to the canoe, find the velocity of the arrow relative to the Earth.

1.16 A train is crossing a bridge at $40\,km\,h^{-1}$ north and a boat passing below the bridge at $20\,km\,h^{-1}$ west. What is the velocity of the train relative to the boat?

1.17 A ship is travelling east at $15\,km\,h^{-1}$ as a man walks across the deck at $5\,km\,h^{-1}$ south. What is the man's velocity relative to the sea?

1.18 A helicopter is rising vertically at $10\,m\,s^{-1}$ when a wheel drops off and reaches the ground 8 seconds later. At what height was the helicopter flying when the wheel left it?

1.19 If a stone is thrown vertically down a well at $5\,m\,s^{-1}$, when will it reach the water surface 60 metres below?

1.20 A circular cycle track has a circumference of 314 metres, with AB as a west-east diameter (Fig. 1–31). A cyclist travels from A to B with a constant speed of $15.7\,m\,s^{-1}$. Find:
(a) the distance he travels,
(b) his displacement,
(c) his average speed,
(d) his average velocity, and
(e) his average acceleration.

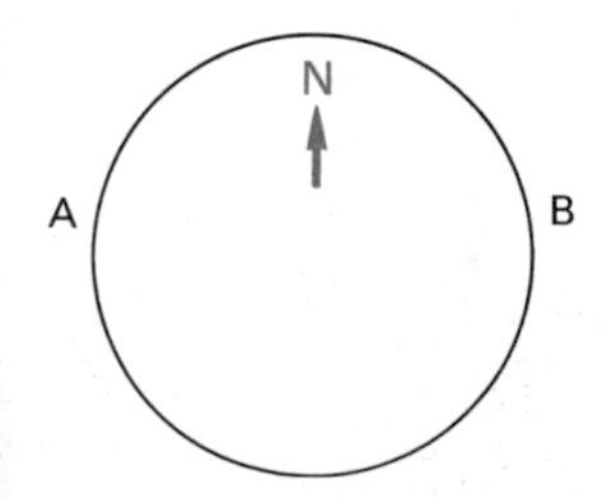

Fig. 1–31

1.21 (a) A plane travelling east at $300\,m\,s^{-1}$ turns northwards in a circular path without altering speed. If after 42.4 seconds it is flying due north, find its change of velocity and its average acceleration.

(b) Imagine that, as the plane is flying east with an air speed of $300\,m\,s^{-1}$, a shot is fired northwards across the cabin at $300\,m\,s^{-1}$ relative to the plane. Find the velocity of the bullet relative to the air outside.

1.22 A body is projected with a horizontal velocity of 10 m/s from the top of a vertical cliff. It strikes the water at a point 30 metres from the foot of the cliff. Estimate:
(a) the time of flight, and
(b) the height of the cliff.
State any assumption made. [S.C.E. (H) 1968]

1.23 (a) The equation below is frequently used in motion calculations.

$$s = ut + \tfrac{1}{2}at^2$$

(i) What fundamental assumption about the motion is inherent in the equation?

(ii) Interpret the expressions ut and $\tfrac{1}{2}at^2$, which represent displacements.

(iii) Derive the equation.

(b) The graph (Fig. 1–32) represents the motion of a ball. Give an interpretation of the motion and predict the next cycle. [S.C.E. (H) 1969]

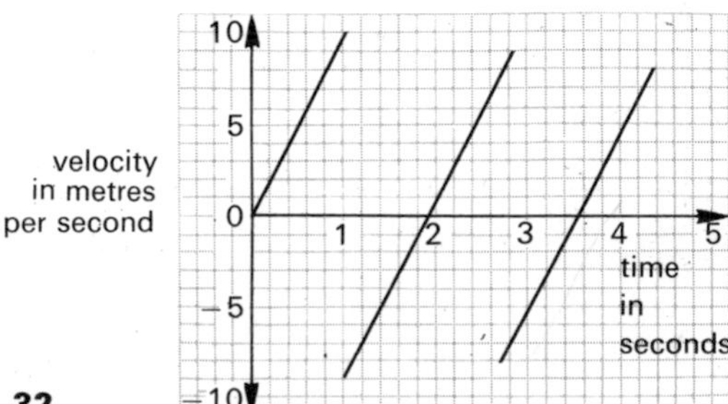

Fig. 1–32

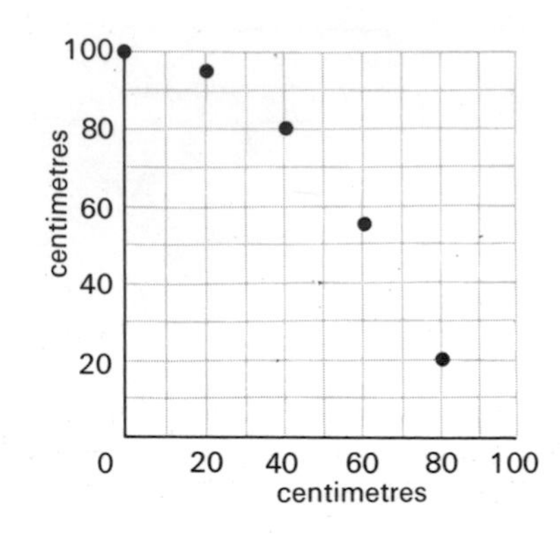

Fig. 1–33

1.24 A ball has rolled off a horizontal table. Fig. 1–33 shows the position of the ball at 1/10 second intervals. Calculate the acceleration due to gravity. [S.C.E. (H) 1970]

1.25 State which of the following situations are possible and give an example of each of these:
(a) a body with acceleration but with zero velocity,
(b) a body with acceleration but with constant speed, and
(c) a body with constant velocity but with increasing speed.
[S.C.E. (H) 1970]

1.26 The graph of velocity against time (Fig. 1–34) represents the motion of a trolley of mass 4 kg constrained to move in a straight line.

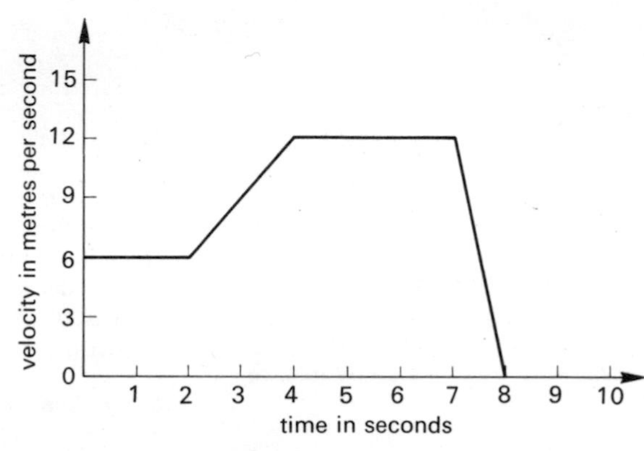

Fig. 1–34

(i) What is the significance of the fact that the graph does not go through the origin?
(ii) What is the acceleration in the 2 s to 4 s time interval?
(iii) What is the magnitude of the force producing this acceleration?
(iv) Use information gained from the above graph to plot a graph of acceleration against time for the whole journey.
(v) Calculate the total distance travelled.
(vi) What is the average velocity for the whole journey?
[S.C.E. (H) 1971]

1.27 An attempt is made to measure the velocity of sound by an echo method (Fig. 1–35). A loudspeaker is made to emit short, regular pulses of sound. A C.R.O., connected to a microphone placed between loudspeaker and reflector displays the trace in Fig. 1–36.

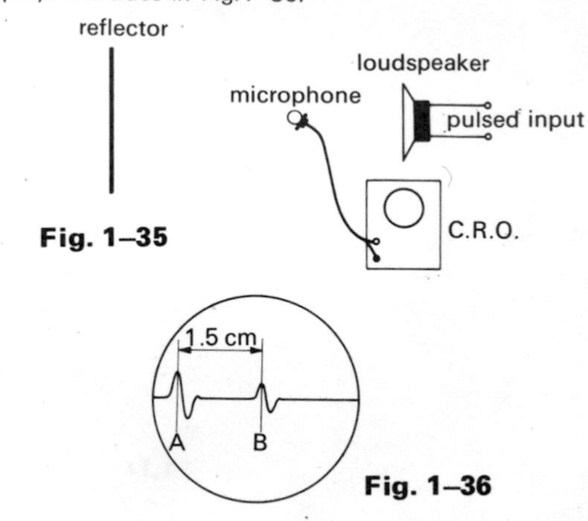

Fig. 1–35

Fig. 1–36

(a) Account for the two peaks at A and B, and explain the difference in their amplitudes.
(b) If the time base were set at 10 milliseconds per centimetre, what would be the time interval between the peaks
(c) Determine the speed of sound, given that the distance between microphone and reflector is 2.5 metres.
[S.C.E. (H) 1971

1.28 A ball is projected at 15 m s^{-1} horizontally from the top of a vertical cliff and reaches the horizontal ground 45 m from the foot of the cliff.
(a) Draw accurate graphs, with the appropriate numerical scales, of
(i) the horizontal speed of the ball against time;
(ii) the vertical speed of the ball against time, for the period from its projection until it hits the ground.
(b) by using a vector diagram or otherwise, find the velocity of the ball 2 s after its projection, giving both the speed and direction. State any assumptions you have made.
[S.C.E. (H) 197

2 Principia

'Nature and Nature's laws lay hid in night.
God said "Let Newton be", and all was light.'
Alexander Pope

'It did not last, the Devil howling "Ho!
Let Einstein be!" restored the status quo.'
Hilaire Belloc

ISAAC NEWTON

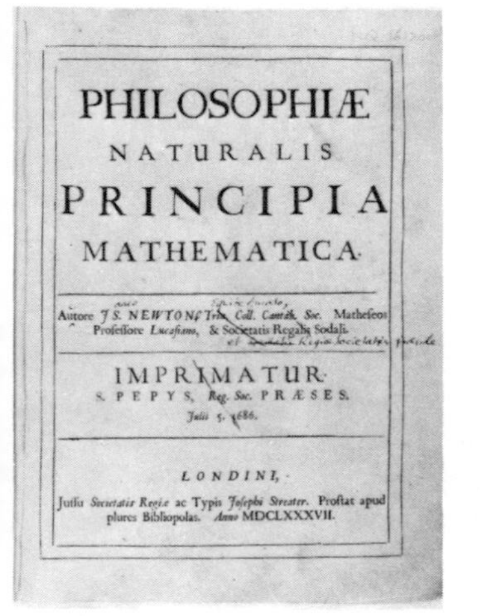

PHILOSOPHIÆ
NATURALIS
PRINCIPIA
MATHEMATICA.

Autore J S. NEWTON, Trin. Coll. Cantab. Soc. Matheseos
Professore Lucasiano, & Societatis Regalis Sodali.

IMPRIMATUR.
S. PEPYS, Reg. Soc. PRÆSES.
Julii 5. 1686.

LONDINI,
Jussu Societatis Regiæ ac Typis Josephi Streater. Prostat apud
plures Bibliopolas. Anno MDCLXXXVII.

Fig. 2–1

'Principles of Natural Philosophy' began a new epoch in the history of physics. If Newton is regarded as the greatest scientific genius the world has known, it is not because he discovered some 'laws of nature' as one might discover an exotic plant or animal but because he invented a way of thinking about *matter in motion* which made sense and produced experimentally verifiable predictions. The Polish astronomer Nicolaus Copernicus replaced the cumbersome Ptolemaic system of epicycles, in which for centuries the Earth had been regarded as the centre of the universe, by an infinitely simpler theory in which the Sun was taken as the centre. Similarly, Newton cut through the confusions of concepts regarding motion and produced a new and amazingly powerful mathematical model.

Newton did not, of course, answer all the questions about motion. Albert Einstein and others have found that his theories do not apply to all kinds of motion, particularly at speeds approaching that of light. But it is a mark of genius to distinguish between questions to which some kind of satisfactory answer is possible and questions which are so fundamental that no explanation in terms of more fundamental concepts is likely at that moment. By attempting to reduce the number of assumptions or axioms to a minimum, Newton produced a mathematical pattern of great beauty and simplicity.

Less than a century earlier, Galileo had pioneered the cause of experimental science. Following in his footsteps, Newton based his work on experiment. The reign of the Greek armchair philosopher was over. Experience was now the sole touchstone of validity.

It was necessary to challenge many of the preconceived ideas of the day. The Greeks had asserted that it was 'in the nature of things' to stay at rest. If a body was moving, it was because something was pushing it. Galileo's experiments provided strong evidence to the contrary. It now seemed that it was 'in the nature of things' to keep moving!

In an age of powerless interplanetary travel it is not difficult for us to believe that a body will keep moving in a straight line if no unbalanced forces are acting on it. It was a different matter three hundred years ago when Galileo came to this conclusion from a series of experiments with inclined planes. Fig. 2–2 shows a multi-flash photo of a puck floating on a near-friction-free cushion of gas so that no unbalanced force is acting on it. The motion is constant speed in a straight line.

You may think that we have simply replaced one assumption, namely that when left to themselves bodies come to rest, with another, that they continue in a state of rest or uniform motion in a straight line. In fact you may feel that the first is more acceptable because it agrees with everyday experience. Two questions, however, must be considered. First, does all the evidence you have available suggest that bodies will in fact come to rest or will they keep going if no unbalanced force acts on them? Laboratory experiments cannot prove Galileo's 'law of inertia' (Newton's First Law), but strong evidence is available.

Fig. 2–2

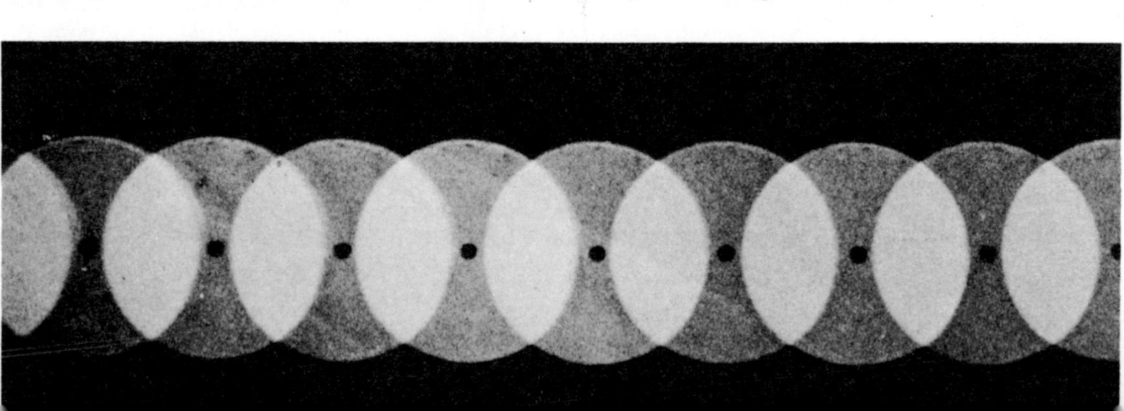

Secondly, why do bodies behave like this? One answer that is sometimes given is 'because of their inertia'. If this answer satisfies you, you have been deluded, for 'inertia' is simply a label to describe this particular property of a body. Another answer might be 'because energy is conserved'. If you accept this, then you could argue that, if there is no change in kinetic energy ($\frac{1}{2}mv^2$) and no change in mass, the speed must be constant. The argument is, however, a circular one. You could equally well cite the inertia of a body as evidence for the conservation of energy.

We accept the notion of inertia because for the past three hundred years it has been found to be extremely *useful* and physicists have found no experimental justification for rejecting it. By accepting it, clear and practical definitions of mass and force are possible and the motions of bodies can be accurately predicted.

frequency

0

number of trolleys (mass)

Fig. 2–5

Mass and Inertia

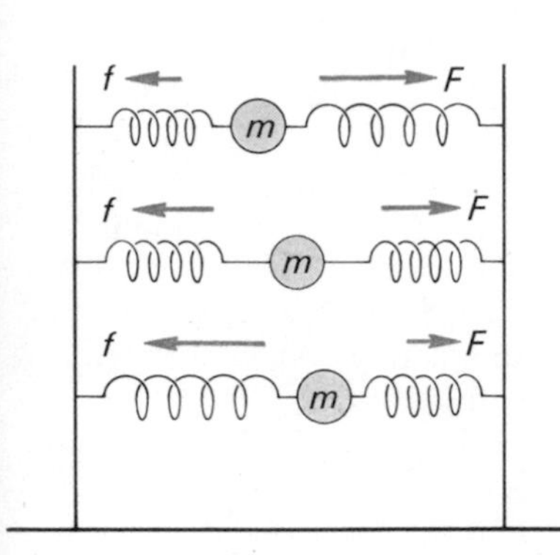

Fig. 2–3

Suppose you were out in interplanetary space, so far from any star or planet that you could ignore the effects of gravitational attraction. Under these conditions a ball of mass m might be attached to two coil springs as shown in Fig. 2–3. If one spring was stretched so that it exerted a force F on the ball and the other stretched only slightly so that it exerted a smaller force f, then there would be an unbalanced force $(F-f)$ acting on the ball. It would therefore accelerate to the right. Soon F would decrease and f increase until the resultant force was zero. There would then be no unbalanced force acting yet the ball would continue to move. As the ball continued to move to the right, an unbalanced force in the opposite direction would be produced and this in turn would cause the ball to decelerate and finally to stop. The whole process would be repeated in reverse and the ball would continue vibrating to and fro at a certain *frequency*.

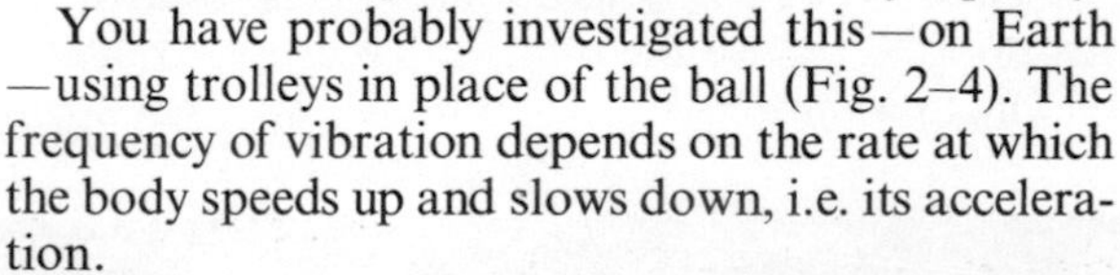

You have probably investigated this—on Earth—using trolleys in place of the ball (Fig. 2–4). The frequency of vibration depends on the rate at which the body speeds up and slows down, i.e. its acceleration.

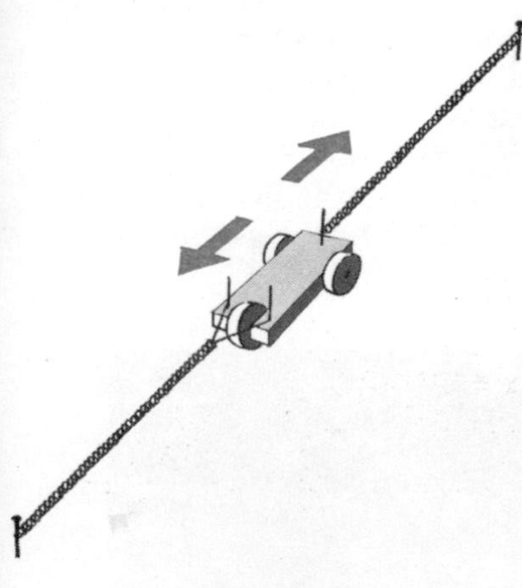

Fig. 2–4

By using the same springs each time, the frequency of vibration could be found for several trolleys stacked on top of one another. The results (Fig. 2–5) show that the frequency and thus the acceleration of the trolleys decrease as the number of trolleys increases. We give the name *inertia* to this 'reluctance' of the trolleys to have their motion altered. Inertia, then, is related to the 'number of trolleys' or the 'amount of matter' or the *mass* which is being moved. A vibrating device such as this or the wig-wag machine (Fig. 2–6) could therefore be used to find the mass of a body. Mass, or inertial mass as it is sometimes called, is simply a measure of the inertia of a body. The unit of mass, the kilogram, is the mass of a certain cylindrical piece of platinum-iridium which is kept in Sèvres near Paris. It is for most purposes equal to the mass of a litre (10^{-3} m^3) of water at 4 °C. Before using an inertial balance to measure mass we would have to calibrate it using 'known masses' based on the standard kilogram.

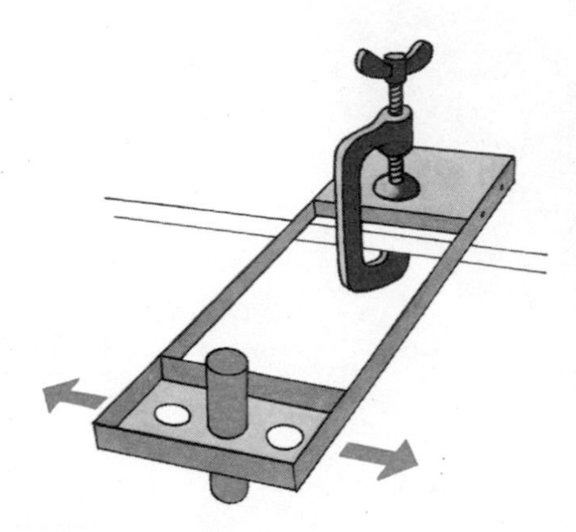

Fig. 2–6

Fig. 2–7

Force

When two identical elastic threads are used side by side to pull a trolley, they exert twice as much force as one thread. Three threads (Fig. 2–7) exert three times the force and so on. The ticker tapes produced show that the acceleration of the trolley is directly proportional to the force.

$$a \propto F \qquad (m \text{ constant})$$

When a given force is used to pull several trolleys, i.e. several units of mass, the acceleration is inversely proportional to the mass.

$$a \propto \frac{1}{m} \qquad (F \text{ constant})$$

If we decide to define the unit of force—the newton—as the force needed to make a mass of one kilogram accelerate at one metre per second, can you show that the two experimental results lead to the following relationship? (2.1)

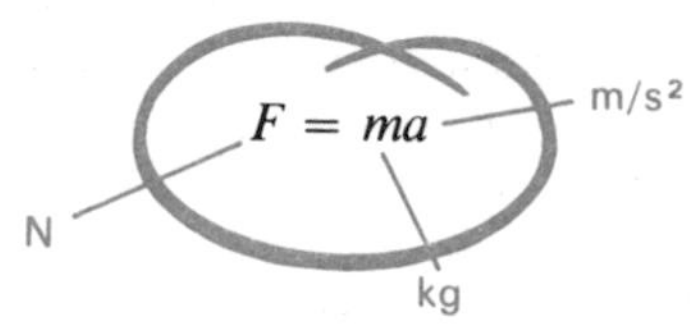

In this equation which relates force and acceleration, the acceleration takes place in the direction of the force. The equation which is, in essence, Newton's Second Law of Motion, makes it possible for us to calibrate a spring balance in newtons. For example, we can find the mass of a loaded trolley by using an inertial balance and 'known masses'. Then, using a ticker timer, we can find the acceleration of the trolley while it is being pulled by a spring balance. Admittedly, the trolley will have to have a fairly large mass and the spring balance will have to be reasonably sensitive . . . but it is possible!

3 cm 5 cm 7 cm

Fig. 2–8

If the 50 Hz ticker tape produced by a loaded trolley of mass 3 kg is represented in Fig. 2–8, what is the force pulling it? (2.2) What are some of the practical snags in this experiment (2.3)

Vectors

Forces can be added vectorially using the tip-to-tail vector rule. Fig. 2–9 shows how two water skiers cause a resultant force R to act on a power boat (Fig. 2–10).

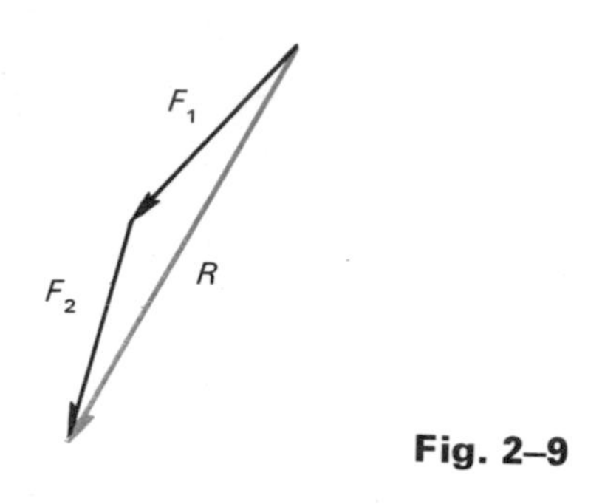

Fig. 2–9

Fig. 2–10

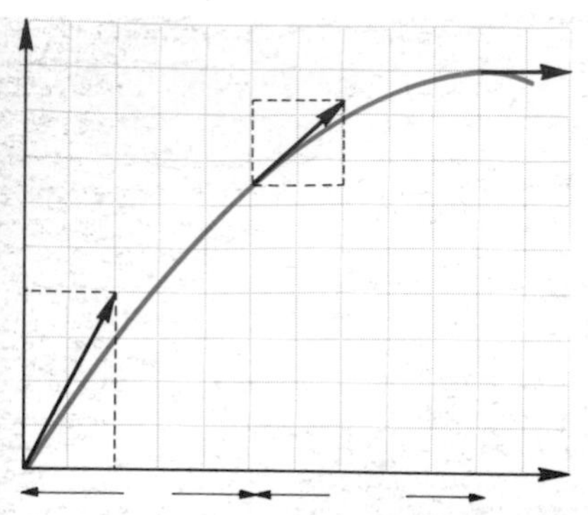

Fig. 2–11

It is also interesting to look at the vector representing the change of velocity when a ball is projected into the air. The direction of the velocity at any point is the direction of the tangent to the path of the ball (Fig. 2–11). If we ignore air resistance, the horizontal component of velocity will remain constant throughout. Taking $\boldsymbol{v}_1$ as the velocity of projection and $\boldsymbol{v}_2$ the velocity after a time Δt, we can calculate the change in velocity $\Delta \boldsymbol{v}$ as shown in Fig. 2–12.

Note that the change in velocity is directed *downwards*. In other words the change in velocity is in the direction of the force acting on the ball. As this change in velocity took place in time Δt, the average downward acceleration will be $\Delta \boldsymbol{v}/\Delta t$. The acceleration is therefore in the same direction as the force, as we might expect from Newton's Second Law.

Fig. 2–12

A similar change in velocity is shown between $\boldsymbol{v}_2$ and $\boldsymbol{v}_3$, which is the velocity when the ball is at its highest point. Again the direction of the change of velocity is downwards. If Δt is the same as before, the length of the $\Delta \boldsymbol{v}$ vector will be the same; that is, the acceleration of gravity $\Delta \boldsymbol{v}/\Delta t$ is constant (Fig. 2–13).

Fig. 2–13

MEASURING MOTION

When Newton set down his Second Law, he spoke of the 'change of motion'. To us today this seems a vague expression, although the meaning was clear to Newton's contemporaries. How can we measure motion?

Your experiments with trolleys showed that in any interaction the total *mass* × *velocity* (labelled *momentum*) remained the same. If momentum is conserved in every interaction, it would seem a reasonable quantity to use to describe motion.

Suppose a driver is pushing his broken-down car along a level road at a steady speed u (Fig. 2–14).

Fig. 2–14

→ v after time t

Fig. 2–15

The force which he exerts on it exactly equals th[e] frictional forces. A volunteer arrives and exerts a[n] additional force F which causes the car to accelerat[e] uniformly to a speed v in time t seconds (Fig. 2–15). The acceleration $= (v-u)/t$. If the car has a mas[s] m, the unbalanced force F is given by

$$F = ma = \frac{m(v-u)}{t} = \frac{mv-mu}{t}$$

The unbalanced force is then equal to the chang[e] of momentum per second; that is

unbalanced force = rate of change of momentum

This is another way of stating Newton's Secon[d] Law. The change of momentum—a vector quan[-]tity—is in the direction of the force.

Newton's Third Law can also be stated in term[s] of momentum. 'Action and reaction are equal an[d] opposite' can be deduced from the experimenta[l] result that momentum is conserved in an inter[-]action between two bodies.

On an air-track vehicle supported on a cushio[n] of air, the friction is negligible. Fig. 2–16 shows th[e] result of a moving vehicle A colliding with a[n] identical stationary one B so that they stick togethe[r] and move off with the same velocity. By measurin[g] the separation of the images before and after th[e] collision, we see that the velocity has been reduce[d] to half. The total momentum is thus conserved.

before	after
mx	$(m+m)\frac{1}{2}x = mx$

If the actual collision takes place in a tim[e] interval Δt and the *average* force acting during th[at] time is F, the product $F\Delta t$ is called the impulse o[f] the force or simply the *impulse*. This impulse give[s] the second vehicle B a velocity of $\frac{1}{2}x$ (i.e. $\Delta v = \frac{1}{2}x$) and *reduces* the velocity of the first vehicle A b[y] $\frac{1}{2}x$ (i.e. $\Delta v' = -\frac{1}{2}x$).

Fig. 2–16

Average force acting on B during collision is

$$F = ma = \frac{m\Delta v}{\Delta t} = \frac{m(\frac{1}{2}x)}{\Delta t} = \frac{\frac{1}{2}mx}{\Delta t}$$

Average force acting on A during collision is

$$F' = ma' = \frac{m\Delta v'}{\Delta t} = \frac{m(-\frac{1}{2}x)}{\Delta t} = \frac{-\frac{1}{2}mx}{\Delta t}$$

Comparing these expressions, we see that $F' = -F$. This means that these forces are the same size but act in opposite directions. It is important to realize that these forces act *on different bodies*.

A exerts a force F on B
B exerts a force $-F$ on A

Remember that Newton's Third Law does not apply to equal and opposite forces acting on *one* body. These are simply balanced forces.

As the time interval Δt for the collision illustrated above is the same for both vehicles, B receives an impulse ($F\Delta t$) which makes it accelerate and A receives an impulse ($-F\Delta t$) of the same size but oppositely directed which slows it down. In each case the impulse is, of course, equal to the change of momentum.

$$F\Delta t = m\Delta v$$

$$F'\Delta t = m\Delta v'$$

If a rocket is accelerating, the force producing its acceleration is due to the 'high speed gas' which it ejects (Fig. 2–17). Electric rockets are being developed for use in spacecraft. In SERT 2 (Space Electric Rocket Test) 33 300 solar cells provide about a kilowatt of power to operate the rocket's

Fig. 2–18

ion motor (Fig. 2–18). Thrust is generated by ionizing mercury vapour. The ions are then accelerated, neutralized and expelled at about 22 000 m/s (50 000 miles per hour). Even so, the thrust produced is only about 30 millinewtons! Ion rockets may be used to make a slight alteration in the position of an orbiting satellite or even to propel spacecraft to distant planets.

A force of reaction, similar to that which drives a rocket, is exerted on the nozzle of a fire hose (Fig. 2–19).

Fig. 2–17

Fig. 2–19

Everything is Squashy

There is no such thing as a perfectly rigid body. Even a fly landing on a steel beam deforms it to some extent.

Would you prefer to jump from a wall 3 metres high into sand or on to a concrete pavement? (2.4) Would you keep your legs straight or bend your knees? (2.5)

To stop your movement a force must be exerted on you. This may be a large force acting for a short time (as with the concrete), or a smaller force acting for a longer time (as in the case of the sand). On landing you would be travelling at about 8 m/s. If it were possible to keep your knees straight and to land on the concrete so that the total deformation (of the concrete, your shoes, and you) was only about 1 cm, you would have to come to rest from 8 m/s in that distance. Your average deceleration would be 3200 m/s^2, which is 320 times the acceleration due to gravity (i.e. 320 g!). If your mass was 50 kg, the concrete would exert a force on you of

$$50 \times 3200 = 160000 \text{ newtons (about 16 tonnes force!)}$$

If, on the other hand, you were to land on sand and sink, say, 20 cm into it, and if you bent your knees so that your body was gradually lowered another 60 cm, your average deceleration would be 40 m/s^2 or 4g. This corresponds to a force of 2000 newtons, which you could withstand without breaking any bones!

From these figures you can also find the duration (Δt) of the impact in each case, and then check that the impulse ($F\Delta t$) is equal to the change of momentum ($m\Delta v$).

mass (kg)	change of velocity (m/s)	average force during impact (N)	duration of impact (s)
50	8	160 000	0.0025
50	8	2 000	0.2

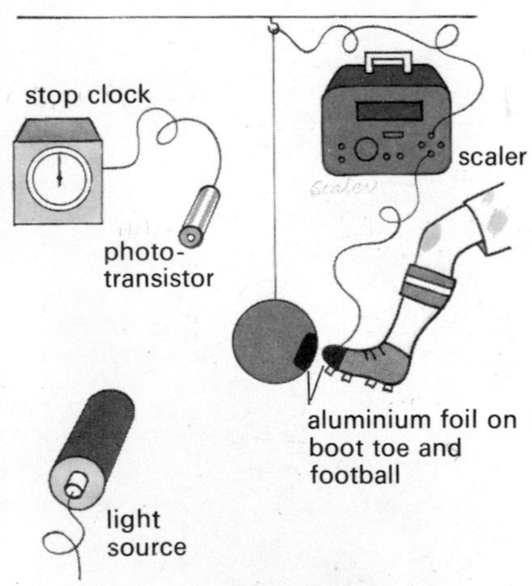

Fig. 2–20

The apparatus illustrated in Fig. 2–20 could be used to find the impulse and the average force exerted when a football is kicked. The ball is suspended by a fine wire which makes contact with a piece of aluminium foil glued to the ball. Another piece of foil is glued to the boot. When the ball is kicked the time of contact Δt is recorded on an electronic clock or scaler.

Another timing mechanism is used to find t time it takes for the ball to pass through a beam light. From this time interval T and the diamet of the ball D, the speed of the ball v can be calc lated, $v = D/T$. If the ball were accelerated fro rest to velocity v in a time Δt, then the increase speed in that time would be $v - 0 = \Delta v$.
If the mass of the football is m, we have the

$$\text{impulse} = F\Delta t = m\Delta v$$

The average force acting during the kick can al be calculated.

$$\text{Average force} = \frac{m\Delta v}{\Delta t}$$

Of course the force is far from constant duri such an experiment. It could vary as shown Fig. 2–21. The impulse or change of momentum then equal to the total area under the graph.

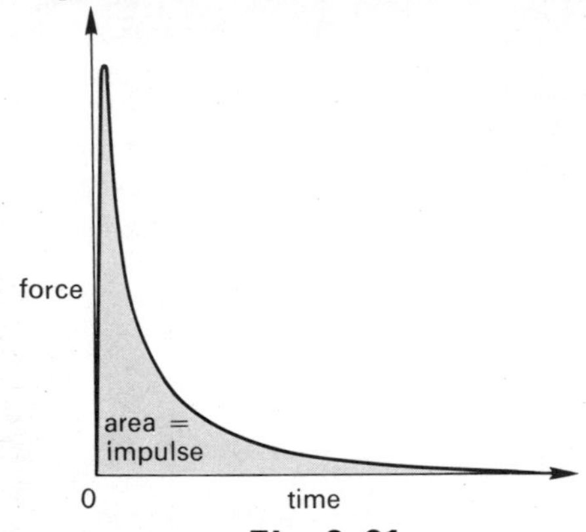

Fig. 2–21

A new type of glass which stretches on imp has been developed in Belgium. If windscree were made of this glass the damage caused collisions could be reduced. Fig. 2–22 shows t effect of a (simulated!) head striking the glass high speed. The time of the impact is increas thus greatly reducing the maximum force exert

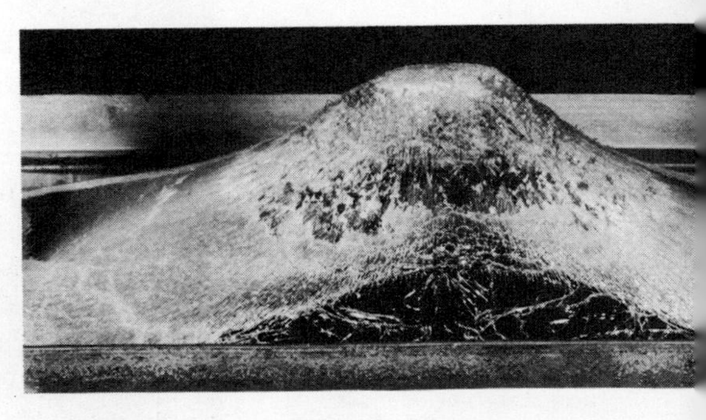
Fig. 2–22

Fig. 2–23

Discuss some other ways in which the force exerted on a driver in a crash can be reduced. (2.6)

IDEAL WORLD

It is sometimes claimed that the physicist has his head in the clouds. He deals with point masses which don't exist, with ideal gases which don't exist, with evacuated space and friction-free motions which are virtually unattainable. Why doesn't he come to terms with the *real* world—the world with air resistance, friction, gravity, humidity and the hundred and one other things that provide excuses for his experiments not working? The simple answer is that the physical world is so complex that it is very difficult if not impossible to make accurate predictions about many events in it. The motions of the stars and planets can be precisely forecast, yet no one would dare predict exactly where a stone thrown down a rocky hillside would eventually come to rest. Physics is, however, in a more fortunate position than many sciences, for it deals with variables that can often be fairly easily controlled.

In the social sciences the task of isolating relevant factors is much more difficult. Sometimes a set of results appears to indicate a causal relationship between two variables. Often, however, other factors come to light which throw doubt on the validity of this relationship. A few years ago, for example, experiments were carried out to see if the accident rates in cities would be affected if drivers used dipped headlights at night. The results looked promising and fewer accidents were recorded. Later the Road Research Laboratory published a report which showed that there were other factors which could have contributed to the reduction in accidents. They were the introduction of radar speed checks, improved street lighting, and very severe weather conditions which reduced speeds during the experimental period. As the dipped headlamps experiment was coupled with a road safety campaign, it was even more difficult to assess the direct effect of using dipped headlamps. In such experiments it is often impossible to ensure that no other changes which might affect the results take place.

Let us return to physics and to the stone on the hillside. If the same stone had been dropped down a well its motion could have been fairly accurately predicted. With an evacuated well the prediction could be even more accurate! By carefully studying all the factors that affect the motion of a falling stone and by altering each in turn, various relationships can be found. When these are known, we can make predictions which will approximate closely to the observed results. Physics is sometimes called an exact science but it would perhaps be better to describe it as dealing with close approximations.

Momentum in the Real World

Our experiments with trolleys, pucks, and air tracks lead us to believe that momentum is conserved in an interaction between two bodies. However, when we look at collisions in the High Street, momentum does not appear to be conserved. When a tanker collides with a café (Fig. 2–24), the café doesn't move off. Where has the momentum gone?

Fig. 2–24

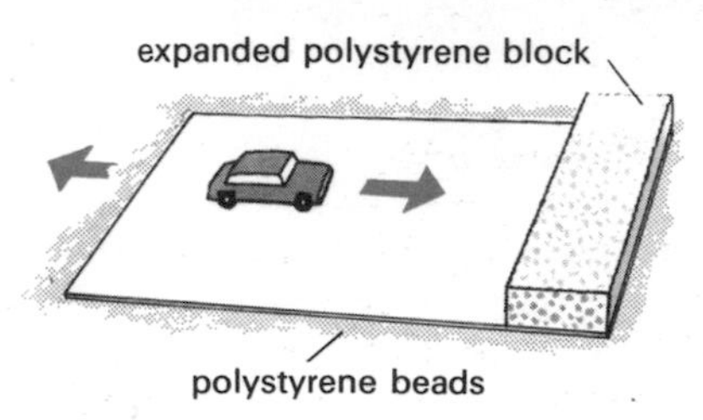

Fig. 2–25

Fig. 2–25 shows an analogue which might strengthen your faith in the laws of physics! A piece of expanded polystyrene is glued to the end of a length of thin cardboard which rests on polystyrene beads. The board moves off to the left when a toy car, with its engine revving, is placed on it. The car then moves off along the board (or the board under the car) until it collides with the polystyrene. The moving board and the car then come to rest as a result of the collision.

If Newton's Third Law is valid, the tanker must have exerted a force on the Earth's surface when it accelerated and then an oppositely directed force on it when it collided with the café. Each collision on the Earth's surface must alter its motion slightly, but of course the mass of the Earth is so great that even the biggest man-made collision is unlikely to make a detectable change in the Earth's motion. You must decide whether the evidence obtained from the experiments you have conducted in the laboratory justifies this faith in the conservation of momentum.

THE EARTH'S FIELD

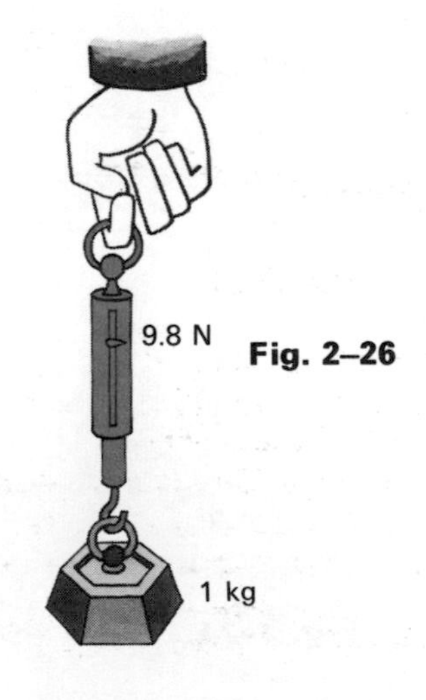

Fig. 2–26

We have seen that the mass of a body can be measured by comparing, on an inertial balance, the inertia of the body with the inertia of a known mass. This measured quantity is sometimes called *inertial mass.* It has nothing to do with gravity and could be measured equally well in a space ship.

A spring balance calibrated to read 1 N when it accelerates a mass of 1 kg at 1 m/s^2 shows that the Earth's gravitational field pulls the 1 kg mass downward with a force of 9.8 newtons (Fig. 2–26). If we now attach additional masses to the spring balance until it reads double the force, i.e. 19.6 newtons, we say that we have attached a mass of 2 kg. Mass measured in this way is sometimes

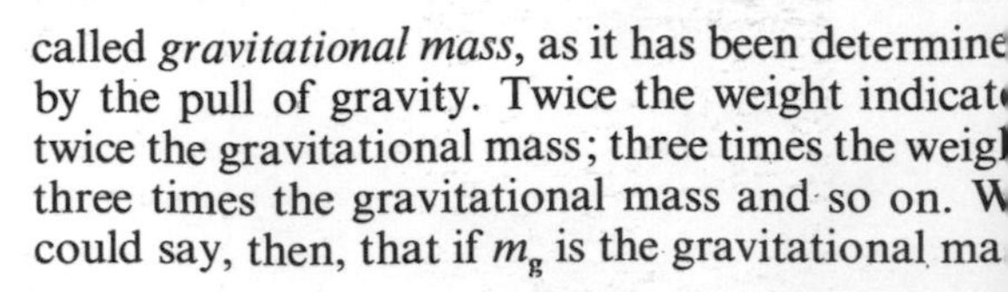

called *gravitational mass*, as it has been determine by the pull of gravity. Twice the weight indicat twice the gravitational mass; three times the weig three times the gravitational mass and so on. W could say, then, that if m_g is the gravitational ma

$$\text{gravitational attraction} = g \times m_g$$

where g is called the *gravitational field strength*, i. the force per unit gravitational mass. Fro Newton's Second Law we know that the acceler tion of a freely falling body is given by

$$a = \frac{F}{m_i}$$

where F is the gravitational attraction and m_i t inertial mass. From these equations we see that

$$a = g\left(\frac{m_g}{m_i}\right)$$

The acceleration of a freely falling body is ther fore directly proportional to its gravitational ma and inversely proportional to its inertial mass.

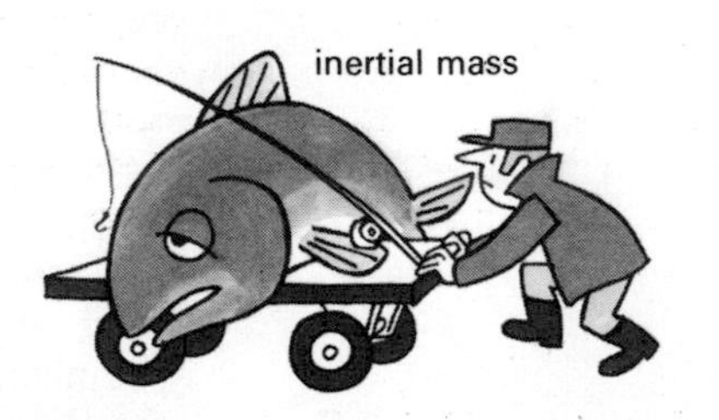

Fig. 2.27

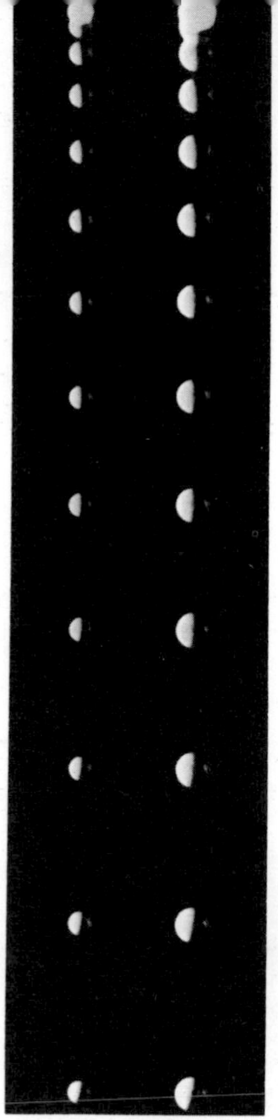

Fig. 2–28

Gravity and Galileo

The Greeks said that it was 'obvious' that heavy bodies fell more quickly than light ones. Nearly two thousand years later Galileo disagreed! His claim might be established theoretically or empirically. You must decide which of the following methods you think more convincing.

First, Galileo's case might be argued logically. If two identical small stones which fall at the same rate are tied together they would then be equivalent to one large stone. They would fall side by side at exactly the same rate whether or not they were tied. A small stone or a large stone will therefore fall at the same rate. You might like to argue the case for a small stone tied to a large stone assuming first that when separate the larger stone falls more quickly.

Alternatively you could conduct the 'Leaning Tower' experiment or a modern version of it. A multi-flash photograph of two spheres, one bigger and heavier than the other, is shown in Fig. 2–28. The two spheres fall together with the same acceleration.

If then all bodies fall with the same acceleration and if

$$a = g\left(\frac{m_g}{m_i}\right)$$

the ratio m_g/m_i must be constant.

By using the same lump of platinum iridium to define the unit of gravitational mass and inertial mass, we make the ratio m_g/m_i equal to 1. Inertial mass and gravitational mass are therefore the same and in future we will refer only to *mass*.

The above equation now becomes $a = g$, where a is the acceleration of gravity, $9.8\,\text{m/s}^2$, and g is the gratvational field strength $9.8\,\text{N/kg}$. *Can you show that the units are also equivalent? (2.7)*

PULLEYS AND PEGS

If a horizontal force of 10 N acts on a mass of 5 kg resting on a smooth horizontal surface, the mass will be given an acceleration of $F/m = 2\,\text{m/s}^2$ (Fig. 2–29). If the mass consisted of, say, two trolleys tied together by a light string as shown in Fig. 2–30, there would be a tension in the string during acceleration. If this tension exerted a force of x newtons on the 3 kg trolley, the unbalanced force causing acceleration would be $(10-x)$ newtons. From Newton's Second Law this force would be equal to the product $m \times a$:

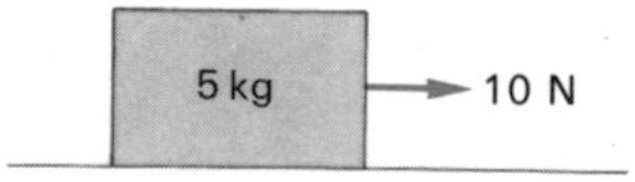

5 kg
10 N

Fig. 2–29

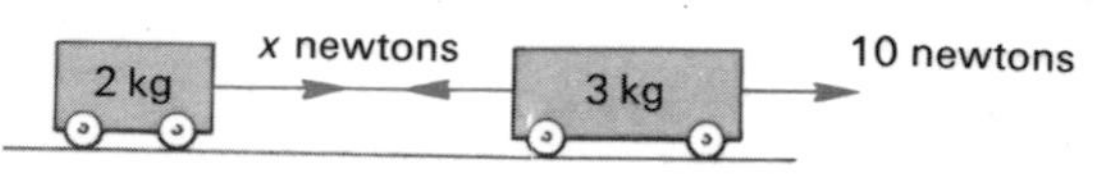

Fig. 2–30

$$10-x = 3a \qquad (1)$$

The tension in the string would also exert a force of x newtons on the 2 kg trolley. As this is the only unbalanced force acting on this trolley, we have

$$x = 2a \qquad (2)$$

From equations (1) and (2) we see that

$$10 = 5a$$

$$\Leftrightarrow \quad a = 2\,\text{m/s}^2$$

The acceleration of the two bodies of total mass 5 kg is therefore the same as the acceleration of the single 5 kg mass. The tension in the string is, of course, much less than 10 newtons. From equation (2) we see that it is 4 newtons.

Now imagine that the string which connects the two trolleys runs over a frictionless pulley and that the string which exerts the accelerating force passes round a smooth peg. The arrangement is shown in Fig. 2–31. For our purposes the 'frictionless pulley' and the 'smooth peg' are identical: they merely change the direction in which the force acts, without altering the *tension* in the strings.

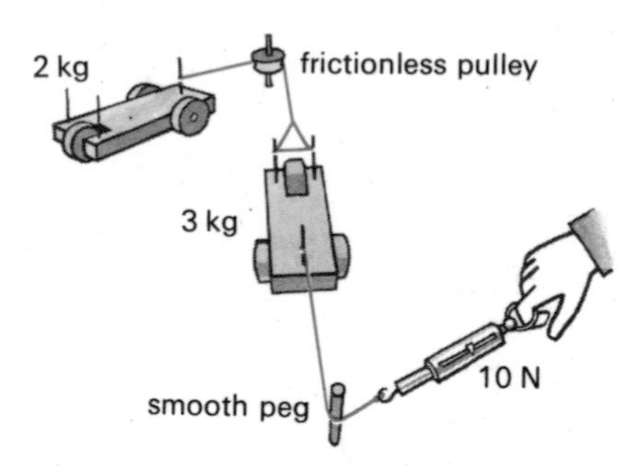

Fig. 2–31

As in the first two cases, the rate of change of speed in the arrangement of Fig. 2–31 is $2\,\text{m/s}^2$. The *direction* in which the accelerating (unbalanced) force acts and the directions in which the two trolleys are accelerated are, however, all different.

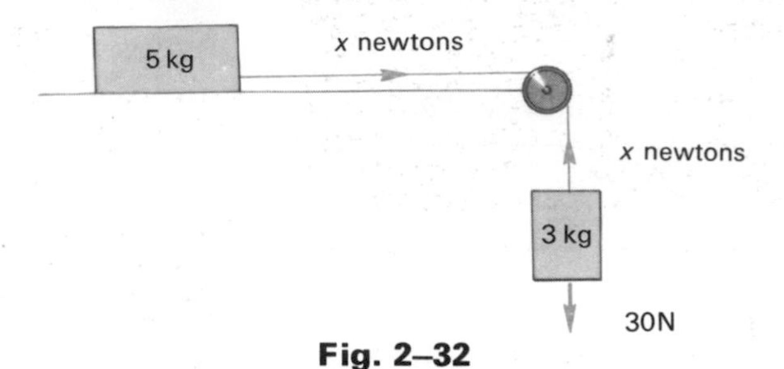

Fig. 2–32

In Fig. 2–32 a 5 kg mass is being pulled along a smooth horizontal surface by a thread which passes over a frictionless pulley. The accelerating force is 30 N produced by the Earth's pull on a 3 kg mass.

First, consider the unbalanced force on each mass.

(a) For the 3 kg mass, the unbalanced force is $(30-x)$ newtons. From Newton's Second Law

$$(30-x) = 3a$$

(b) For the 5 kg mass, the unbalanced force is x newtons

$$x = 5a$$

From these equations we have

$$30 = 8a$$

$$\Leftrightarrow \quad a = 3.75 \text{ m/s}^2$$

SUMMARY

Newton's Laws of Motion.

Second Law $a = \dfrac{F}{m}$

First Law is a special case of the Second Law when $F = 0$, $a = 0$; that is, the body continues at a constant speed in a straight line.

Third Law If $F_1 = -F_2$, then, from

$$F = ma = \frac{m\Delta v}{\Delta t}$$

we have $m_1\Delta v_1 = -m_2\Delta v_2$ after time interval Δt.
Momentum mv is conserved.

Impulse $F\Delta t = m\Delta v$ = change of momentum.

Both impulse and momentum are vector quantities.

PROBLEMS

2.8 What are the readings X and Y on the spring balan shown in Fig. 2–33?

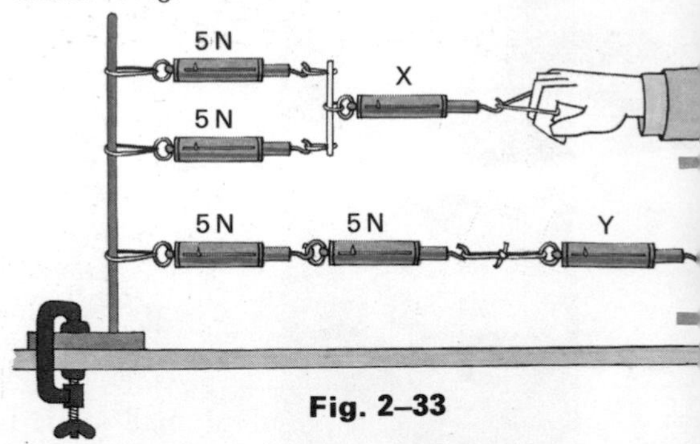

Fig. 2–33

Which of the two situations is represented by the vector Fig. 2–34?

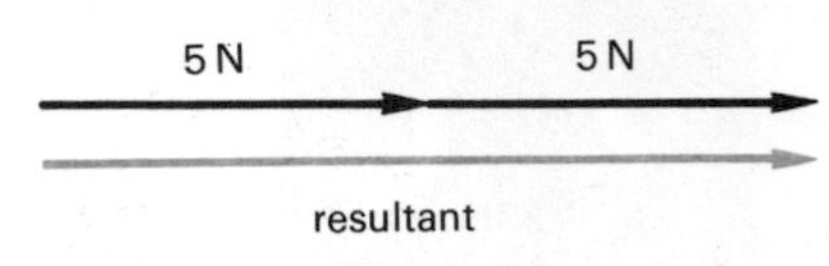

Fig. 2–34

Is the resultant equal to X or Y?

2.9 Discuss any difference in behaviour when a shee paper and a ball bearing of the same mass are dropped r the Earth's surface and near the Moon's surface.

2.10 A girl on skis has a total mass of 50 kg. Sh travelling at 21 m s^{-1} when she strikes a snowdrift and co to rest in 3 seconds. What is the average force exerted her? How far does she travel in the 3 seconds?

2.11 In 1966 the mass of the *Agena* rocket was founc accelerating it with a push from a *Gemini* spacecraft. thrust produced by *Gemini's* motor was known to be 89 and when they fired for 7 seconds *Agena* and *Ge* speeded up together by 0.93 m s^{-1}. If *Gemini's* mass 3360 kg, find the mass of *Agena*.

2.12 A toy rocket has a mass of 0.15 kg. It ej 30 grams of fluid per second at a velocity of 100 m If the rocket is fired horizontally, find its *initial* accelerat What would be its *initial* acceleration if fired vertic upwards?

2.13 When a ball is thrown vertically upwards in a vac it takes the same time to rise as to fall. If it were throwr in air there would be some drag. Discuss whether or not ball would now rise and fall in equal times.

2.14 The value of g varies slightly over the Earth's surf Discuss how the range obtained in the long jump or putt will be affected by competing in different parts of world.

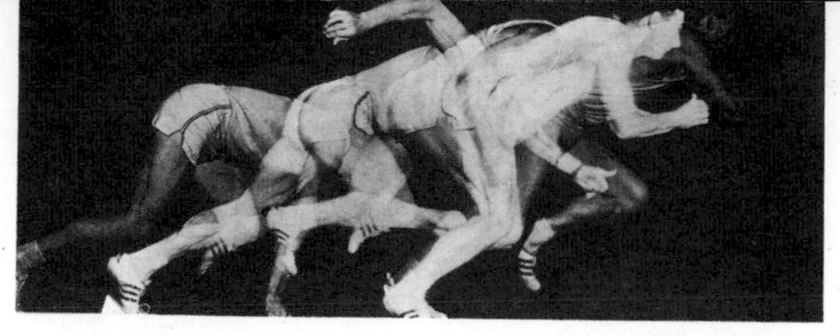

Fig. 2–35

2.15 Discuss why the body must be in a slanting position for a sprint start (Fig.2–35).

2.16 A man stands on bathroom scales placed on the floor of a lift. When the lift is at rest the scales read 80 kg and when it is accelerating uniformly upwards the reading is 96 kg. Find the acceleration of the lift. If the man throws a ball vertically upwards at 4 m s^{-1} with respect to this accelerating lift, how long will it take to return to his hands?

2.17 A car accelerates uniformly from rest to 30 m s^{-1} (67 miles h^{-1}) in 10 seconds. If the driver has a mass of 50 kg, estimate the force his back will exert on the seat.

2.18 A uniform bar AB is 5 m long and has a mass of 10 kg. It is hinged to a wall at A and held in a horizontal position by a rope tied to B. Find the tension in the rope which makes an angle of 150° with the bar.

2.19 Show that when a ball is projected horizontally in the Earth's field its path is parabolic, i.e. the equation is of the form $y = kx^2$.

2.20 A plane flying at 300 m s^{-1} over the sea drops a flare from a height of 2 km. Neglecting air resistance, find the time taken for the flare to reach the water. If P is a point on the water surface immediately below the point of release, how far from P will the flare land?

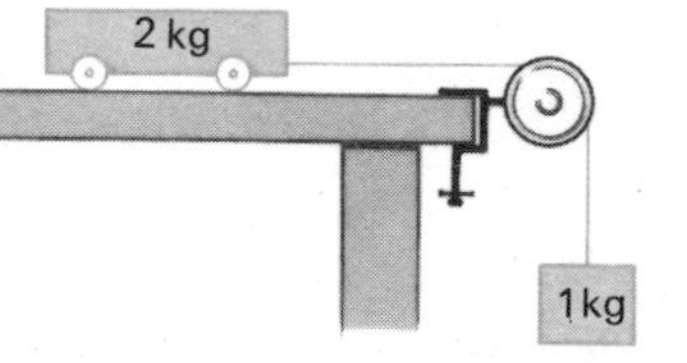

Fig. 2–36

2.21 A 2 kg trolley is placed on a friction-compensated runway. A thread tied to the trolley runs over a frictionless pulley and is then attached to a 1 kg mass (Fig.2–36). What will be the acceleration of the trolley?

If a 2 kg mass were attached to a 2 kg trolley, what would be the acceleration? If a 100 kg mass were attached to a 0.1 kg trolley, what would be the approximate acceleration?

2.22 A string 5 metres long is tied to two nails 4 metres apart on a horizontal beam. A 10 kg mass is then tied to the string 1 metre from one end. Find the tension in each part of the string.

2.23 A 1 kg block of wood slides at constant speed down a friction-compensated runway which is 2 metres long and raised 800 mm at one end. Find the frictional force between the block and the runway.

2.24 A truck is pulled at a constant speed along a railway line by a horizontal rope which lies at an angle of 20° to the direction of motion. If the tension in the rope is 600 N, find the work it does in moving the truck 100 metres along the line.

2.25 What force is needed to accelerate a 600 kg Mini uniformly from rest to 20 m s^{-1} in 15 seconds? How far will it travel in that time?

If the clutch and brake pedals are then suddenly operated so that the car comes to rest in 60 m, what is the effective force of friction between the road and the wheels?

2.26 Benjamin Franklin once objected to the idea that light consisted of particles on the grounds that such a particle travelling at 3×10^8 m s^{-1} would have the same impact as a 10 kg cannon ball fired at 100 m s^{-1}. If this were true, what would be the mass of one of the particles?

2.27 Fig. 2–37 shows the result of a collision between two vehicles of mass 2 units and 3 units. They collide and move off again in opposite directions after the collision. Measure the distance gone in (say) 5 flash intervals and calculate, in arbitrary units, the momentum of each vehicle before and after the collision. Compare the total momentum before and after the collision. Then compare the *change* of momentum for each vehicle as a result of the impulses each receives. Remember to take direction into account. Finally, compare the average force acting on one vehicle with that acting on the other during the collision lasting Δt units of time.

Fig. 2–37

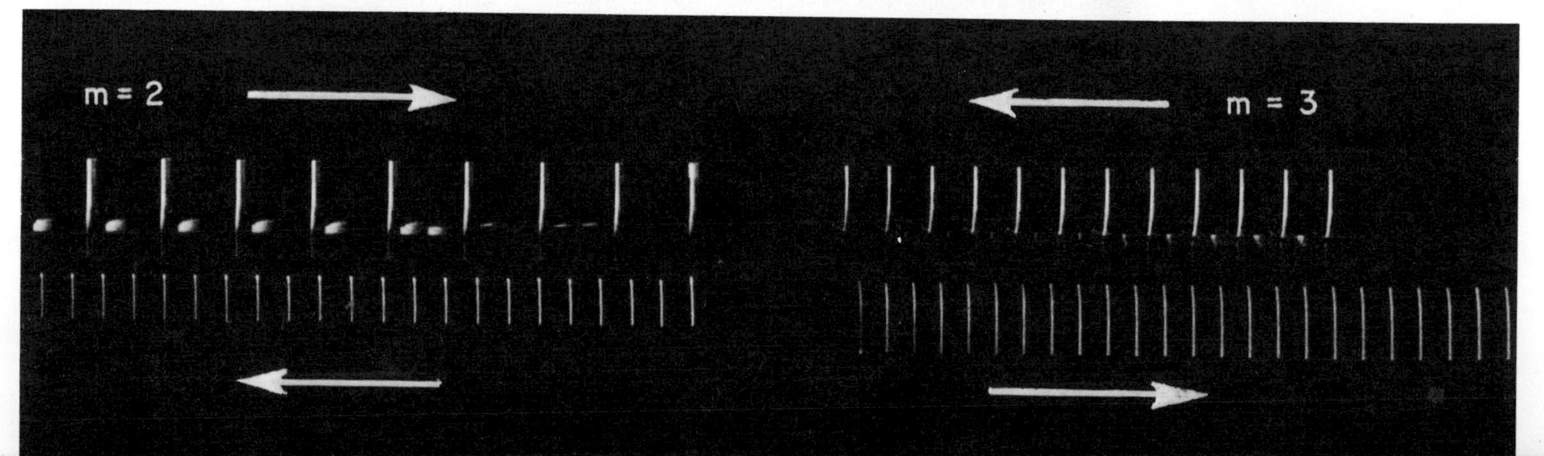

Fig. 2–38

2.28 Three of the four jets in a rotary garden sprinkler are sealed and the volume of water per second coming from the fourth jet is found to be 50 cm³ (Fig. 2–38). If the area of the jet is 10 mm², what is the speed of the water and the minimum force F needed to prevent rotation? (Density of water = 10^3 kg m^{-3}.)

How might you measure this force?

2.29 Find the impulse when a cricket ball of mass 0.16 kg travelling at 30 m s^{-1} is hit back towards the bowler at 20 m s^{-1}. If the impact lasts for 50 ms, find the average force applied.

2.30 A fireman plays a horizontal jet of water on to a wall. It strikes the wall at 5 m s^{-1} and bounces back at 1 m s^{-1}. If the mass of water hitting the wall every second is, on average, 20 kg, find the force exerted on the wall.

Fig. 2–39

2.31 A shell lying on a frozen loch explodes into three parts which fly off along the friction-free surface of the ice. Two of the pieces fly off at right angles to each other, a 1 kg piece at 8 m s^{-1} and a 400 g piece at 25 m s^{-1}. Find the mass of the other piece if its speed is 10 m s^{-1}. Show on a diagram the direction of the third piece.

2.32 A fly, flying horizontally, bumps into a fast moving train coming in the opposite direction (Fig. 2–39). The fly is stopped and its direction reversed. At one instant the fly must be stationary with respect to the Earth. At that instant it is in contact with the front of the train, so that the train must also be stationary with respect to the Earth. The fly has stopped the train. Do you agree? If not, why not?

F			
Δt	1	0.1	0.01

Fig. 2–40

2.33 A boy of mass 50 kg jumps in the air and lands with a speed of 4 m s^{-1}. If he bends his knees, he comes to rest in 1 second (Δt). Find his deceleration. What is the average force F acting on him? If he had bent his knees only very slightly, he might have come to rest in 0.1 seconds (Δt). Find the deceleration and force now. Finally, find the deceleration and force acting if he had kept his legs straight and come to rest in 0.01 seconds (Δt). Copy out the table shown in Fig. 2–40 and complete it. Can you discover a way of combining F and Δt so that the result is the same each time? What is this quantity called?

2.34 A 500 g football reaches 15 m s^{-1} by being in contact with a player's boot for 0.02 seconds. What was the average force exerted on the ball? What is the impulse (i.e. $F \times \Delta t$)?

2.35 How long will it take a force of 10 N to stop a mass of 2.5 kg which is moving at 20 m/s?

2.36 If a 1000 kg car accelerates at 2 m s^{-2}, what is the average force acting on it? If, after accelerating for 10 seconds, it crashes into a wall and comes to rest in 0.5 seconds, find the average force of resistance and the impulse.

2.37 A string is attached to a trolley and the other en after passing over a pulley, is tied to a metal block of mass (Fig. 2–41).

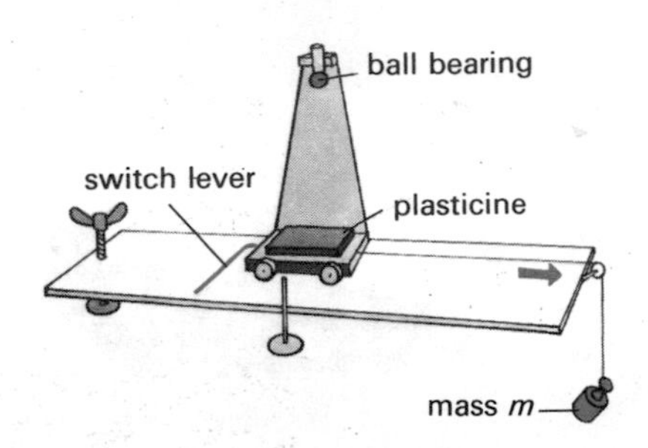

Fig. 2–41

As the trolley moves along a friction-compensated benc the ball-bearing is released and strikes the plasticine 10 c from a mark vertically below the electro-magnet. If the b falls 45 cm vertically, what is the acceleration of the trolle If the trolley and fittings have a mass of 1.8 kg, find t tension in the string and the mass m.

2.38 Two trolleys of mass 10 kg and 30 kg are placed friction-compensated slopes as shown in Fig. 2–42.

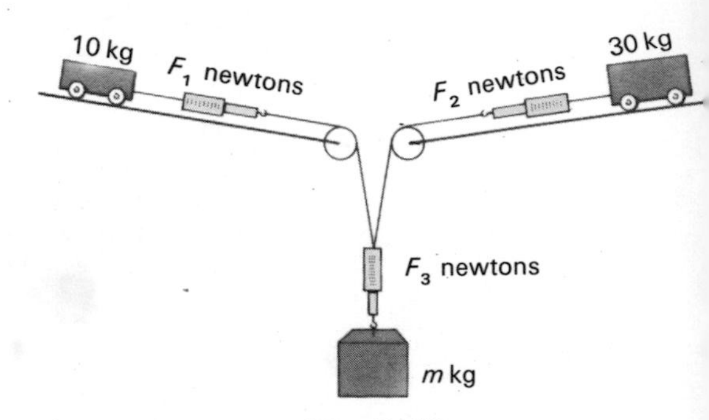

Fig. 2–42

They are attached to a mass of m kg by threads contain spring balances. When released the mass drops 1 metre 2 seconds. If the mass of the strings and balances can neglected, find the acceleration of the trolleys and readings on all three balances as the mass is dropping. W is the upward force acting on the mass? What is the dow ward force acting on the mass? What is the unbalan force acting on the mass? It is this unbalanced force wh causes the mass to accelerate. Using Newton's Second L find the value of m.

2.39 An unbalanced force of 10 N gives a mass of an acceleration of 10 m s^{-2}. It also produces an accel tion of 20 m s^{-1} when applied to a mass of y kg. If x ar are joined together, what acceleration would be produ by the same force?

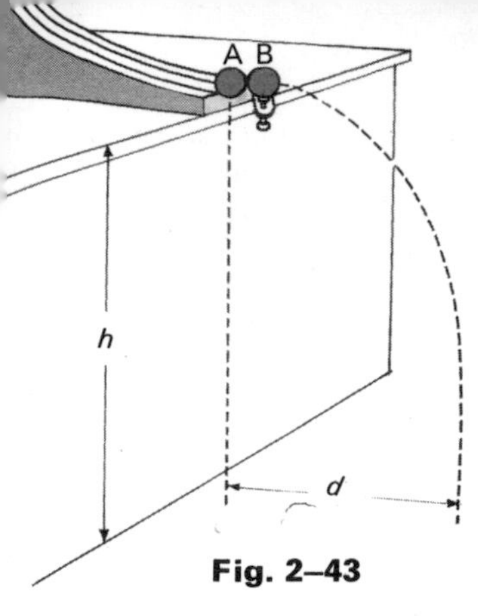

Fig. 2–43

2.40 In the experiment illustrated in Fig. 2–43, two identical ball bearings of mass 30 g are used. A rolls down the slope, strikes B, which is stationary, and then drops vertically to the floor. If h is 1.25 m and d is 1.5 m, find:
(a) the time taken for A to fall,
(b) the velocity with which B leaves A,
(c) the momentum of B as it leaves A,
(d) the impulse during the impact,
(e) the average acceleration of the ball if the impact lasts 50 microseconds,
(f) the average force acting during the impact, and
(g) the velocity of B just before it lands.
Give its magnitude and direction.

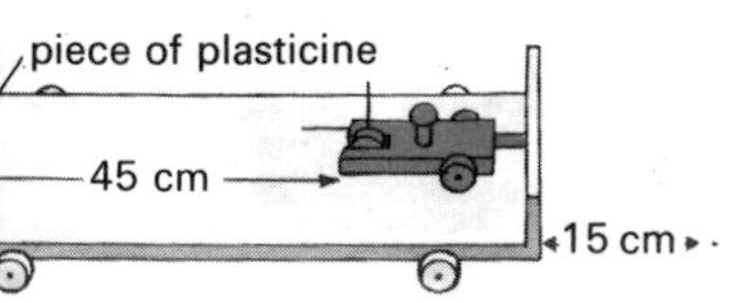

Fig. 2–44

2.41 A trolley fitted with a spring-loader plunger is placed at one end of a long trolley as shown in Fig. 2–44. A needle in the small trolley can penetrate a lump of plasticine fixed to the opposite end of the long trolley. When the plunger is released the long trolley moves 15 cm to the right. The distance from the needle to the plasticine is 45 cm. If the small trolley has a mass of 750 g, what is the mass of the long trolley? You may ignore frictional forces.

Think very carefully; this is not quite as simple as it looks!

2.42 An unbalanced force is applied for 2 seconds to a mass of 4 kg originally at rest. If the mass moves 10 metres in that time, find:
(a) the acceleration of the mass,
(b) the speed at the end of the two seconds,
(c) the impulse applied to the body, and
(d) the change of momentum.

Fig. 2–45

2.43 A boy and a girl of equal mass hang on either end of a rope which passes over a frictionless pulley (Fig. 2–45). If the boy starts to climb the rope who will reach the pulley first? Explain the answer you give.

2.44 A 1 kg trolley moving at 2 m s^{-1} collides and sticks to another 1 kg trolley originally at rest. At what speed do they move off?

The experiment is now repeated with a powerful magnet fixed to one end of each trolley so that they accelerate towards each other. How will this affect (a) their relative velocity before impact, (b) the total momentum after impact?

2.45 A boy of mass 50 kg is travelling at 3 m s^{-1} on a playground trolley of mass 20 kg. The boy jumps off the back of the trolley and in doing so applies a force to it for 0.15 seconds, thus giving it a forward impulse of 15 N s. Find:
(a) the average force applied to the trolley by the boy,
(b) the final speed of the trolley, and
(c) the horizontal velocity with which the boy lands.

2.46 A ball of mass 0.6 kg travelling at 10 m s^{-1} (east) strikes a wall perpendicularly and rebounds at 9 m s^{-1} (west) The impact lasts 0.05 seconds.
(a) What was the original momentum of the ball?
(b) What was the change of momentum of the ball?
(c) What was the impulse on the wall?
(d) What was the average force acting on the wall?
(e) What was the average force acting on the ball?

2.47 A football of mass 0.5 kg reaches a velocity of 20 m s^{-1} (east) after a kick lasting 0.02 seconds. Find:
(a) the impulse,
(b) the change of momentum of the ball,
(c) the average force exerted on the ball, and
(d) the displacement of the ball in 3 seconds.

What assumption must you make to answer the last question? Do you think the error would be insignificant or fairly large? Under what conditions could the error be very large?

2.48 Normally, the fuel in a rocket accounts for nearly all its mass. The payload can be compared to the shell of an egg and the fuel to the entire contents. The dummy *Apollo* capsule plus the shell of the second-stage rocket had a mass of 8×10^3 kg, whereas the combined lift-off mass of the capsule and *Saturn* two-stage rocket was 5.7×10^5 kg. If the thrust generated by the booster rocket was 6.7×10^6 N, find:
(a) the unbalanced force acting on the rocket and
(b) the vertical acceleration of the rocket initially. Neglect air resistance—but *not* gravity!

2.49 A skier of mass 60 kg runs into a bank of snow and is brought to rest in 2 seconds. If he was travelling at 12 m s^{-1} find:
(a) the average force exerted by the snow on the skier,
(b) the average force exerted by the skier on the snow, and
(c) the impulse.

2.50 A car of mass 1500 kg has a maximum braking force of 7000 N. Find:
(a) the maximum deceleration possible,
(b) the distance gone before stopping from speeds of 4.5 m s^{-1} (10 miles h^{-1}), 18 m s^{-1} (40 miles h^{-1}) and 45 m s^{-1} (100 miles h^{-1}).

2.51 A water skier is being towed by a boat moving at a constant speed of 10 m s^{-1}. The skier is moving on a straight track parallel to the boat's path but displaced from it so that the tow-rope makes an angle of 30° with the skier's forward direction. If the tension in the rope is 100 N, find the size of the force retarding the skier's forward motion.

Why is the tension in the rope greater than this force?
[S.C.E. (H) 1967]

2.52 A block of wood rests on a 'frictionless' horizontal surface. A rifle bullet is fired horizontally into the block and becomes embedded in it. What apparatus would you use and what measurements would you make to determine as accurately as possible the common velocity of the block and the bullet?

Using the following data, estimate the initial velocity of the bullet.

Mass of bullet = 0.010 kg
Mass of wooden block = 0.50 kg
Common velocity of block and bullet = 2.0 $m s^{-1}$.

[S.C.E. (H) 1968]

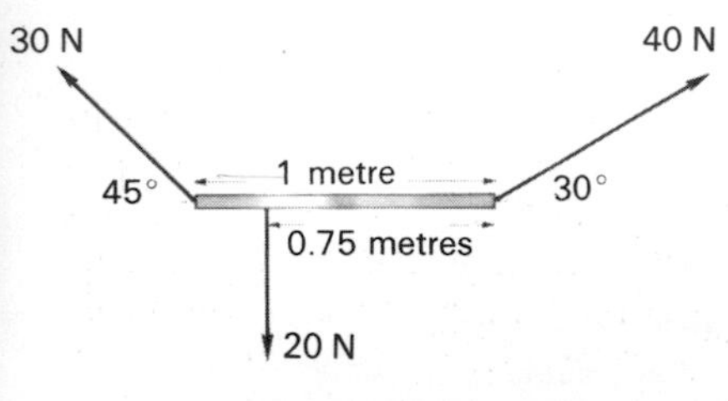

Fig. 2–46

2.53 The diagram (Fig. 2–46) represents a uniform rod of weight 10 N subject to the forces shown, all of which are in the same vertical plane.

State whether the rod is subject to (a) a resultant force and (b) a turning effect. In each case give *one* reason for your conclusion. [S.C.E. (H) 1969]

2.54 A mass of 1.0 kg hangs from a spring balance which is attached to the roof of a lift. The balance has a scale marked from 0 to 25 N.
(a) What will the balance read when the lift is at rest?
(b) How will this reading be affected when the lift is moving
(i) upwards with constant velocity;
(ii) downwards with constant velocity;
(iii) upwards with constant acceleration;
(iv) downwards with constant acceleration?

[S.C.E. (H) 1970]

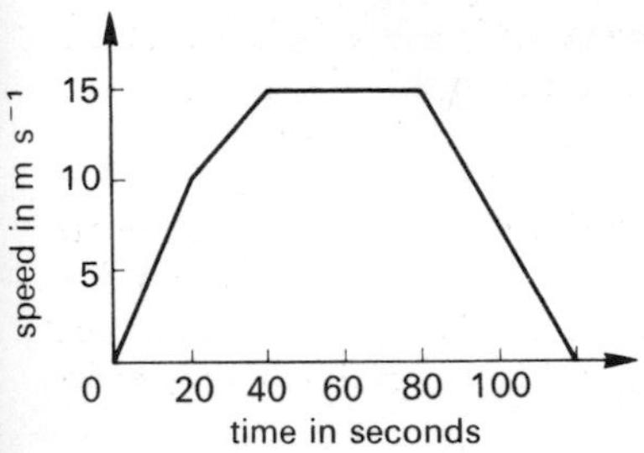

Fig. 2–47

2.55 The graph (Fig. 2–47) shows the speed of an electric train of mass 2.0×10^5 kg as it makes a journey between two stations on a horizontal track.
(a) Using suitable scales, draw accurate graphs to show how the following quantities vary during the journey:
(i) acceleration
(ii) accelerating force;
(iii) the total force applied to the train, assuming that the resistance due to friction, etc., amounts to 1.0×10^4 N and is independent of speed.
(b) Calculate the power output in kilowatts of the electric motors at the end of the 30th second.

[S.C.E. (H) 1970]

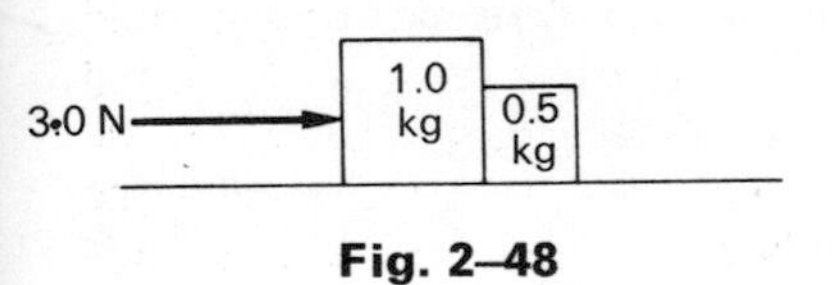

Fig. 2–48

2.56 (a) Two blocks of mass 1.0 kg and 0.50 kg rest on a horizontal surface as shown in Fig. 2–48. A force of 3.0 N then acts on the 1.0 kg block.

Neglecting friction, find:
(i) the acceleration of the blocks and
(ii) the horizontal force exerted on the 0.50 kg block by the 1.0 kg block.

If the 3.0 N force acted instead on the 0.50 kg block and in the opposite direction, explain whether the acceleration would be the same and find the horizontal force on the 1.0 kg block. Suggest a method of measuring the forces between the blocks.
(b) A car A, of mass 800 kg, travelling at 24 $m s^{-1}$, collides at right angles with a lorry B, of mass 4000 kg, travelling at 8 $m s^{-1}$ (Fig. 2–49).
Assuming that the two vehicles interlock on colliding, calculate the magnitude and direction of the velocity of the wreckage immediately after the collision. [S.C.E. (H) 1970]

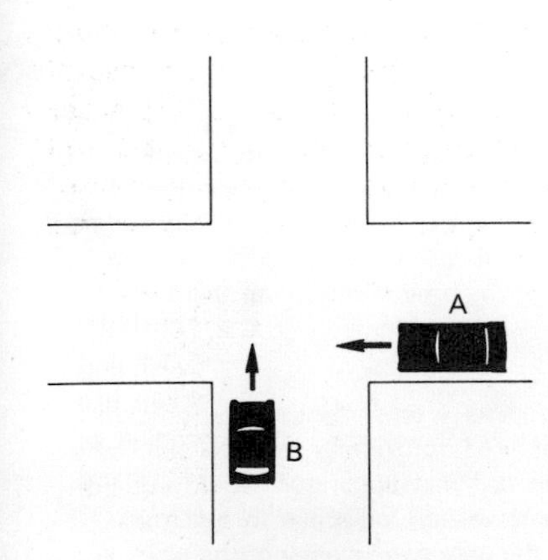

Fig. 2–49

2.57 The sketch (Fig. 2–50) is a copy of part of a photo-graph showing a ball-bearing falling through glycerine. The photograph was obtained by taking a series of exposures at regular time intervals.

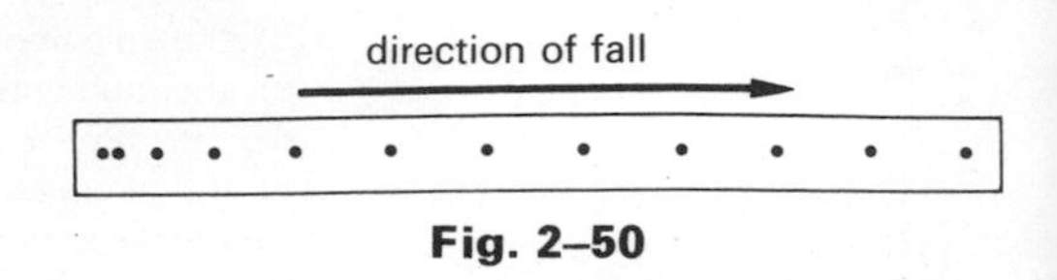

Fig. 2–50

(a) Explain the motion of the ball-bearing in terms of the forces acting on it.
(b) The ball-bearing is now dropped through the same depth of water. In what respects will its motion differ from that in (a)? [S.C.E. (H) 197

2.58 A ball is suspended inside a cylindrical can by string as shown in the sketch (Fig. 2–51).
The string is severed at X.

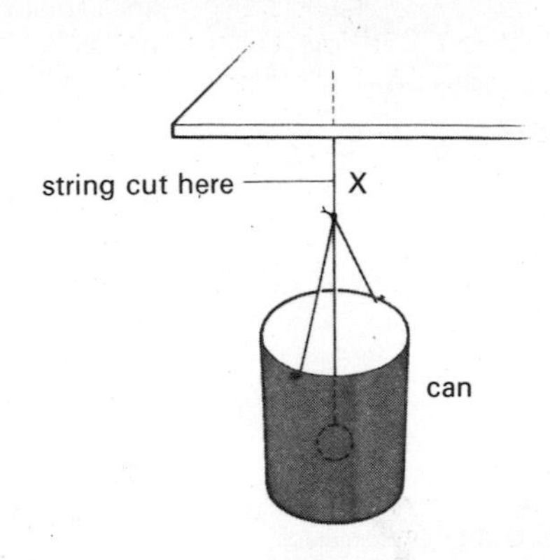

Fig. 2–51

(a) Ignoring air resistance, discuss and compare the subsequent motion of the ball and the cylinder as they fall.
(b) How does your answer to (a) help to explain the 'weightlessness' experienced by astronauts circling the Earth? [S.C.E. (H) 197

2.59 In a laboratory experiment, a 2 kg block of wood was pulled along a bench top by a horizontal force of 6 N. A paper tape attached to the block was drawn through a vibrator which made dots on it at equal time intervals. Part of the tape is shown in Fig. 2–52.

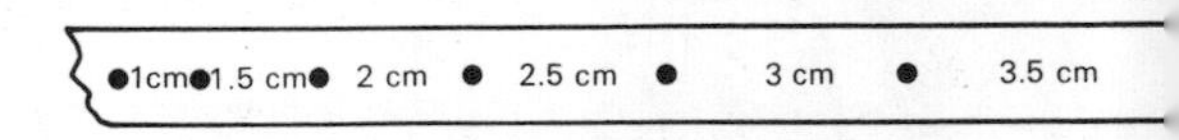

Fig. 2–52

If the vibrator made 20 dots per second, calculate:
(i) the acceleration of the block in m/s^2 and
(ii) the frictional force acting on the block.

[S.C.E. (H) 19

3 Energy

Fig. 3–1

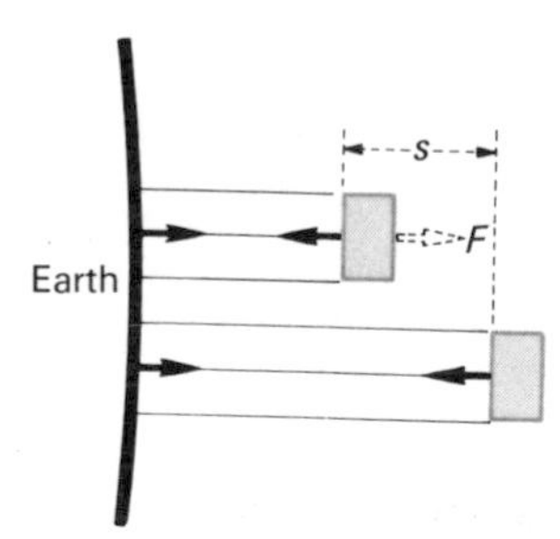

Fig. 3–2

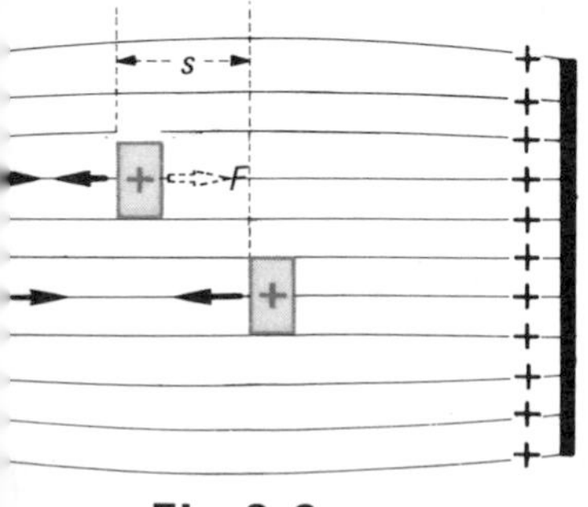

Fig. 3–3

Awake or asleep we are continually changing energy stored in our bodies to other forms. The air around us is normally colder than our skin temperature and so we lose energy by warming the surrounding air. Whenever we do manual work we are again transforming energy by moving objects against some resistance or other. We refer to the first form of energy transfer as *heat*. It can be measured by the product $cm\Delta T$, where c is the specific heat capacity, m the mass and ΔT the temperature change. The second we call *work* and it is measured by multiplying the force exerted by the distance moved in the direction of the force, i.e. Fs.

Fortunately most people enjoy their food—even school meals—and so have no trouble replacing this energy which is being continually transferred. Food, then, is our most important source of energy and its production, particularly in the developing countries, still constitutes one of the world's major problems. On average each of us takes in and gives out about one hundred joules per second. That is, we transfer energy at about the same rate as a 100 W lamp!

In the technologically developed countries other forms of energy are being used in ever increasing amounts. At present these countries are using fuels such as gas, electricity, oil, coal, and nuclear fuels at an average rate of about 5000 joules per second (5 kW) per head of population and there is a real danger that many of the fuels, particularly the fossil fuels, will soon be exhausted. New fuels are desperately needed and scientists are looking to atomic, geothermic, and solar energy as possible future sources. All our present needs could be supplied by the sun if only the energy could be harnessed. It is estimated that on average the sun supplies the required 5 kW to every three square metres of the Earth's surface.

STORING ENERGY

Small quantities of energy can be readily stored but there is no known way at present of storing the kind of quantities that are required for national demands. Pumped storage systems can be used to store some of the surplus energy available from a power station when the demand is smaller than the output. The energy is stored in the Earth's gravitational field by pumping water up to a reservoir, such as in the Cruachan scheme shown in Fig. 3–1.

When a body of mass m is raised at a constant speed through a small distance s by a force F $(=mg)$, the work done (energy transfer) is Fs. This energy, potential energy (P.E. or E_p), is stored in the Earth's gravitational field (Fig. 3–2).

$$E_p = Fs = mgs$$

This represents the *increase* in energy relative to an arbitrary starting level. If we choose the Earth's surface as our starting level, the distance s becomes the height h above the Earth's surface and we have

$$E_p = mgh$$

If the air resistance is ignored, the body when released would accelerate towards the Earth and have a kinetic energy $= mgh$ when it reached its original position.

Energy can also be stored in an electric field (Fig. 3–3). If a force F moves a positively charged body at a constant speed through a distance s, the work done is Fs. This energy is stored in the electric field. If the body were released and accelerates towards its starting point it would have a kinetic energy $= Fs$ when it returned to its original position.

Magnetic fields will also store energy (Fig. 3–4). The work done moving a north pole at a constant speed through a distance s would be Fs. Again the energy would be stored in the magnetic field and would be available to produce an equal amount of kinetic energy if the pole were released. For this

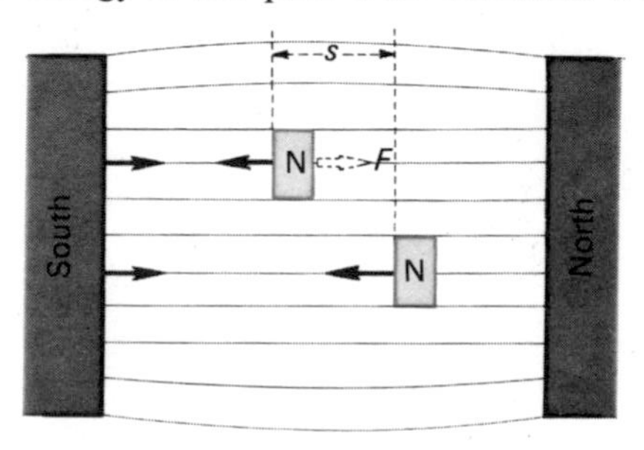

Fig. 3–4

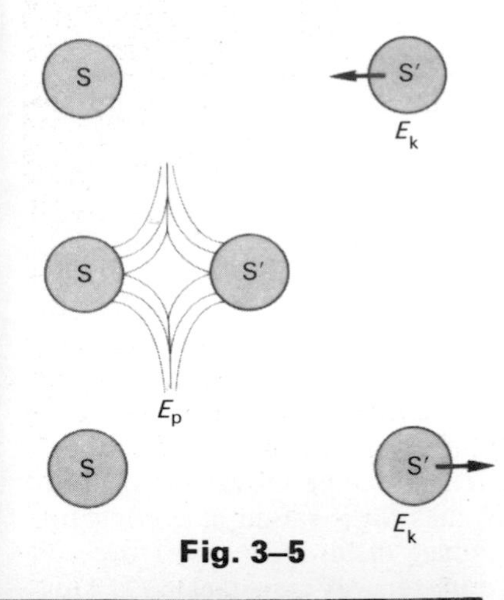

Fig. 3–5

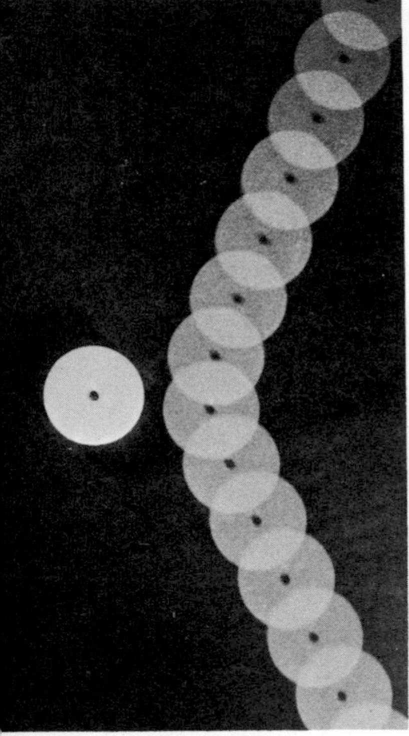

Fig. 3–6

'thought experiment' the north pole could be one end of a very long bar magnet. The south pole would then be so far away that we could ignore its effect.

The energy associated with a collision can be stored in a magnetic field. Fig. 3–5 represents a frictionless puck S′ moving towards a fixed puck S. The upper faces of the pucks have south poles and will therefore repel each other. This repulsive force eventually brings S′ to rest. At this instant the energy is stored in the magnetic field between the pucks. The force of repulsion soon pushes S′ away so that it gains the same amount of kinetic energy as before. As there is no loss of kinetic energy (because there is no friction) this is called an *elastic collision*. Fig. 3–6 shows another elastic collision.

In gravitational, electric, and magnetic fields it is possible to store energy and to get the same amount of energy back without loss, when required. There is, however, unfortunately no practical way of storing large quantities of energy in these fields, apart from the pumped storage system.

Springs

Watches, clocks, and clockwork toys have energy stored in springs. All the energy is not, however, recovered, as the temperature of the spring rises when it is stretched, bent or squeezed. This rise in temperature warms the surroundings; that is, some of the energy is transferred. This loss of energy is normally very small and can usually be ignored.

As the force needed to stretch a spring increases linearly with the extension (Hooke's Law), the work done in stretching a spring can be represented by the area under the curve shown in Fig. 3–7.

$$\text{Work done} = \text{energy stored}$$
$$= \tfrac{1}{2}(\text{maximum force} \times \text{extension})$$

When a body is moved in a uniform field such as the Earth's gravitational field, the force is, however, constant, so that the work done is simply force × distance (Fig. 3–8).

KINETIC ENERGY

When you exert a force on a body, for example, a curling stone (Fig. 3–9), there are three ways in which the work you do can be transferred. First, as we have seen, you could lift the stone and store the energy, as gravitational potential energy (*mgh*). This energy can be easily reclaimed by allowing the stone to return under gravity to its original position.

Secondly, you could push the curling stone very slowly along a level carpeted floor so that it stopped the moment you stopped pushing it. As the stone is not raised it has no potential energy. The stone is again at rest and therefore has no kinetic energy. All the work that you have done has slightly raised the temperature of the stone and the carpet. As the temperature rise is very small, it is not a practical proposition to use it to do other useful work, and very soon the heat will be 'lost' to the surroundings.

Thirdly, you can use the stone for the purpose for which it was designed! If we assume that the surface of the ice is completely friction-free, the stone will continue moving at the speed reached when it leaves your hand. If you have exerted a

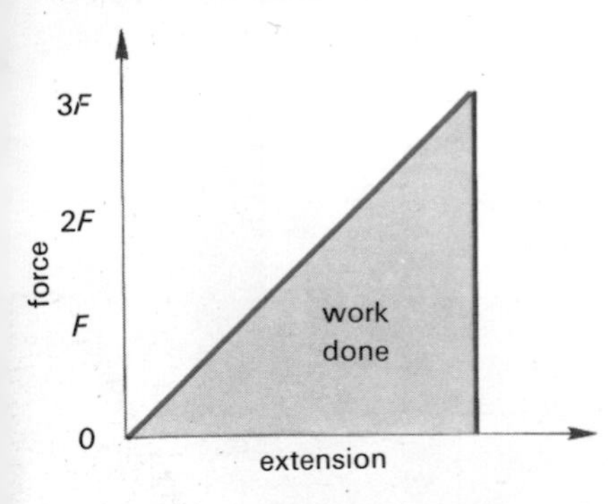

Fig. 3–7

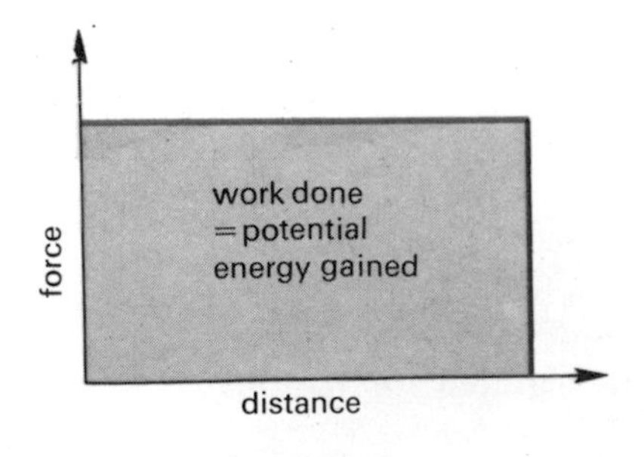

Fig. 3–8

Fig. 3–9

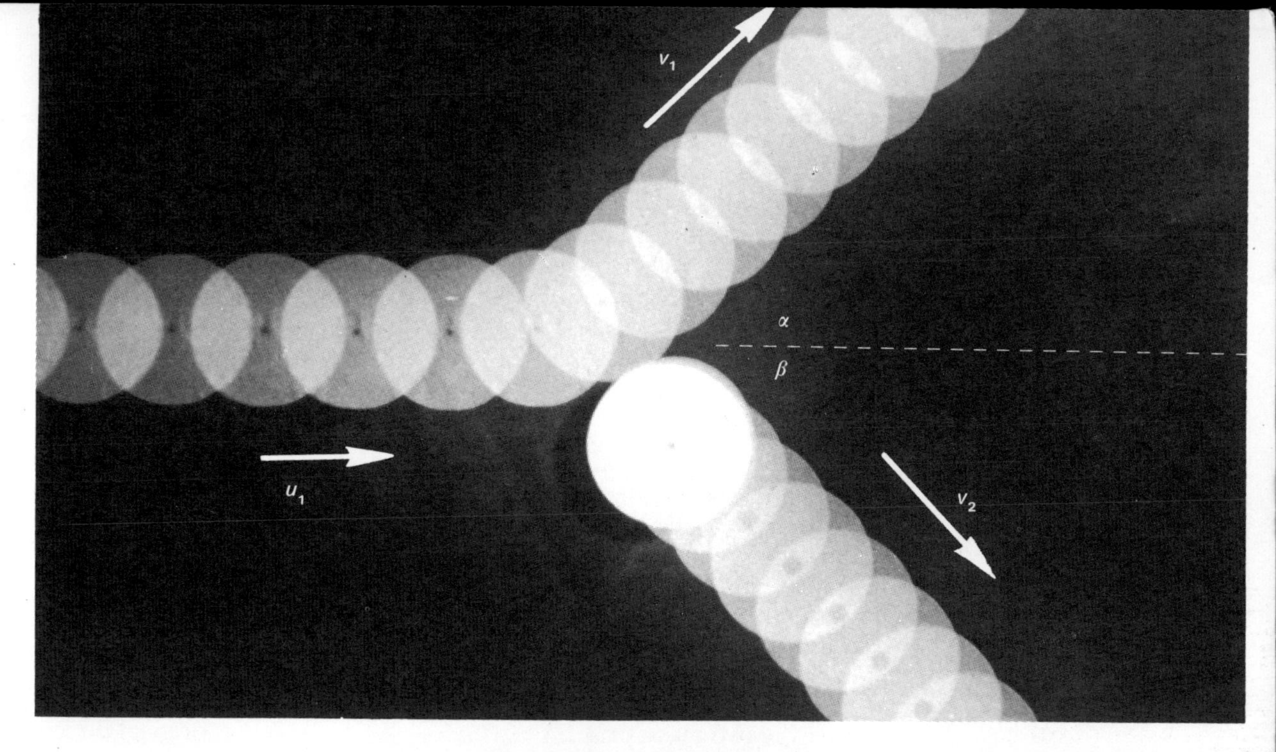

Fig. 3–11

force F for a distance s a stone of mass m will have a kinetic energy of $\frac{1}{2}mv^2$.

$$F = ma$$

$$Fs = mas$$

but $$v^2 = 2as$$

thus $$Fs = \tfrac{1}{2}mv^2$$

In practice two or more methods of energy transfer could be in operation at the same time. In curling, for example, the ice may not be absolutely smooth or even absolutely level, so that there could be small changes of potential energy as the stone moves over the ice surface. Then the free movement of the stone depends on the ice melting slightly as a result of the temperature rise produced by friction. Completely friction-free motion is therefore impossible. In spite of these complications valuable predictions can be made provided we consider all the factors involved and then make allowances for them.

COLLISIONS

Curling stones are made from a hard mineral such as Ailsa Craig granite. Round their circumference the stones have a rough 'striking band' which reduces slipping when two stones collide. These two factors reduce the energy lost during a collision to a minimum and thus produce a near-elastic collision. Before considering an actual collision between curling stones, we will again look at a few ideal cases.

Elastic Collisions

If all the kinetic energy (E_k) lost during impact is regained when the bodies move apart, the collision is said to be *elastic*. Fig. 3–10 shows the result of such a collision between two air-track vehicles of equal mass m, one initially at rest. The separation of the images, in mm, is proportional to the velocity v.

	Before the collision		*After the collision*
Momentum	$m10.5$	=	$m10.5$
E_k	$\frac{1}{2}m(10.5)^2$	=	$\frac{1}{2}m(10.5)^2$

Fig. 3–10

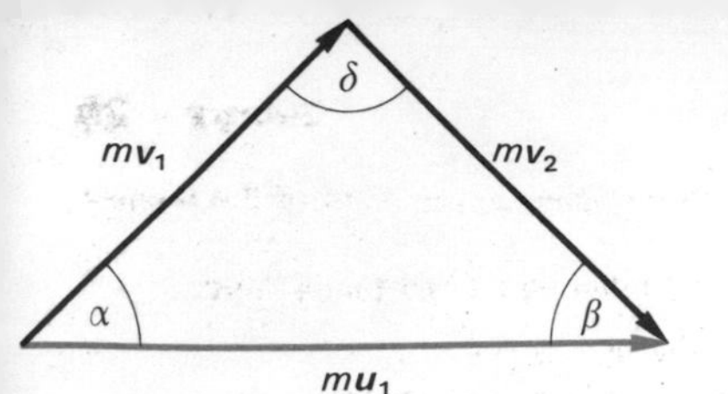

Fig. 3–12

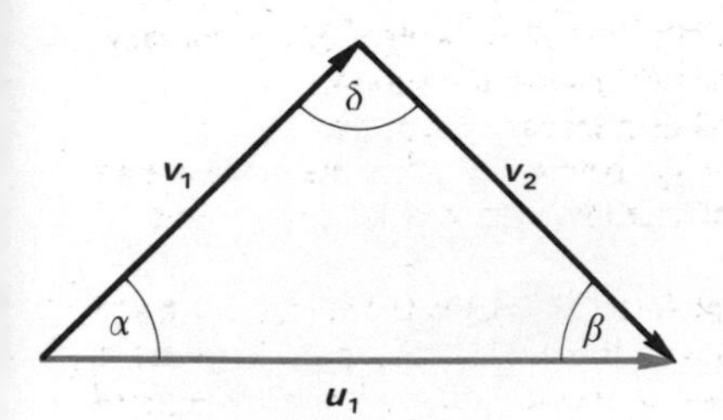

Fig. 3–13

Non-elastic Collisions

When one vehicle collides inelastically with another of equal mass so that they stick together and move off at the same speed, we have the result shown in Fig. 2–15 (page 14).

	Before the collision		*After the collision*
Momentum	$m11$	=	$2m(5.5) = m11$
E_k	$\frac{1}{2}m(11)^2$		$\frac{1}{2}2m(5.5)^2$
	$m60.5$		$m30.25$

Here momentum has been conserved but kinetic energy has been 'lost'. In this particular case the kinetic energy after the collision is 50% of the original kinetic energy. This happens when a moving mass collides with an equal stationary mass and they move off together. In general the loss of kinetic energy can have any value up to nearly 100%. When a very small mass collides with a very large mass nearly all the kinetic energy is lost. It is transformed to internal energy as the temperatures of the two masses rise.

Elastic Collisions in Two Dimensions

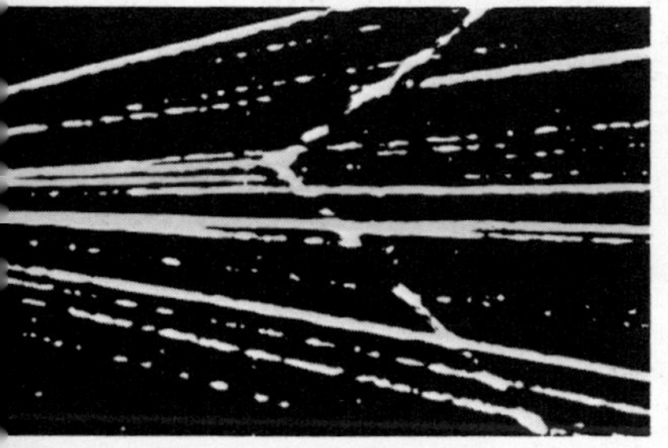
Fig. 3–14

A moving magnetic ice puck collides with an identical stationary one to produce the result shown in Fig. 3–11. Analysis shows that the angle between the paths after the collision is very nearly 90°. Analyse this picture and see if (1) momentum is conserved in the original direction and (2) the total kinetic energy is the same before and after the collision. If momentum is conserved, a vector diagram can be drawn (Fig. 3–12). In this special case the colliding masses are equal and so we can represent the *velocities* by a similar vector triangle (Fig. 3–13).

	Before the collision		*After the collision*	
Momentum	$m\mathbf{u}_1$	=	$m\mathbf{v}_1 + m\mathbf{v}_2$	
thus	$\mathbf{u}_1$	=	$\mathbf{v}_1 + \mathbf{v}_2$	(1)
E_k	$\frac{1}{2}mu_1^2$	=	$\frac{1}{2}mv_1^2 + \frac{1}{2}mv_2^2$	
thus	u_1^2	=	$v_1^2 + v_2^2$	(2)

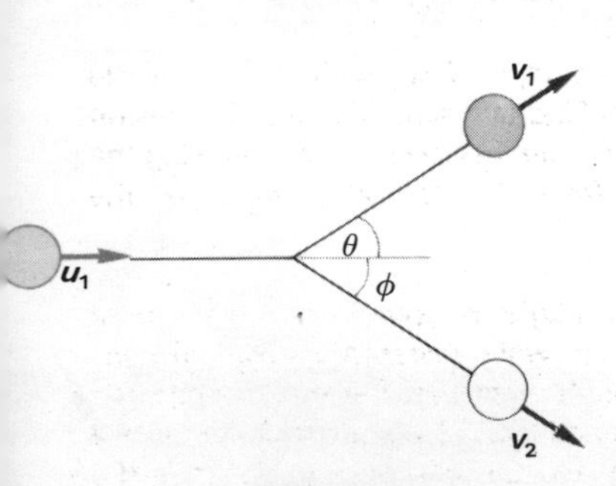

Fig. 3–15

The vector diagram (Fig. 3–13) can represent the vector equation (1) *and* the scalar equation (2) only if δ and thus $(\alpha+\beta)$ are 90°. Thus for *equal masses* a 90° angle implies both conservation of momentum and conservation of energy; that is, the collision is elastic.

Fig. 3–14 shows a collision between an alpha particle and a helium nucleus. We cannot be certain that the particles are moving exactly in the plane of the photograph but evidence from many similar photographs suggests that in this case they are moving very nearly in this plane. *How would you interpret this cloud chamber photograph? (3.1)*

The case of a collision between two similar curling stones is more complex (Fig. 3–15). If we ignore friction, we can assume that momentum is conserved, but some loss of energy will have to be considered.

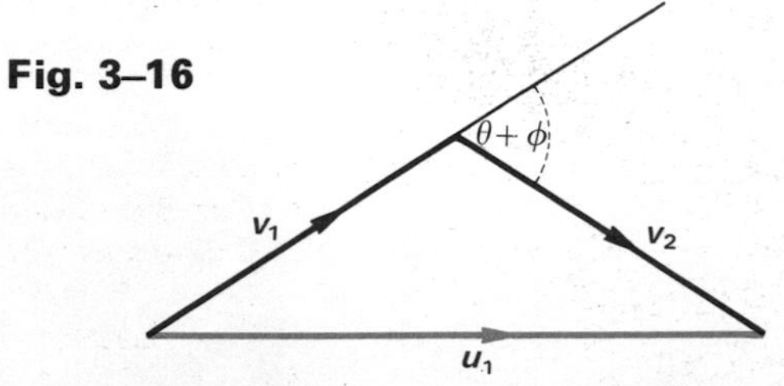

Fig. 3–16

Assuming conservation of momentum and taking components parallel and perpendicular to the original direction of the curling stone

$$mu_1 = mv_1 \cos\theta + mv_2 \cos\phi$$

$$mv_1 \sin\theta = mv_2 \sin\phi$$

But, as $(\theta+\phi)$ is found by experiment to be less than 90° (Fig. 3–16).

$$u_1^2 > v_1^2 + v_2^2$$

and thus $$\tfrac{1}{2}mu_1^2 > \tfrac{1}{2}mv_1^2 + \tfrac{1}{2}mv_2^2$$

That is, there must have been a loss of kinetic energy during the collision.

SUMMARY

Energy transfer

Work = force × distance in the direction of the force

P.E. = $mg \times h$
= potential energy (E_p) of mass lifted in the Earth's field

Heat = $cm\Delta T$ or ml
(c = specific heat capacity, l = specific latent heat)

Kinetic energy (E_k) = $\frac{1}{2}mv^2$

In elastic collisions, momentum and kinetic energy are both conserved. In inelastic collisions, momentum is conserved, but there is a loss of kinetic energy.

PROBLEMS

Fig. 3–17

3.2 A wooden cube with 1 m sides has a mass of 700 kg. What is the minimum amount of work which has to be done to overturn it?

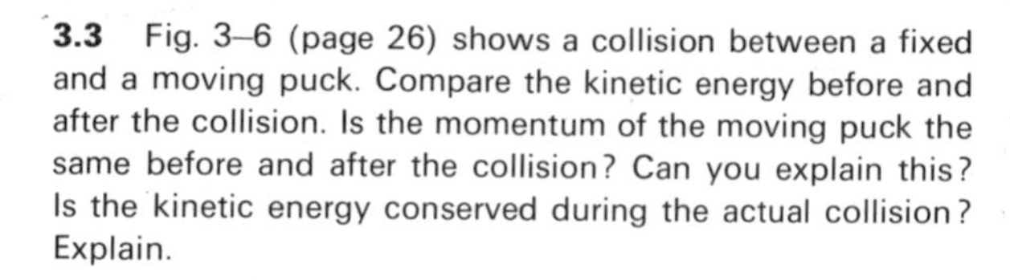

3.3 Fig. 3–6 (page 26) shows a collision between a fixed and a moving puck. Compare the kinetic energy before and after the collision. Is the momentum of the moving puck the same before and after the collision? Can you explain this? Is the kinetic energy conserved during the actual collision? Explain.

3.4 A 50 g light bulb is dropped from the window of a block of flats. If it is travelling at 30 m s^{-1} when it strikes the ground 80 m below, how much energy was transferred as heat during its fall?

3.5 An elastic spring is compressed 100 mm by a maximum force of 20 N. When released it accelerates a 2 kg trolley from rest on a friction-compensated runway. Find the final speed of the trolley.

3.6 If a million kilograms of water per second flow over a waterfall 40 m high, what power is available? What normally happens to this power?

3.7 What force is needed to stop a 2 kg trolley moving at 5 m/s (a) in 1 second, (b) in 1 metre?

A body of mass m is moving with a speed v. Show that the force needed to bring it to rest in x metres is always $\frac{1}{2}v$ times the force needed to stop it in x seconds.

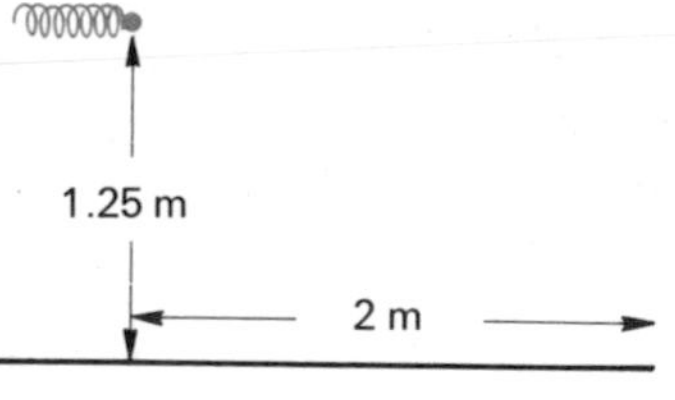

Fig. 3–18

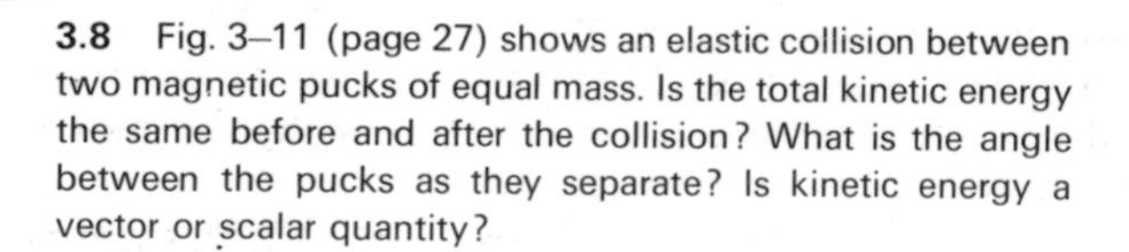

3.8 Fig. 3–11 (page 27) shows an elastic collision between two magnetic pucks of equal mass. Is the total kinetic energy the same before and after the collision? What is the angle between the pucks as they separate? Is kinetic energy a vector or scalar quantity?

3.9 In a dry ice puck experiment a moving puck collides with a stationary puck of equal mass and, as they separate, their tracks make an angle of 60°. If the moving puck had a speed of four units before the collision and three units after the collision, find the final speed of the puck originally at rest. What percentage of the original kinetic energy of the puck remains as kinetic energy of the pucks after the collision? What has happened to the rest?

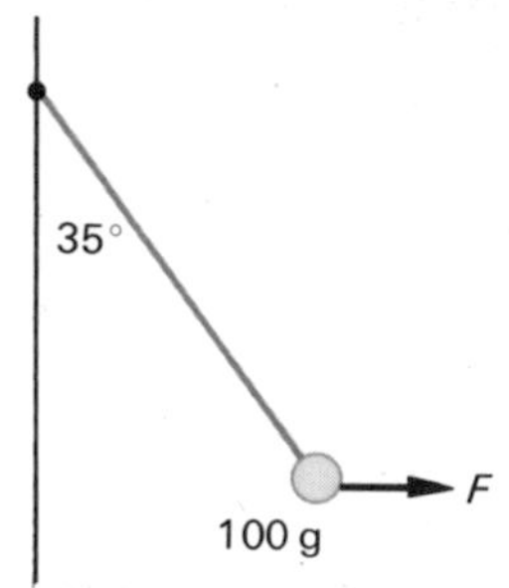
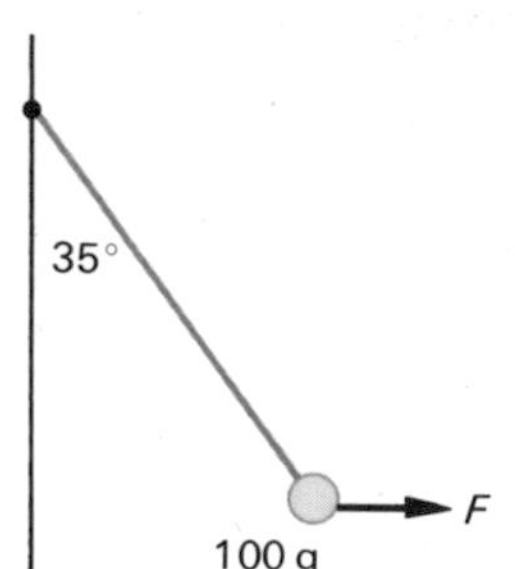

Fig. 3–19

3.10 A cable car is supported on an overhead cable and pulled uphill by a second cable (Fig. 3–17). If the incline is 1 in 4 (i.e. it rises 1 m for every 4 m of cable), find the force exerted by the second cable when the car is moving up at a constant speed. The mass of the cable car and trolley is one tonne (i.e. 10^3 kg) and the frictional forces on the trolley 500 N.

3.11 A ball of mass 0.04 kg is projected horizontally by a compressed spring. When released the ball is in contact with the spring as it moves forward 0.02 m. The ball is released from a height of 1.25 m and falls 2 m in front of the release point (Fig. 3–18).

Taking g as 10 m s^{-2} and ignoring air resistance, find:

(a) the time taken for the ball to fall,
(b) the velocity of the ball when it leaves the spring,
(c) the velocity of the ball just before it lands,
(d) the momentum of the ball as it leaves the spring,
(e) the impulse on the ball,
(f) the (average) acceleration of the ball while being pushed by the spring,
(g) the average force acting on the ball as it accelerates,
(h) the time during which the ball is accelerated horizontally,
(i) the K.E. of the ball as it leaves the spring,
(j) the P.E. of the ball as it leaves the spring,
(k) the K.E. of the ball just before it touches the ground, and
(l) the P.E. of the ball just before it touches the ground.

3.12 A hammer head (mass 0.6 kg) is travelling at 4 m s^{-1} just before it strikes a nail. If the nail (of negligible mass) is driven 2 cm into a piece of wood, find the kinetic energy of the hammer head and hence the average resistance. Can you find the resistive force by another method?

3.13 A bullet of mass 0.02 kg is fired horizontally into a sandbag of mass 7.98 kg. The sandbag is suspended as a pendulum and is seen to rise 0.08 metres (measured vertically). If the bullet remains embedded in the sandbag,
(a) find the speed of the bullet,
(b) find the kinetic energy before and after impact,
(c) how do you explain the results you obtain?
(d) what percentage of the original mechanical energy remained in this form after the impact?

3.14 In a particle accelerator 3×10^4 protons are accelerated from rest to 2×10^7 m s^{-1} in one second. If each particle has a mass of 1.7×10^{-27} kg, find the total kinetic energy of the particles.

3.15 Einstein's equation $E = mc^2$ gives the energy equivalent (in joules) locked up in m kilograms of matter. If the velocity of light c is taken as 3×10^8 m s^{-1}, find the energy equivalent of 1 gram of matter. How much would this amount of energy cost from the Electricity Board if it charges 1 p per kW h?

3.16 A pendulum bob of mass 100 g is pulled horizontally by a force F (Fig. 3–19).
When the string makes an angle of 35° with the vertical the system is in equilibrium. Find the value of F and the tension of the string. How much work has been done in moving the bob to this position from the vertical position of the pendulum?

3.17 A ball of mass 0.1 kg is projected horizontally by a compressed spring which, when released, remains in contact with the ball through 30 mm. If the ball is projected from a height of 1.25 m and lands 3 m in front of the release point, find the impulse and the time it lasts. What is the kinetic energy of the ball just before it lands?

3.18 A pile driver of mass 50 kg is dropped from a height of 5 m on to a stake. If there is no rebound and the stake is driven 200 mm into the ground, find:
(a) the impulse of the force,
(b) the average deceleration of the pile driver,
(c) the kinetic energy of the pile driver just before it strikes the stake, and
(d) the average force exerted on the stake.

Discuss any assumptions you have made in answering these questions.

3.19 A 100 gram marble travelling $1\,m\,s^{-1}$ strikes a stationary 25 gram marble head on (Fig. 3–20).

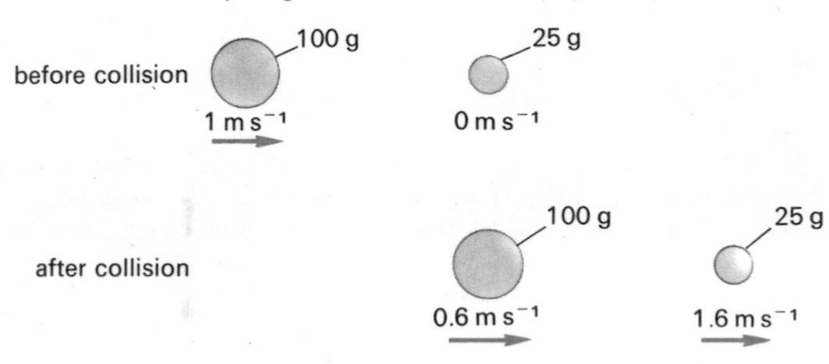

Fig. 3–20

What was their relative velocity before the collision?

After the impact the larger marble is slowed down to $0.6\,m\,s^{-1}$ and the small one shoots off at $1.6\,m\,s^{-1}$. What is their relative velocity now? Is it a velocity of approach or separation? What was the total momentum before and after the collision? What kind of collision is this?

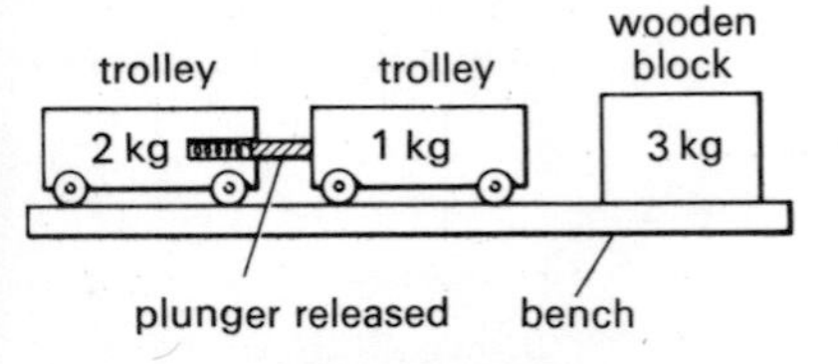

Fig. 3–21

3.20 The principle of conservation of momentum and the principle of conservation of energy are widely applied in physics. It is important to remember that momentum is a vector whereas energy is not.
(a) What is meant by this distinction?
(b) Two free-running trolleys of mass 1 kg and 2 kg respectively were placed in contact with one another on a horizontal bench (Fig. 3–21).

They were exploded apart by the sudden release of a spring-loaded plunger attached to the 2 kg trolley.
(i) Briefly discuss the forms in which the energy of the system existed just before and just after the release of the plunger.
(ii) Describe in some detail how the energy of one of the trolleys could have been measured after the release of the plunger.
(iii) Given that the 2 kg trolley moved off at $2\,m\,s^{-1}$ after the explosion, find the velocity of the 1 kg trolley.
(iv) Given that the plunger was in contact with the 1 kg trolley for 50 milliseconds during the explosion, find the average force exerted by the plunger on the trolley.
(v) Just after the explosion the 1 kg trolley collided with and adhered to the 3 kg wooden block. The combination moved 10 cm from the point of impact. Find the initial velocity of the combination and the average force of sliding friction between the block and the bench.
(vi) Suggest a method by which the duration of conta[ct] between the plunger and the 1 kg trolley could ha[ve] been measured. [S.C.E. (H) 196]

3.21 The sketch (Fig. 3–22) represents schematically [a] possible arrangement for bench testing an engine in order [to] find its efficiency.

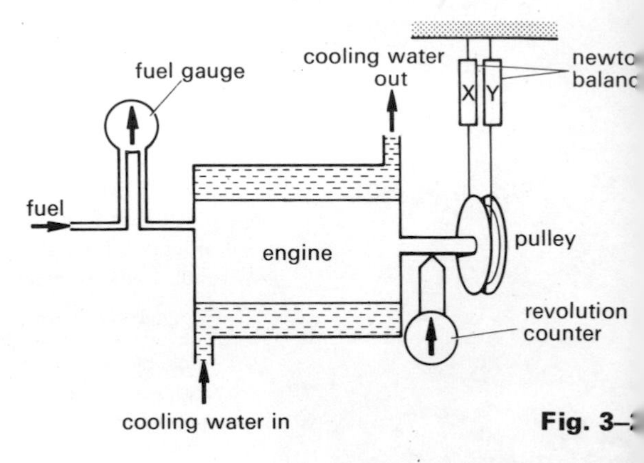

Fig. 3–

The following observations were made in a 5 minute test ru[n]

Average reading on balance X	=	500 N
Average reading on balance Y	=	1000 N
Circumference of pulley	=	0.5 m
Total number of revolutions	=	6000
Total fuel consumed	=	0.5 kg
Calorific value of the fuel	=	$3\times10^{7}\,J\,kg^{-1}$

(a) Calculate the power developed by the engine.
(b) Calculate the rate at which energy is being supplied [to] the engine and hence its efficiency.
(c) Given that the inlet water is at 20 °C and the out[let] water is at 80 °C and assuming that all the heat produc[ed] in the engine is removed by the water in the cooling syste[m] calculate the rate at which water flows through the syste[m] (Specific heat capacity of water is $4200\,J\,kg^{-1}\,°C^{-1}$
(d) Show how you would modify the above experiment [to] investigate the efficiency of an electric motor. State [the] readings you would take and how you would use them. [S.C.E. (H) 196]

3.22 A catapult is extended a distance of 0.50 m. The fo[rce] required to hold it in this extended position is 160 N.
(a) How much potential energy is stored in the catapult
(b) With what velocity will a stone of mass 0.05 kg lea[ve] the catapult?
(c) State *one* important assumption you have made. [S.C.E. (H) 197]

3.23 A golf ball is dropped on to a horizontal concr[ete] floor. It rebounds to three-quarters of its original heig[ht]. State in detail what energy changes take place. [S.C.E. (H) 197]

3.24 Estimate the time required to raise the temperat[ure] of 0.75 kg of water from 20 °C to 100 °C using an elec[tric] kettle rated at 1.50 kW. Mention any assumptions you ma[ke].

Fig. 3–24

Fig. 3–25

Fig. 3–26

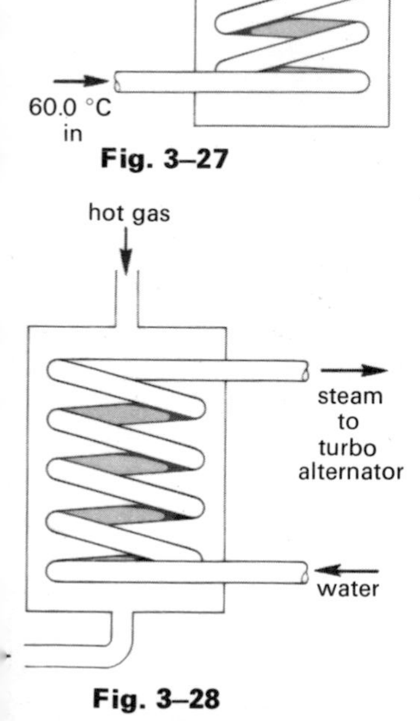

Fig. 3–27

Fig. 3–28

and explain whether the time actually required will be greater or less than your estimate. (Specific heat capacity of water = 4.2×10^3 J kg^{-1} K^{-1}.) [S.C.E. (H) 1970]

3.25 A jet of steam at 100 °C is directed into a bucket containing 10.0 kg of water at 10 °C. Two minutes later the temperature of the water is 100 °C. How much steam is condensed per minute? (Specific latent heat of vaporization of water = 2.26×10^6 J kg^{-1}. [S.C.E. (H) 1970]

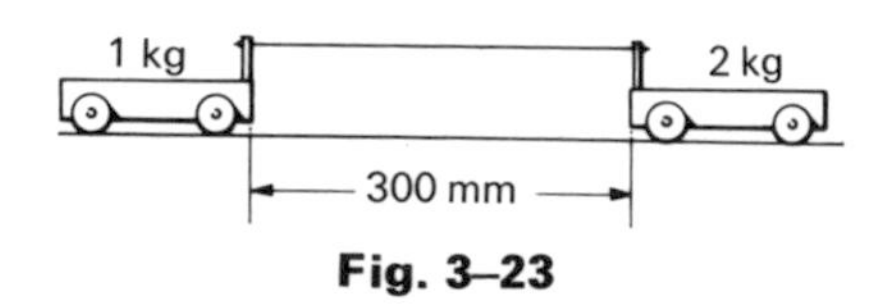

Fig. 3–23

3.26 The sketch (Fig. 3–23) shows two free-running trolleys on a horizontal surface. They are connected by an elastic cord, pulled apart to the positions shown and then released.

(a) Compare the unbalanced forces on the trolleys throughout their motion.

(b) Explain where you would expect the trolleys to meet.

(c) What would be the ratio of their kinetic energies just before impact? [S.C.E. (H) 1971]

3.27 One million kilogrammes of water per minute flow through the turbines of a hydroelectric power station after falling through a height of 50 m. The turbines develop a power of 5 MW.

(a) Calculate the efficiency of this energy conversion.

(b) Apart from energy losses due to friction, what other factor is likely to account for the efficiency being considerably less than 100%? [S.C.E. (H) 1971]

3.28 Running water is used to remove excess heat at the rate of 70 J s^{-1} from the oil-diffusion stage of a high-vacuum pump.

(a) Estimate at what rate the water flows if it enters the pump at 20 °C and leaves it at 22 °C.

(b) If the water were cut off, the temperature would not rise indefinitely. Suggest a reason why this should be so. [S.C.E. (H) 1971]

3.29 A lead pellet of mass 1.0 g is fired from an air gun with a muzzle velocity of 120 m s^{-1}. This is sufficient to cause it to pass completely through the block of plasticine, fixed on a vehicle at rest on a horizontal surface (Fig. 3–24). The vehicle and plasticine have a total mass of 0.5 kg and are found to acquire a velocity of 0.2 m s^{-1}.

(i) Outline briefly the method you would use to measure the velocity acquired by the vehicle. State the measurements you would make and explain how you would use them in calculating the velocity.

(ii) With what velocity does the pellet leave the plasticine?

(iii) What is the kinetic energy lost by the pellet?

(iv) Estimate the kinetic energy gained by the vehicle plus plasticine.

(v) Compare your answers to (iii) and (iv) and account for the difference.

(vi) Assuming that the pellet leaves the plasticine horizontally at a height of 5 cm above the horizontal surface, estimate how much further it travels before striking the surface. [S.C.E. (H) 1971]

3.30 An object A collides obliquely with an identical object B which is at rest. The velocity of each is found after the collision and is indicated in Fig. 3–25.

(a) What kinetic energy is lost as a result of the collision if A and B each has a mass of 0.20 kg?

(b) Suggest a laboratory method for investigating oblique collisions of the type described above.

(c) What difference would you expect in the results if the collision were elastic? [S.C.E. (H) 1972]

3.31 In a laboratory experiment a vehicle on a track is set in motion by a catapult incorporating a number of identical elastic cords. Fig. 3–26 shows the arrangement of the catapult system.

(a) What practical steps would you take to ensure that friction is made negligible and how would you check that this has been done?

(b) The speed achieved by the vehicle when catapulted by different numbers of elastic cords is recorded in the table below.

Number of cords N	Speed in m s^{-1} v
1	0.41
2	0.57
3	0.69
4	0·80

(i) From theoretical considerations, what relationship would you expect between N and v?

(ii) Show whether the results above verify this relationship.

(c) State any *two* practical factors likely to lead to discrepancies in the results. [S.C.E (H) 1972]

3.32 (a) Hot water from a boiler is pumped through a coiled tube inside a hot water storage tank (Fig. 3–27). The water enters the coil at a temperature of 60.0 °C and comes out at 55.0 °C. How much water must flow through the coil per second if the rate of heating is to be the same as that provided by a 3.0 kW immersion heater?

(b) Hot gases from a gas-cooled nuclear reactor (Fig. 3–28) are made to turn water into high pressure steam in a heat exchanger.

(i) List the data required to calculate the heat transferred from the gas each second.

(ii) What design features of the heat exchanger would lead to good heat transfer? [S.C.E. (H) 1972]

4 Pressure

Fig. 4–1

A skater glides gracefully over a near friction-free surface because a thin layer of ice melts temporarily as the skate passes over it. The ice melts partly because of the tremendous pressure exerted by the skate but mainly because of frictional forces. The pressure exerted by a ski is considerably less as the area of contact may be a thousand times greater. Frictional forces are then responsible for the melting process.

As pressure is measured by the force exerted on unit area, it would seem reasonable to assume that pressure, like force, ought to be a vector quantity. Before rushing to this conclusion we must, however, look carefully at the exact definition of pressure. It is not simply force per unit area. In fact the direction of the required force is built-in to the definition. Pressure is concerned with the force or component of the force acting at right angles to a given area. If one leg of the lunar module exerts a force F on an area A of the Moon's surface (Fig. 4–2) the pressure is *not* F/A. It is $F \cos\theta/A$ where θ is the angle the leg makes with the vertical (Fig. 4–3).

$$\text{Pressure} = \frac{\text{force normal to an area}}{\text{area}}$$

Pressure indicates the magnitude of this normal force per unit area and no direction such as north or east need be stated. Pressure is therefore a scalar quantity. Our main concern in this chapter will, however, be with the pressure exerted by a fluid at rest. Simple kinetic theory interprets this pressure in terms of continuous bombardment by the molecules of the fluid. The force exerted on a given area is then equal to the rate of change of momentum of the molecules during the bombardment, and the direction of the force is always *normal* to this area. Brownian motion strongly suggests that molecules in a fluid move in a random way and so we might expect the pressure resulting from molecular bombardment to be, on average, the same in all directions (Fig. 4–4). Evidence from Pascal's syringe

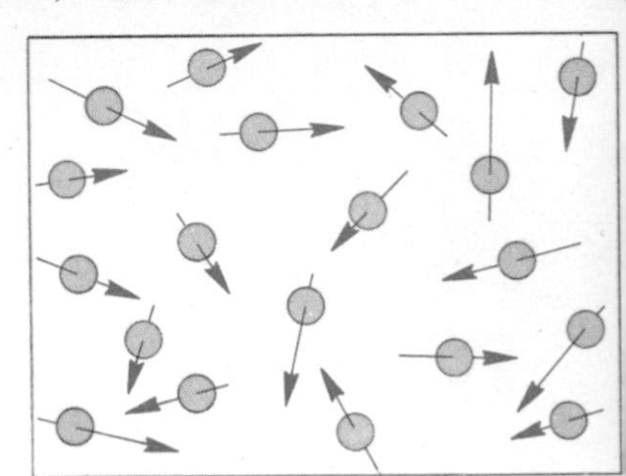

Fig. 4–4

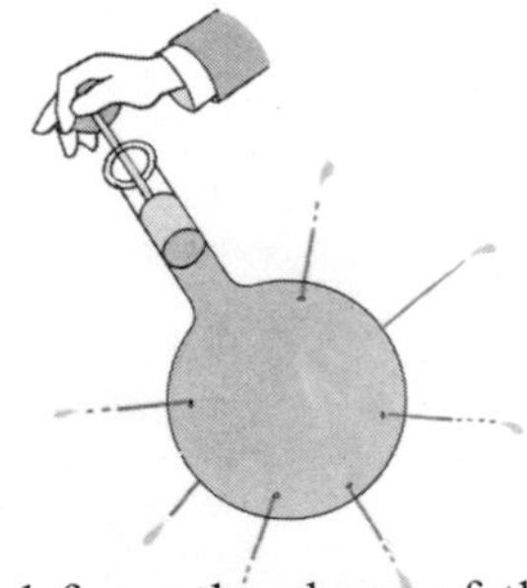

Fig. 4–5

(Fig. 4–5) and from the shape of the bulb being produced by the Caithness glass-blower (Fig. 4–6) confirms this assumption.

We have already met the idea of instantaneous speed, for example, the reading on a car's speedometer, as a limiting case of distance/time when the time interval shrinks to a very small quantity but does not disappear completely. Speed = $\Delta s/\Delta t$

Fig. 4–2

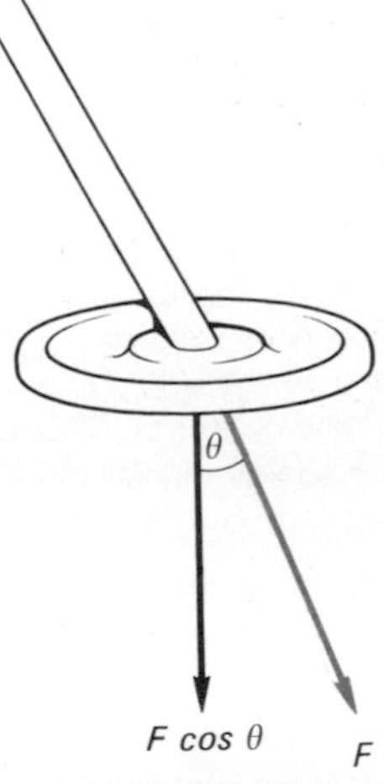

Fig. 4–3

Fig. 4–6

Fig. 4–7

Fig. 4–8

There is a similar concept relating to pressure, namely, 'pressure at a point'. Here the area shrinks but does not disappear. Pressure at a point is thus $\Delta F/\Delta A$ where ΔA is very very small.

PRESSURE IN A LIQUID

As every diver knows, the pressure exerted by water increases with depth. The future of underwater exploration depends on a fuller understanding of the effect of pressure on the human body. In the underwater habitat 10 metres below the surface of the Mediterranean (Fig. 4–7), the pressure is 2 atmospheres. With Aqua-Lung equipment (Fig. 4–8) it is possible to go down to depths of about 50 metres for short periods. As the pressure of the air breathed must also increase with depth, more and more nitrogen is dissolved in the blood stream. This forms bubbles in the diver's blood when he returns to the surface, causing severe pain ('diver's bends') or even death. In an emergency the diver can be placed in a decompression chamber where the pressure is *gradually* reduced over a long period, thus preventing formation of bubbles. If a diver is going to operate at depths greater than 50 metres he uses a cylinder containing a mixture of about 5% oxygen and 95% helium, instead of air.

Naval tests suggest that free swimming at depths of 500 metres and pressures of 50 atmospheres may be possible using neon mixed with oxygen.

The American Bathyscaphe, shown in Fig. 4–9, has been built to explore depths well beyond the reach of any diver. It is designed to withstand tremendous pressures of about 1000 atmospheres exerted 10 kilometres below the surface of the Pacific Ocean.

Fig. 4–9

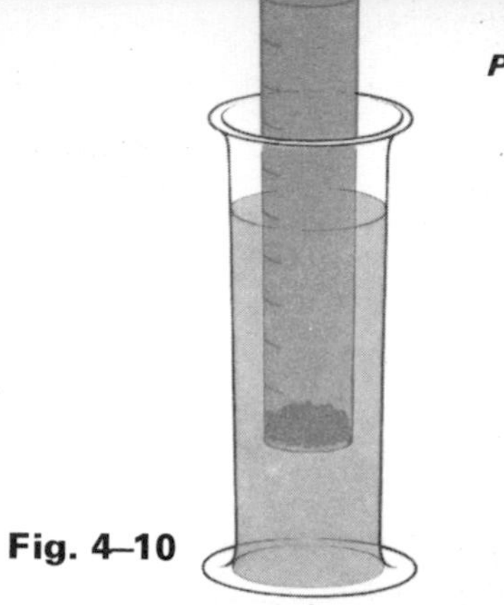

Fig. 4–10

A simple experiment in which a floating tube is loaded with lead shot can be used to investigate the variation of pressure with depth in a liquid (Fig. 4–10). The effect of density on pressure can be

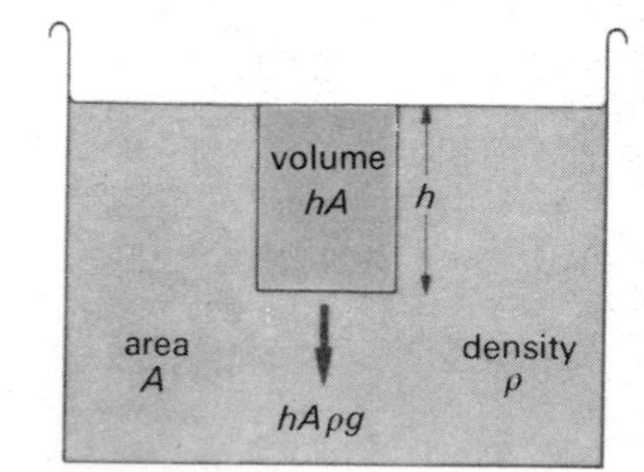

Fig 4–11

studied with the same apparatus. The results of such experiments might have been predicted in the following way. Consider a cylindrical volume of liquid of length h and area A (Fig. 4–11). Its volume is $h \times A$. If the density of the liquid is ρ, the mass (m = volume × density) of this volume is $hA\rho$ and its weight (downward force) is $mg = hA\rho g$.

$$\text{The pressure at area } A = \frac{\text{force normal to area } A}{\text{area } A}$$

$$= \frac{hA\rho g}{A}$$

$$= h\rho g$$

As the area A can be as small as we wish to make it, this expression represents pressure at a point in a liquid.

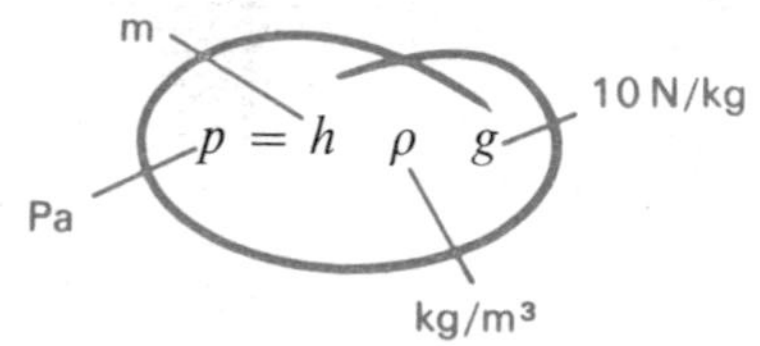

Pressure is thus directly proportional to the depth (or height h) and the density (ρ) of the fluid. It is the same in all directions. The unit of pressure is the pascal (Pa). 1 pascal is 1 newton per square metre.

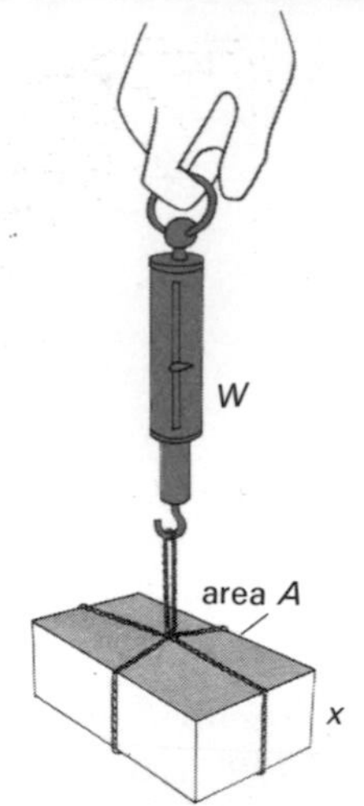

Fig. 4–12

Here is a simple way of testing the relationship which we have just derived theoretically. To do this let us consider what *should* happen to a brick if we suspend it in a bucket of water and assume that the pressure acting on every point on the brick is given by $h\rho g$.

In air the brick has a weight W and the volume is xA (Fig. 4–12). If it is lowered into a bucket full of water, the same volume of water, xA, will spill over. As the pressure will be the same at all points on the same level, the forces acting on opposite sides of the brick will be equal in magnitude but act in opposite directions. The net sideways force on the brick will therefore be zero.

However, the top and bottom of the brick are at different depths and will therefore have different forces acting on them. If the top of the brick is at a depth h below the water surface (Fig. 4–13), then

downward force on the upper surface f

$$= \text{pressure} \times \text{area}$$

$$= h\rho g A$$

upward force on the lower surface F

$$= (h+x)\rho g A$$

$$\text{difference in forces } (F-f) = x\rho g A$$

This difference between the force on the bottom and the force on the top of the brick $(F-f)$ is the net force acting upward. This is called the *upthrust*, and so

$$\text{upthrust } U = xA\rho g$$

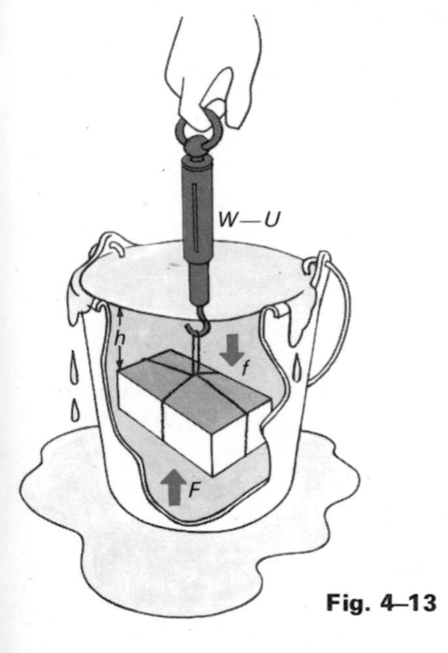

Fig. 4–13

As the volume of the water spilled is xA and its density is ρ, the mass of water spilled will be $xA\rho$ and therefore its weight mg is $xA\rho g$. That is, the weight of liquid displaced $= xA\rho g$. As both the upthrust and the weight of the liquid displaced are equal to $xA\rho g$, they must be equal to each other.

$$\text{Upthrust} = \text{weight of liquid displaced}$$

What is this famous statement called? (4.1) How would you measure the upthrust and the weight of the liquid displaced, and thus check the validity of the expression $p = h\rho g$? *(4.2) What would happen if, as a wooden brick was being lowered into the liquid, the upthrust equalled the weight of the brick before the brick was totally submerged? (4.3)*

Fig. 4–14

Fig. 4–14 shows an easy way to lift an 18-stone police sergeant! America's astronauts train in large water tanks where the weightlessness of space can be simulated (Fig. 4–15). *Describe the conditions necessary to produce this effect. (4.4)*

Fig. 4–15

PRESSURE GAUGES

The Bourdon pressure gauge (Fig. 4–16) is really a sophisticated development of a popular toy (Fig. 4–17). Before it can be used to measure pressure the gauge must first be calibrated by comparing its readings with those of a standard instrument such as a manometer.

Fig. 4–16

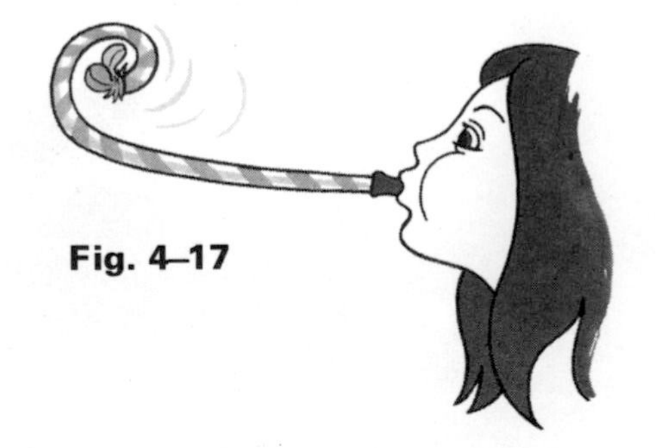

Fig. 4–17

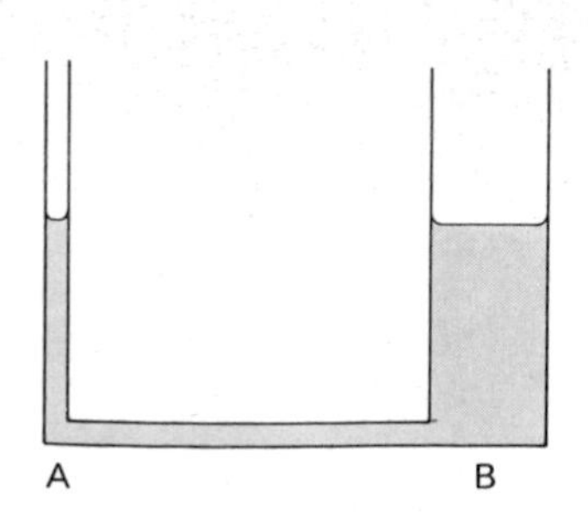

Fig. 4–18

If the pressure in a liquid depends on the depth and the density, the pressure in any given liquid will vary only with the depth. At any two points on the same horizontal level in a liquid, the pressures will therefore be the same. It is sometimes interesting to ask what might happen if nature behaved differently! Suppose, for example, that when water is poured into a U-tube with unequal arms the pressure at B (Fig. 4–18) is greater than at A. Water would then flow along the horizontal tube from B to A until the pressures at A and B were equal. People often think that the levels shown in Fig. 4–19 represent the equilibrium state. *What quantities are they then confusing? (4.5)*

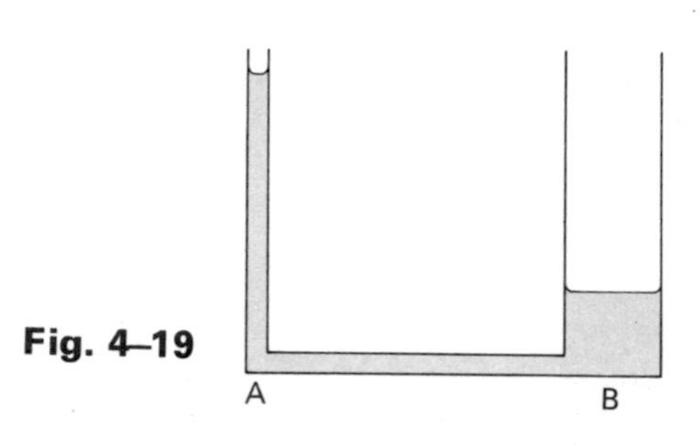

Fig. 4–19

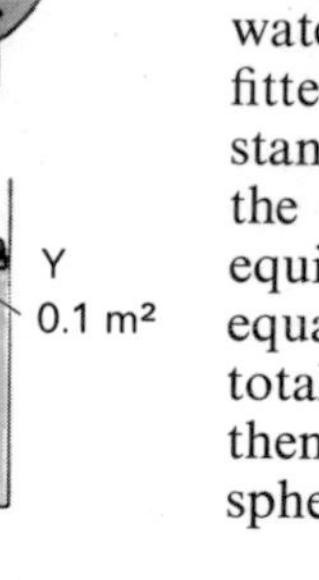

Fig. 4–20

Experiments show that Fig. 4–18 represents the correct equilibrium state. The thickness of the tubes makes no difference to the pressures at A and B. Of course it would be possible to push the liquid up the narrow tube by exerting a force on the surface of the liquid in the wide tube. If a giant water manometer was constructed and a piston fitted to the right-hand tube (Fig. 4–20), you could stand on the piston and estimate your weight from the rise in water level in the other tube. In the equilibrium position the pressures at X and Y are equal. If the area of the piston is $0.1\ \mathrm{m}^2$ and the total weight of the piston and you is W newtons, then, considering pressures in excess of atmospheric pressure, the

$$\text{pressure at Y} = \frac{\text{force}}{\text{area}} = \frac{W}{0.1} = 10W\ \text{Pa}$$

As the density of water is $10^3\ \mathrm{kg/m^3}$ then

$$\text{pressure at X} = h\rho g = 0.7 \times 10^3 \times 10$$
$$= 7 \times 10^3\ \text{Pa}$$

As these pressures are equal, $W = 700$ N.

How would the height of the liquid column be affected if (a) both tubes had an inner cross-sectional area of $0.1\,m^2$ and (b) the liquid used was oil which is less dense than water? (4.6)

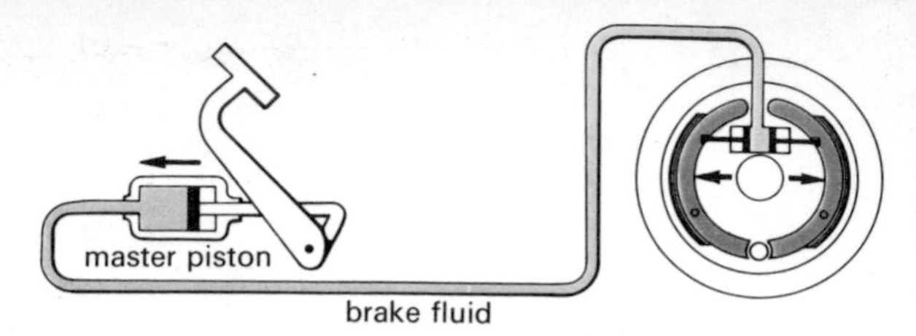

Fig. 4–21

The transmission of pressure in a fluid is used to operate the hydraulic brakes in a car. The principle is illustrated in Fig. 4–21. One valuable feature of such a system is that the pressure transmitted to each of the four wheel cylinders is the same, thus producing the same braking force on each wheel.

The Manometer

A manometer is normally used to measure gas pressures. For example, if a water manometer is

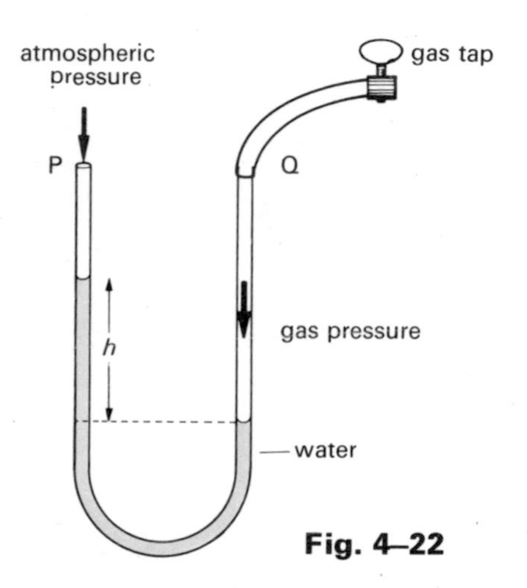

Fig. 4–22

connected to a gas supply as shown in Fig. 4–22 the *difference* between the pressures at P and Q is indicated by the head of water h. As such instruments have been widely used for measuring pressure, it is still common to see pressures expressed in 'millimetres of water' or 'millimetres of mercury'. A manometer can be calibrated in pascals by using the expression developed earlier:

$$p = h\rho g$$

If, for example, a manometer, filled with oil of density $800\ \mathrm{kg/m^3}$, was used to measure gas pressure, it might indicate a head h of 140 mm. The pressure, *in excess of atmospheric pressure*, would then be

$$p = h\rho g = 0.14 \times 800 \times 10 = 1120\ \text{Pa} = 1.12\ \text{kPa}$$

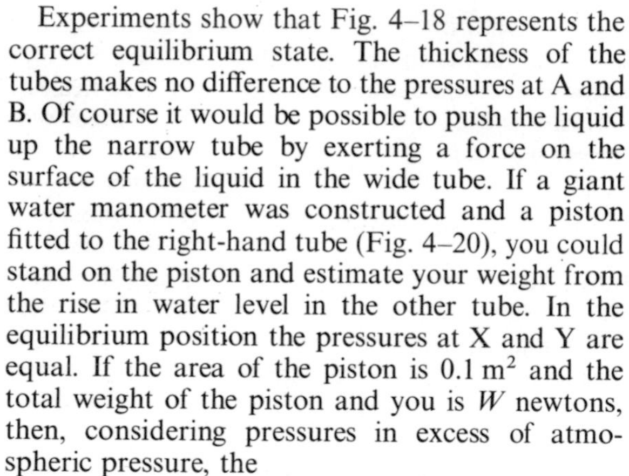

Fig. 4–23

ATMOSPHERIC PRESSURE

Unless special steps were taken, the air pressure in *Concorde* (Fig. 4–23) flying at its cruising height of 18 km would be about 7 kPa instead of the usual pressure on the Earth's surface of 100 kPa. As breathing would be impossible under these conditions, air is pumped into the plane to maintain a reasonable, but slightly less than normal, air pressure. A pressure of 80 kPa equivalent to the air pressure at a height of 2 km is used. If we think of the atmosphere as a 'blanket of air' held to the Earth by gravitational attraction we can see that the nearer we come to the Earth the greater the mass of air there is above us. The weight of air pushing down on every square metre, i.e. the pressure, is therefore greatest on the surface of the Earth. The variation of pressure with height is shown on Fig. 4–24.

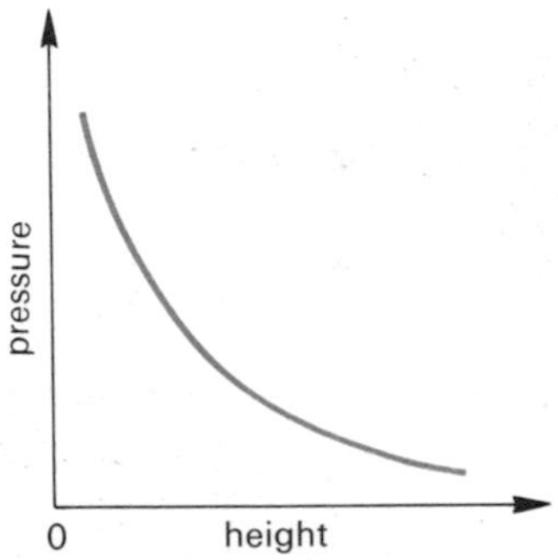

pressure

0 height

Fig. 4–24

The difference between the atmospheric pressure exerted on the top and the bottom of the hot air balloon shown in Fig. 4–25 results in enough upthrust to lift the balloon and its passengers. A 700 kW propane gas burner is used in this modern version of a two hundred year old sport.

Fig. 4–25

Eighteen Palms

To measure the atmospheric pressure we could evacuate one side of a mercury manometer (Fig. 4–26). Alternatively we could remove the air from a long glass tube standing in a basin of mercury (Fig. 4–27). In each case the air pressure in the evacuated tube would be zero, so that the head of mercury h would indicate the atmospheric pressure.

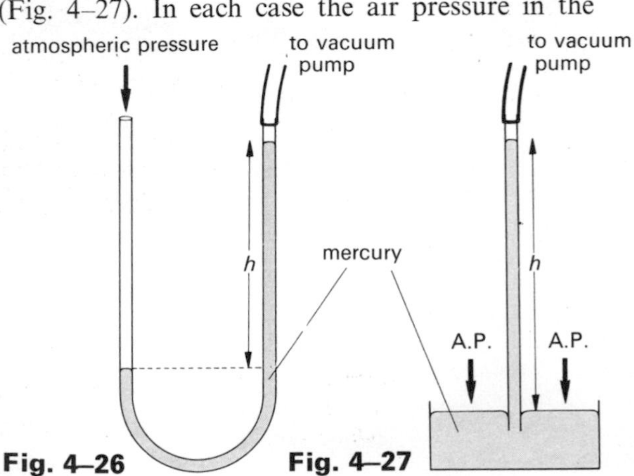

Fig. 4–26 Fig. 4–27

This experiment was first conducted over three hundred years ago—by accident! The Grand Duke of Tuscany, so the story goes, gave orders to his men to dig a well in the palace grounds. Unluckily they found that they had to dig to a depth of about 13 metres before coming to water. When they tried to pump water up out of the well they found that no matter how energetically they worked the pump no water came out.

The Duke sent for Galileo who discovered that the water rose in the pipe to a height of 'eighteen palms'—about 10 metres. Galileo commented that although 'Nature abhors a vacuum' its horror ends when the water has risen eighteen palms inside one. He then left to one of his research students, Evangelista Torricelli, the task of investigating the problem. This Torricelli did by using mercury in place of water for his simple barometer.

Although such an instrument can be made quite simply, several precautions must be taken. A long *clean* tube sealed at one end is needed and some *clean* mercury. First you fill the tube with mercury and put your finger over the open end. By inverting the tube several times you can collect air bubbles in the tube. Once you are satisfied that there are no air bubbles left in the tube, top it up, and, with your finger over the open end, insert the tube in a basin of mercury. When you remove your finger the mercury drops leaving a vacuum (sometimes called the Torricellian vacuum) above it. The head h then indicates the atmospheric pressure (Fig. 4–28). This simple barometer is, in effect, a manometer with the reservoir acting as the other tube.

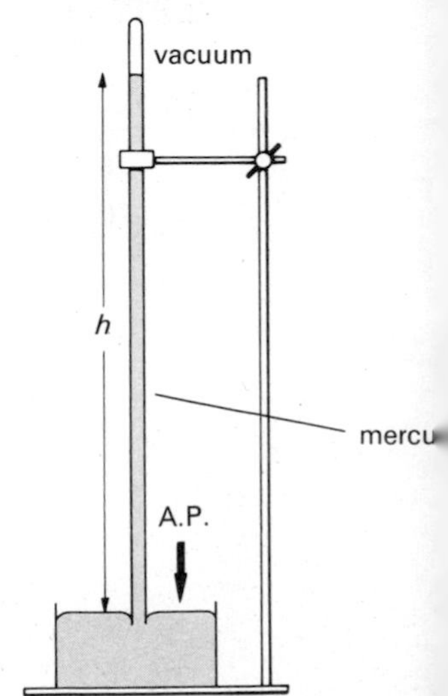

Fig. 4–28

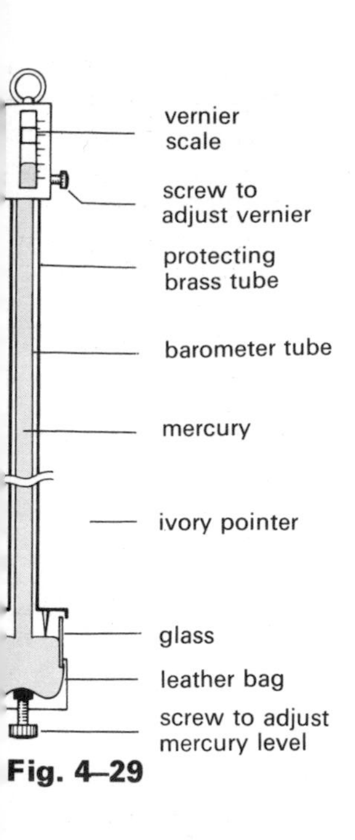

Fig. 4–29

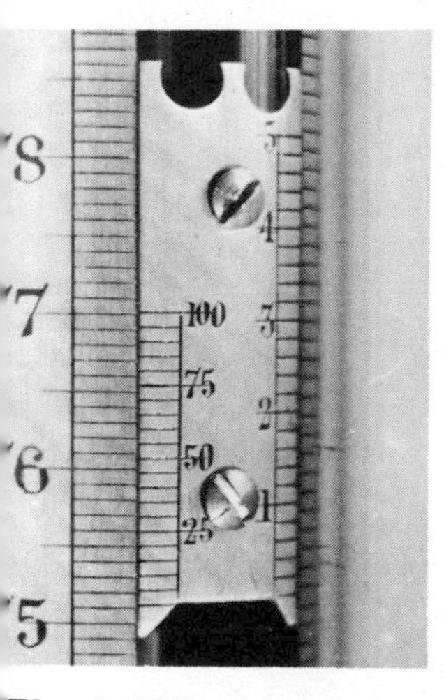
Fig. 4–30

Fig. 4–31

Standard atmospheric pressure sustains a mercury column of 760 mm; thus

$$\begin{aligned}\text{Standard atmospheric pressure} \\ &= h\rho g \\ &= \frac{760}{1000} \times 13\,600 \times 9.8 \\ &= 10^5 \text{ Pa approx.}\end{aligned}$$

In the Fortin barometer (Fig. 4–29)—a practical form of simple barometer—the zero of the scale is indicated by the point of an ivory tooth. The level of the mercury, which is contained in a leather bag, can be adjusted until the surface is in contact with the tooth. A vernier scale allows very accurate readings to be taken. In Fig. 4–30 the mercury stands at 75.12(5) cm.

Although they are very accurate, mercury barometers are expensive and awkward to use. Consequently barographs, altimeters, and gauges for measuring blood pressure often operate on the same principle as the aneroid barometer (Fig. 4–31). A small evacuated box is fitted with a flexible lid which moves slightly as the pressure varies. The movement is magnified by a series of levers which then operate a pointer.

Fig. 4–32 shows how the pressure of the atmosphere assists the installation of double glazing in the Edinburgh University Library. Four large 'suction' pads hold the glass.

For immunization from certain diseases the day of frightening 'jabs' could be over. A high pressure jet injector (Fig. 4–33) uses compressed air to shoot the liquid drug, in the form of a high speed spray, painlessly through the skin.

FLUIDICS

The rapid development of solid state electronics has sometimes given the impression that hydraulic and pneumatic devices are obsolete. Fluidics is, however, a relatively new science dealing with the control of machines by fluid operated devices. A large quantity of fluid can be controlled by means of a much smaller variable flow. It is claimed that fluid amplifiers have certain advantages over electronic amplifiers. They are said to be more reliable and robust, have a longer life and are not affected by explosive or radioactive environments. Fig. 4–34 shows a fluidic control system operating a plug-making machine.

Fig. 4–32

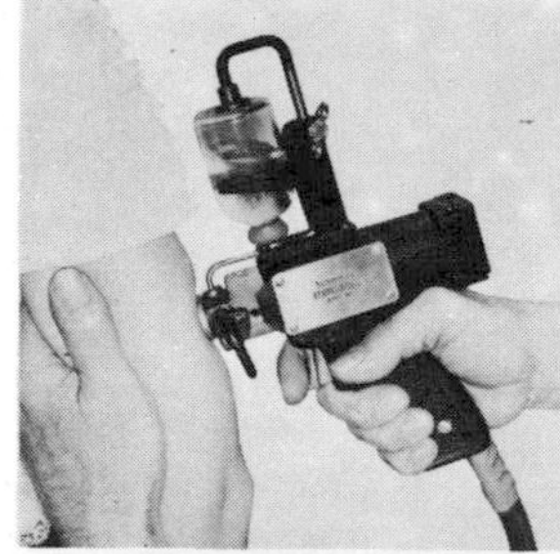
Fig. 4–33

Fig. 4–34

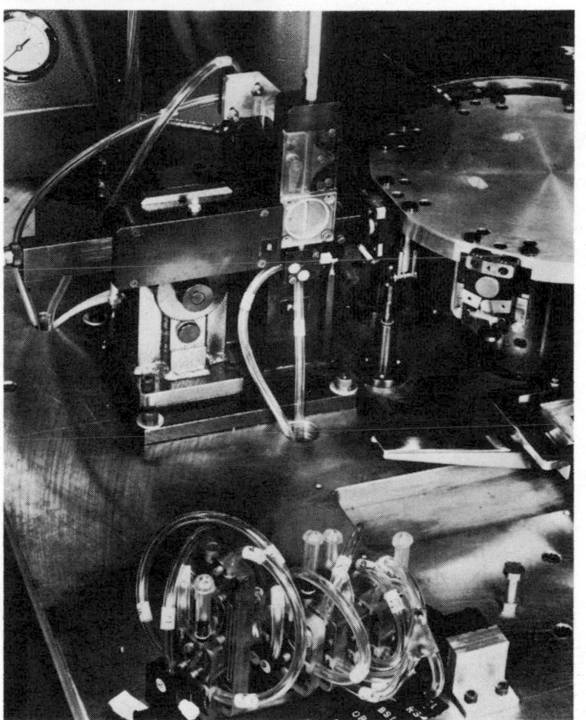

SUMMARY

Pressure $= \dfrac{\text{force}}{\text{area}}$ (a scalar quantity)

Pressure in a fluid $= h\rho g$

Standard atmospheric pressure $= 1.013 \times 10^5$

$\approx 10^5$ Pa (i.e. N/m^2)

PROBLEMS

4.7 What is the pressure at the point of a drawing pin (area 10^{-8} m^2), if a force of 50 N is applied?

4.8 Find the pressure due to a 400 mm column of olive oil of density 900 kg m^{-3}

4.9 If the base of a tank of water measures 3 m × 6 m, find the force exerted on the base when the tank is filled with water to a depth of 1.5 m. (Density of water = 10^3 kg m^{-3}.)

Water is taken from a tank via a tap 18.5 m below the base of the tank. Find the water pressure at the tap.

4.10 What is the height of the mercury in a barometer at sea level when the atmospheric pressure is 1015 millibars? (1 bar = 10^5 Pa.)

If the same barometer reads 720 mm on top of a mountain, how high is the mountain, assuming the *average* density of air is 1.2 kg m^{-3}?

Fig. 4–35

4.11 A diving bell which is 7 m high is lowered into fresh water (Fig. 4–35). The water rises 3 m inside the bell. How far is the top of the bell below the surface? (Atmospheric pressure = 10^5 Pa.)

4.12 Explain why it is so easy to float on the very salty Dead Sea.

4.13 Draw a graph showing how you would expect the pressure around a diver to vary with depth below the surface of the sea. Sketch a second graph showing how you would expect the atmospheric pressure around a space ship to vary as it entered the Earth's atmosphere from space. That is, plot atmospheric pressure against 'depth' below the upper limit of the atmosphere.

4.14 If the combined mass of the car, platform, and piston in Fig. 4–36 is 1000 kg, what will be the air pressure, in atmospheres, in the air reservoir when the car is being raised? What would be the required air pressure, if the area of the reservoir was reduced to 0.5 m^2?

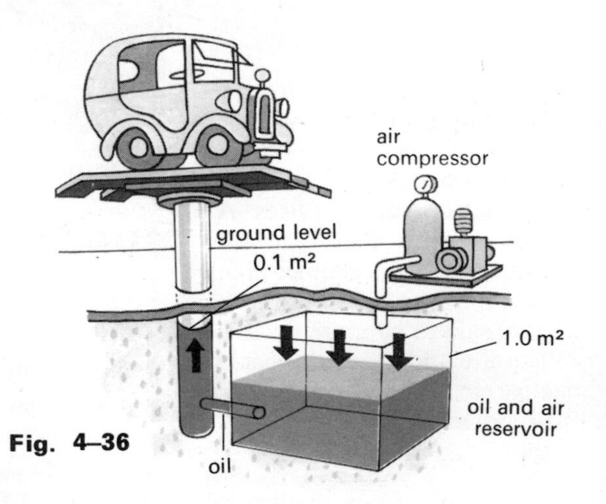

Fig. 4–36

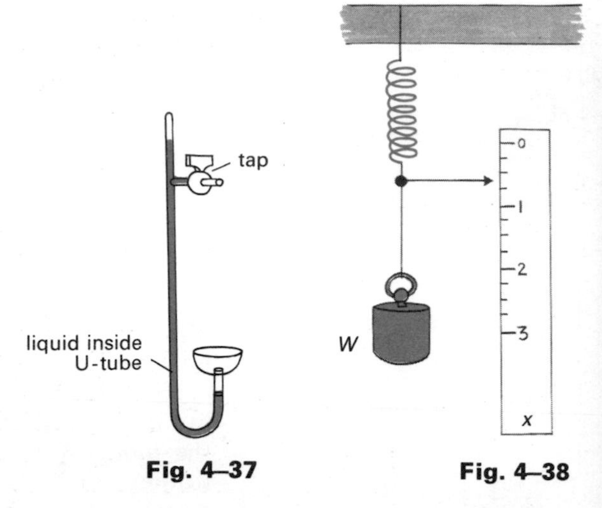

Fig. 4–37

Fig. 4–38

4.15 A barometer is made from a U-tube as shown in Fig. 4–37. What will happen if the tap is slowly opened?

4.16 A pointer attached to a spring reads zero when there is no load on the spring. Draw a graph showing the extension of the spring x plotted against the weight W when masses are attached as shown in Fig. 4–38).

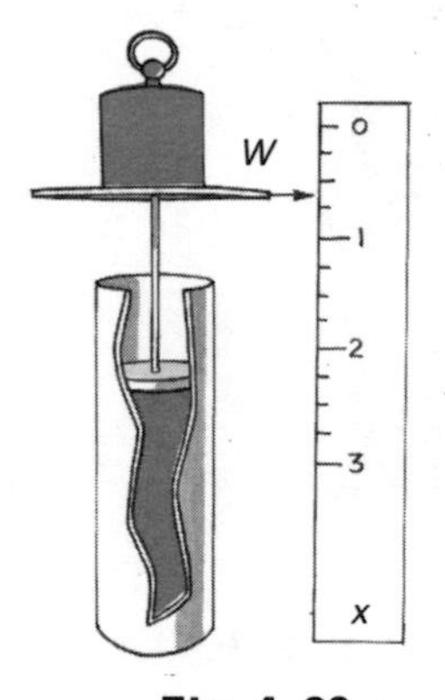

Fig. 4–39

4.17 A cylinder is fitted with a weightless piston and plate and the plate is opposite the zero mark on the scale when there are no masses on it. Draw a rough graph showing the displacement of the piston x with weight W when masses are placed on the plate (Fig. 4–39).

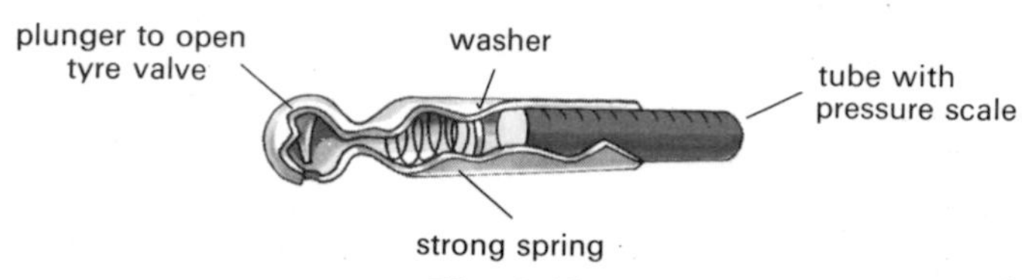

Fig. 4.40

4.18 Would you expect the scale on the tyre pressure gauge shown in Fig. 4–40 to be linear? Explain your answer.

4.19 Jack and Jill decide, for a change, to measure variations in atmospheric pressure as they climb a not very high hill. Jack makes a pressure gauge by sealing one end of a capillary tube and introducing a mercury thread as shown in Fig. 4–41(a).
Jill claims that small pressure differences will be detected more readily if, instead of sealing the end of the capillary tube, it is fitted into a flask as shown in Fig. 4–41(b). Discuss Jill's claim, mentioning any advantages or disadvantages you see to this kind of pressure gauge.

4.20 The diver in Fig. 4–42 is trying to raise a piece of metal from the sea bed. Explain his use of the two containers.

Fig. 4–41

Fig. 4–42

5 Particles

It seems to me that God, in the beginning, formed Matter in solid, hard, massy, impenetrable, moveable Particles.' *Isaac Newton*

Most of us are not satisfied by an investigation which shows *what* happens: we want to know *why* it happens. We want some kind of 'explanation'; we want in some sense to 'understand' the event. But there is no standard recipe for understanding, no single explanation which will satisfy everyone. You might invoke Newton's Third Law to explain the appearance of a dent in the off-side wing after borrowing the family car. Even if your father is a scientist, he is hardly likely to be impressed by such an explanation. The writer of *Genesis* and the writers of physics texts are talking about different things when they 'explain' the rainbow. In these examples it is not difficult to distinguish between the *motive*, in human terms, and the *mechanism*, in scientific terms, and few of us would confuse the issues involved.

It is sometimes said that a 'why' question refers to motives and a 'how' question to scientific descriptions. This is, however, an oversimplification. When people ask a 'why' question they often expect the answer to describe some phenomenon in terms of something with which they are more familiar. Moreover, one 'why' question usually leads to another which in turn leads to another and so on until we are forced to answer 'because it is in the nature of things' or else refer the questioner to a department of theology!

Now consider a more philosophical question. To what extent can we expect to understand a complex system in terms of the behaviour of its constituent parts? Certainly an explanation of the working of, say, a machine may be simply a description of the way in which its various parts behave. Yet the whole is *not* merely the sum of the parts in every case. The properties of water are not, for example, revealed by looking at the properties of its constituents—hydrogen and oxygen.

In this chapter we are going to look at attempts that have been made first to study the microscopic and sub-microscopic particles that make up matter and then, secondly, to see if their behaviour can be used to explain the bulk properties of solids, liquids, and gases. You must judge whether or n such explanations are helpful.

ATOMS

'This process, as I might point out, is illustrated by an image of it that is continually taking place before our very eyes. Observe what happens when sunbeams are admitted into a building and shed light on its shadowy places. You will see a multitude of tiny particles mingling in a multitude of ways in the empty space within the light of the beam, as though contending in everlasting conflict, rushing into battle rank upon rank with never a moment's pause in a rapid sequence of unions and dis-unions.

From this you may picture what it is for the atoms to be perpetually tossed about in the illimitable void. To some extent a small thing may afford an illustration and an imperfect image of great things. Besides, there is a further reason why you should give your mind to these particles that are seen dancing in a sunbeam: their dancing is an actual imitation of underlying movements of matter that are hidden from our sight. There you will see many particles under the impact of invisible blows changing their course and driven back upon their tracks, this way and that, in all directions. You must understand that they all derive this restlessness from the atom.

It originates with the atoms, which move themselves. Then those small compound bodies that are least removed from the impetus of the atoms are set in motion by the impact of their invisible blows and in turn cannon against slightly larger bodies. So the movement mounts up from the atoms and gradually emerges to the level of our senses, so that those bodies are in motion that we see in sunbeams, moved by blows that remain invisible.'

If this passage rings an amazingly modern b you may be interested to know that it was writt over 2000 years ago by Lucretius and that ato had been 'invented' hundreds of years before th

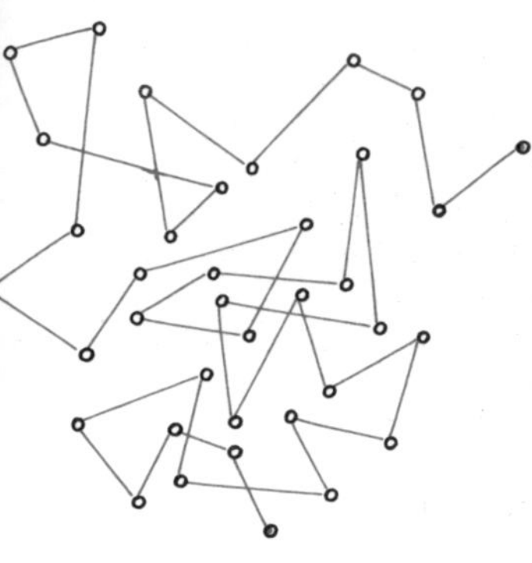
Fig. 5–1

Although we now know that convection currents cause the dust particles to dance around in the sunlight, the above passage might well have been a description of the *Brownian motion* you have studied under the microscope.

Fig. 5–1 represents the zigzag path of a smoke particle under haphazard bombardment by invisible air molecules. The effects of air molecules bombarding a tiny mirror can also be shown with

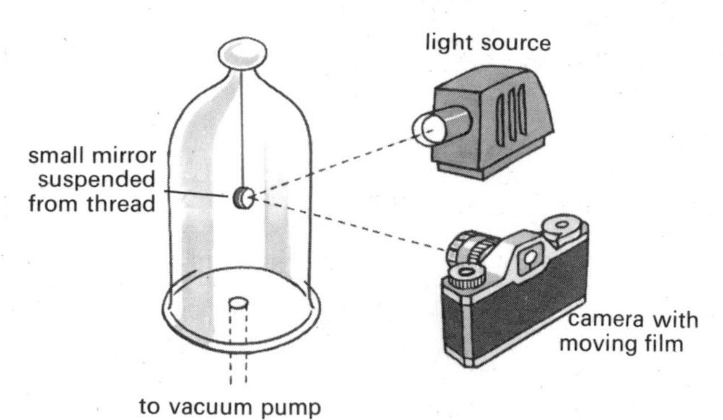

Fig. 5–2

the apparatus illustrated in Fig. 5–2. The mirror is freely suspended in air at a very low pressure. A beam of light reflected from the mirror on to a moving film in a recording camera will produce an erratic pattern similar to that shown in Fig. 5–3.

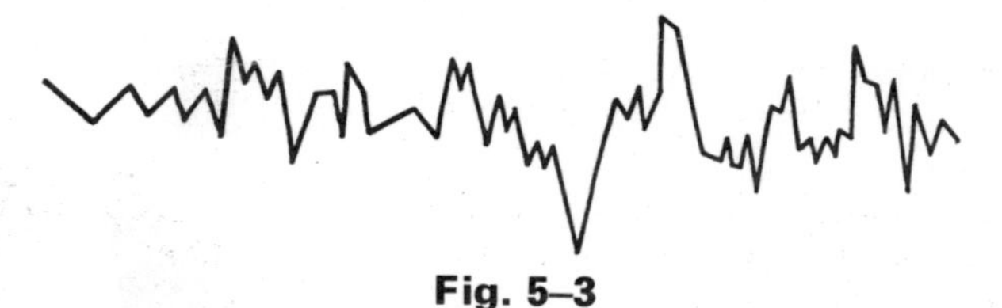
Fig. 5–3

Although the idea of small indivisible chunks of matter had been in existence since about 400 B.C., it was not until the beginning of the nineteenth century that atoms became scientifically respectable. The work of John Dalton and Amedeo Avogadro was largely responsible for this. A distinction was beginning to emerge between atoms and molecules. Avogadro was able to explain that, at standard temperature and pressure,

He suggested that, at a specific temperature and pressure, a given volume of *any* gas contains the same number of particles (molecules).

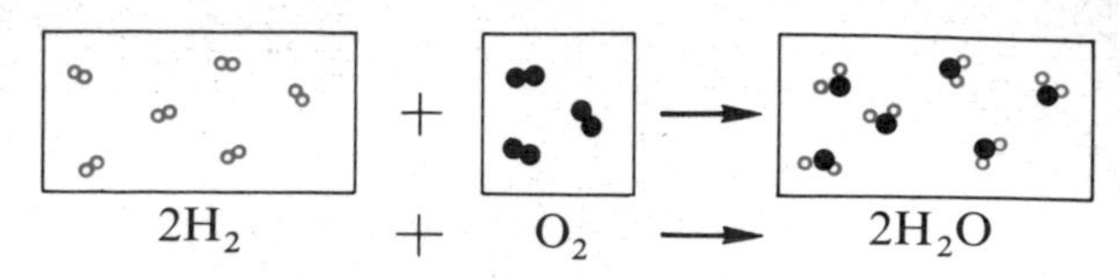

The atomic model gained popularity during the nineteenth century but it was not until 1899 that the first experimental estimate was made of molecular dimensions. In that year Lord Rayleigh estimated the thickness of a film of oleic acid floating on water. Your own oil-film experiment (Fig. 5–4) should have given a similar result: namely that, if the film is one molecule thick, the length of the molecule must be of the order of 1 nanometre ($1\ \text{nm} = 10^{-9}\ \text{m}$).

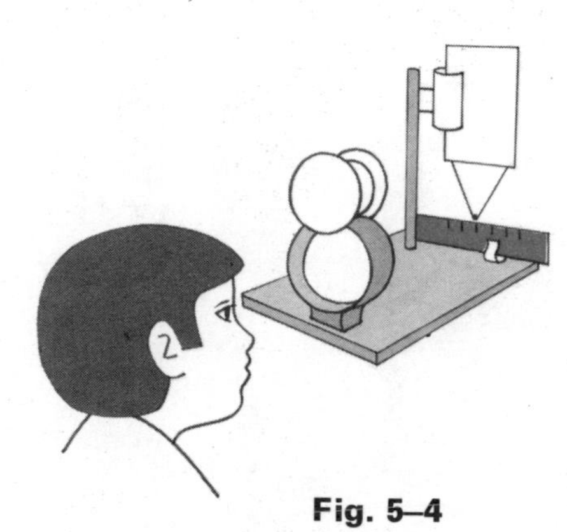
Fig. 5–4

Evidence supporting the existence of small particles of matter is now overwhelming. Fig. 5–5 represents the atoms on the tip of a fine tungsten needle—magnified a million times. An electron-

Fig. 5–5

Fig. 5–6

Fig. 5–9

Fig. 5–7

microscope photograph of a protein crystal is shown in Fig. 5–6. The large molecules can be clearly seen. They are magnified about 100 000 times. Again the growth of regular crystals such as those of copper sulphate (Fig. 5–7) could be accounted for by assuming that they are built up by layer upon layer of identical particles. Fig. 5–8 shows a shadow picture of a crystal lattice. It was produced by bombarding the crystal with a beam of protons.

If bromine vapour enters an air filled jar, it very gradually diffuses through it. If the air and the bromine consist of large numbers of randomly moving particles often in collision, this diffusion would make sense.

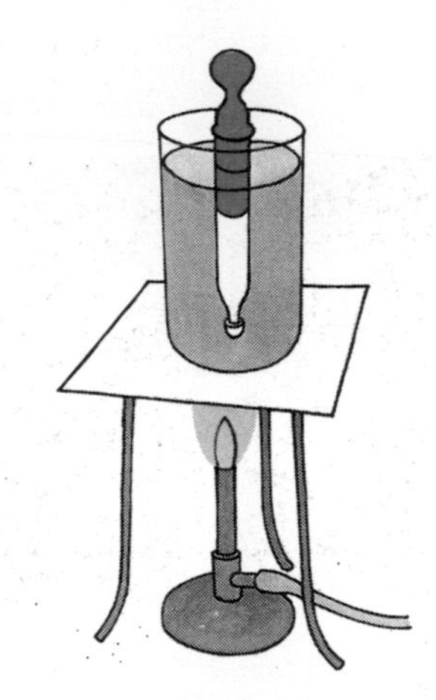

Fig. 5–10

Fig. 5–8

CHANGE OF STATE

When you melt a solid such as wax or lead or ice, the liquid formed has very nearly the same volume. However, experiments show that when a solid or a liquid changes to gas a dramatic change takes place. When dry ice changes to carbon dioxide gas (Fig. 5–9) or water to steam (Fig. 5–10), the volume increases to about a thousand times its original volume.

These experiments suggest that, if we imagine the molecules in a solid or liquid to be more or less 'in contact' with each other, then in the gaseous state at room temperature and pressure they would be around $1000^{\frac{1}{3}} = 10$ molecular diameters apart.

In terms of our atomic model we imagine the atoms in a solid vibrating to and fro about their mean position in the crystal lattice (Fig. 5–11). To overcome the attraction between the atoms or molecules energy must be supplied. This is the latent heat of fusion. This energy gives the molecules more freedom to move around in the liquid state (Fig. 5–12).

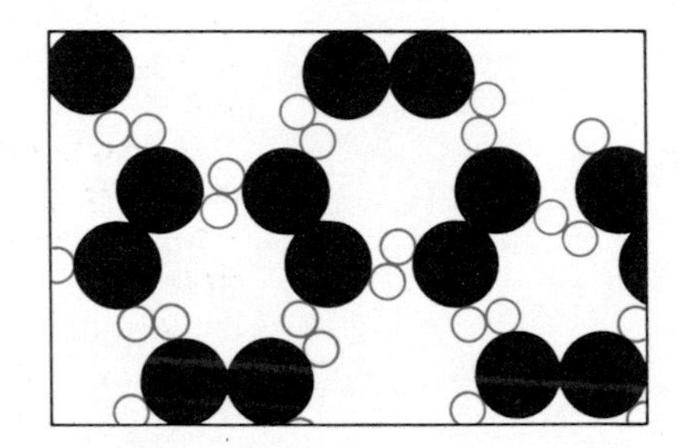

Fig. 5–11

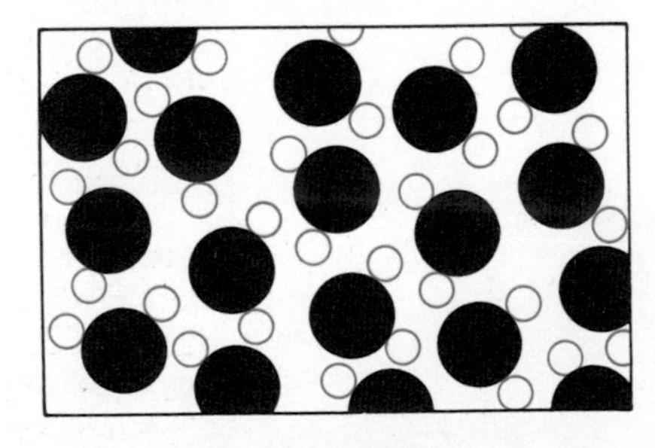

Fig. 5–12

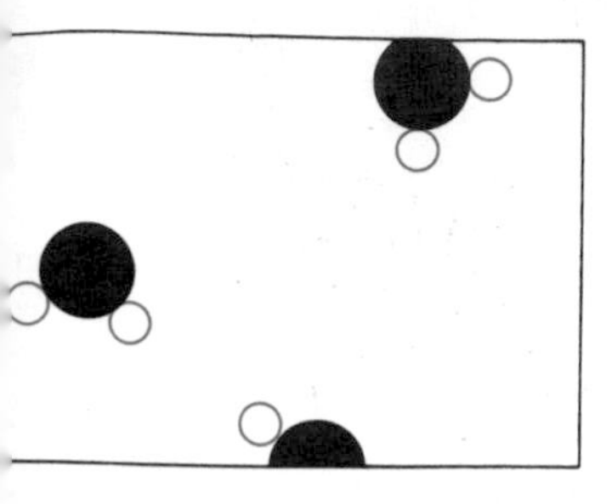

Fig. 5–13

If we continue supplying energy to the liquid, some of the molecules will eventually leave the liquid completely and fly around on their own (Fig. 5–13).

The energy needed to give them this additional freedom is called the latent heat of vaporization. The internal energy of a gas includes the kinetic energy of the molecules. It increases with temperature.

With the help of a computer it is possible to simulate the conditions of a small number of atoms in the three different states. The paths of the 'atoms' displayed on an oscilloscope are represented in Fig. 5–14.

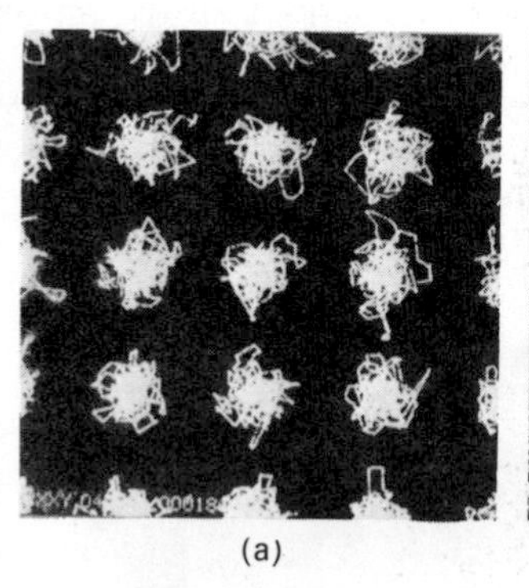

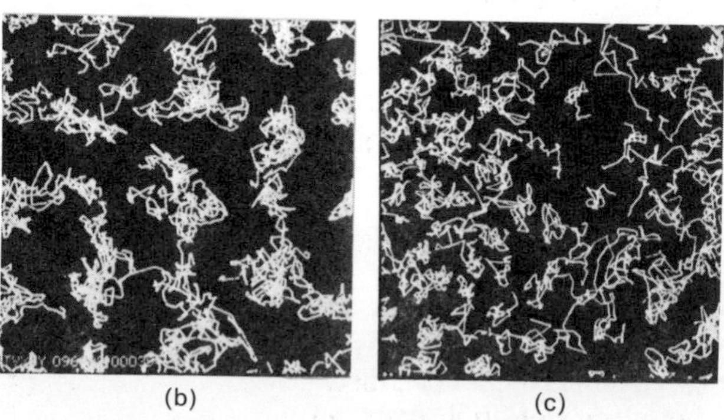

(a) (b) (c)

Fig. 5–14

In the simulated solid (a) the atoms vibrate about one position in the lattice. In the 'liquid' state (b) the atoms move around but in a restricted way, and in the 'gas' state (c) there is a completely random movement.

To sum up, we can think of the three states of matter (Figs. 5–11, 5–12, and 5–13) as results of a struggle between inter-molecular forces of attraction trying to hold the particles together, and increased internal energy trying to make them vibrate more violently or rush off energetically on their own. As the internal energy increases, a solid changes to a liquid and then to a gas.

Internal energy is the *total* energy in a substance. It includes potential energy gained during a change of state and kinetic energy gained when the temperature is raised. The internal energy of a substance can be increased, firstly, by doing work on the substance. For example, the internal energy of the air in a bicycle pump is increased if the air is compressed quickly. The internal energy of the brake pads in a car is increased if the brakes are applied when the car is moving. Secondly, the internal energy of a substance is increased when it is brought into contact with a hotter substance so that there is a transfer of energy (heat).

GAS LAWS

Small metal spheres jostling in a glass tube are sometimes used as an analogue of gas molecules (Fig. 5–15). If we increase the pressure by increasing the weight of the piston, the volume decreases. This certainly is in line with the results of our Boyle's Law experiment (Fig. 5–16), in which the pressure varies inversely with the volume

$$p \propto \frac{1}{V} \text{ (temperature constant)}$$

Fig. 5–15

Fig. 5–16

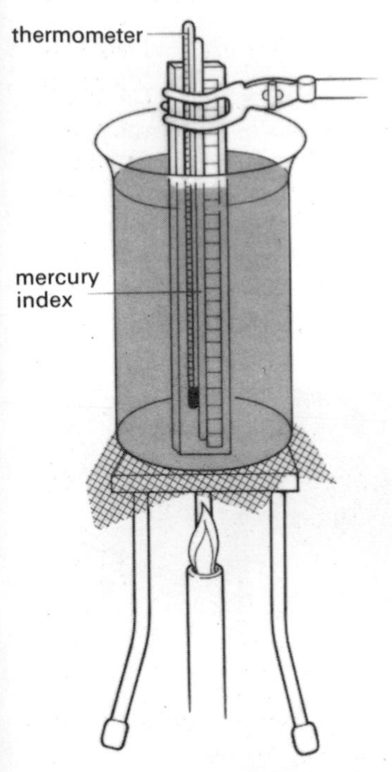

Fig. 5–17

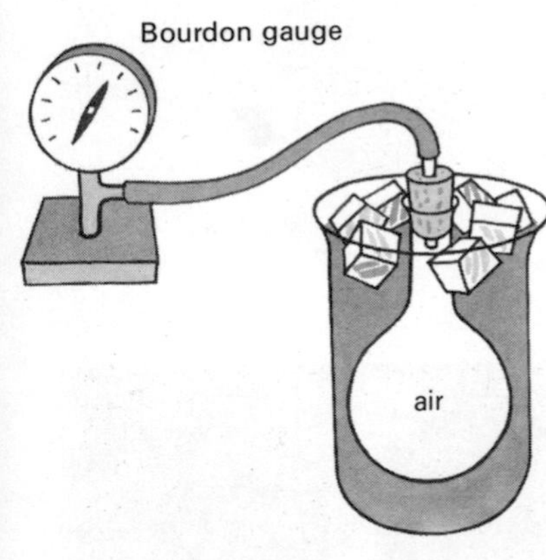

Fig. 5–18

Fig. 5–19

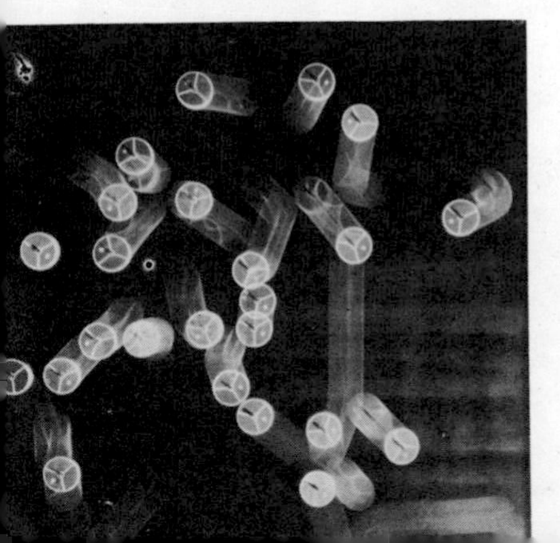

If we now return to our analogue and increase the speed of the vibrating spheres, the piston rises; that is, the volume increases. The Charles' Law experiment (Fig. 5–17) gives results which are consistent with this.

$$V \propto T \quad \text{(pressure constant)}$$

Finally, we could keep the piston in the same place (constant volume) as we increase the speed of the vibrating spheres (increased temperature). To do this we find that we must add masses to the piston to increase its weight (greater pressure). The pressure law experiment (Fig. 5–18) gives similar results.

$$p \propto T \quad \text{(volume constant)}$$

It seems, then, that at least in a qualitative way the molecular model developed from the vibrating spheres analogue predicts the same kind of results as we obtain from experiment.

MATHEMATICS FOR THE MILLION

The Greeks, as we have seen, pictured a gas as millions of tiny particles flying around in all directions at high speeds, but it was not until some two thousand years later that the theory was developed by Jakob Bernouilli in the eighteenth century and James Clerk Maxwell in the nineteenth century to produce what we now call the kinetic theory of gases.

Real gases are extremely complicated, particularly when they have molecules consisting of a number of different atoms. So in trying to build our mathematical model we will have to simplify and, of course, let's face it, falsify the situation to some extent. Some of the assumptions we will have to make can be illustrated by the two-dimensional mechanical analogue illustrated in Fig. 5–19. A number of identical pucks are floating on a cushion of air on a horizontal table and a vibrating wire round the sides keeps them moving hither and thither. A camera shutter is then opened just before cutting off the air supply, so that the length of the track gives an indication of the velocity of each puck before it is suddenly stopped by friction. You can see that the pucks were moving at many different speeds in many different directions. On average there will be the same number of collisions per second with each wall; that is, the motion is random. We could extend the analogy by imagining that the pucks are magnets and that the sides of the table are lined with magnets so that all the collisions would then be elastic. Finally, if we imagine tha there are no frictional forces, the pucks could g on moving indefinitely without the help of th vibrating wire.

In our simplified theory for an 'ideal' gas we wi assume that

(a) the gas consists of a large number of identic molecules each of mass m,
(b) the molecules are very small compared to th spaces between them,
(c) the molecules exert no forces on one anothe
(d) the molecules are in continuous rando motion, and
(e) all collisions with the walls of the container a elastic; that is, there is no energy transfer betwee the molecules and the walls.

The calculations we are going to make are base on Newton's Second Law of Motion; so let us sta with a simple example:

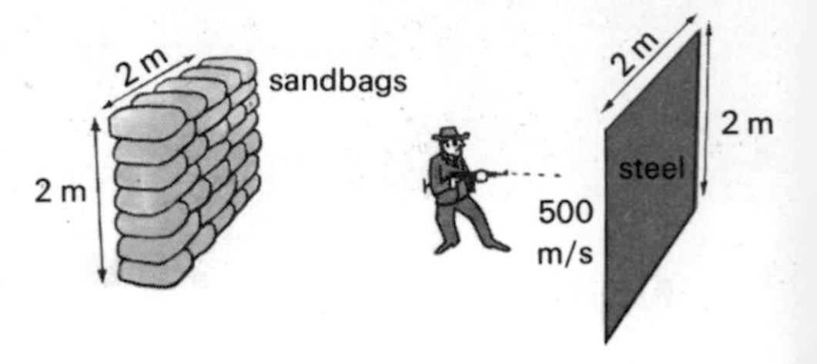

Fig. 5–20

A man fires a spherical rubber bullet (of ma 0.01 kg) directly at a steel wall (2 m × 2 m), as show in Fig. 5–20. If the bullet bounces back at the sa speed (500 m/s) then, taking the initial direction the bullet as the positive direction, we have

momentum of bullet before striking wall =

$$mv = 0.01 \times 500 = 5\ \text{kg m/s}$$

momentum of bullet after striking wall =

change of momentum of bullet =

$$-mv-(+mv) = -5-5 = -10\ \text{kg m/s}$$

If the bullet misses the man on its return journ and comes to rest in a bed of sandbags (al 2 m × 2 m), the change of momentum of the bul is only -5 kg m/s.

Imagine now that a continuous stream of bull is fired from the gun at a rate of 20 per secon Then the total change of momentum of the bull at the steel wall every second is as follows:

total change of momentum of the bullet

$$= 20 \times \Delta(mv)$$
$$= 20 \times (-10)$$
$$= -200 \text{ kg m/s}$$

Now, as the force is equal to the rate of change of momentum, we have

$$\text{force} = \frac{\text{change of momentum}}{\text{time}}$$
$$= -200 \text{ kg m/s per second}$$
$$= -200 \text{ kg m/s}^2$$
$$= -200 \text{ N}$$

The minus sign indicates that the force acts in the negative direction, i.e. wall to man. It is the force exerted by the wall on the bullet. By Newton's Third Law the force exerted by the bullet on the wall will be +200 N. As this is a positive quantity, the direction of the force is positive, i.e. man to wall. The average pressure at the steel wall can then be found by dividing the force by the area.

$$\text{Pressure} = \frac{\text{force}}{\text{area}} = \frac{200}{4} = 50 \text{ Pa}$$

The pressure at the wall of sandbags

$$= \frac{100}{4} = 25 \text{ Pa}$$

You can see that the pressure at the steel wall due to elastic collisions is twice as great as the pressure at the sandbags due to inelastic collisions. This fact can be illustrated by dropping a cricket ball and then a lump of plasticine of the same mass on to a steel plate on a compression balance (Fig. 5–21).

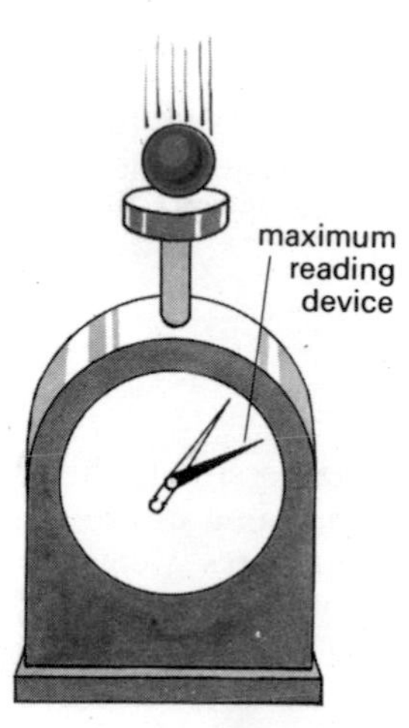

Fig. 5–21

Imagine that a rectangular box of length l, breadth b, and height h contains gas molecules flying around in all directions. We will assume that on average as many fly up and down (direction h) as in and out (direction b) or to and fro (direction l).

To calculate the force exerted on the container by this gas, we will start by thinking about a single molecule of mass m moving to and fro between the shaded walls (Fig. 5–22).

Fig. 5–22

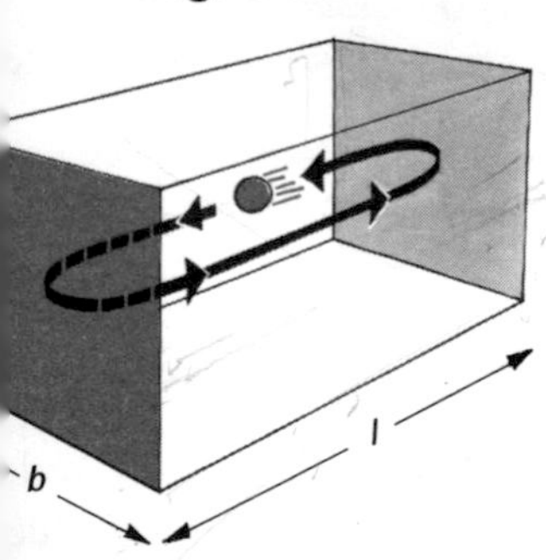

Molecular speed is sometimes represented by c. To avoid confusion with the speed of light we will use v. If then the molecule moves with a constant speed v and takes a time Δt to move from one wall to the other and back again, then

$$v = \frac{\text{distance}}{\text{time}} = \frac{2l}{\Delta t}$$

Therefore $$\Delta t = \frac{2l}{v}$$

As the molecule moves towards the red wall its momentum is mv and as it leaves its momentum will be $-mv$, so that the

$$\text{change of momentum} = -mv - mv$$
$$= -2mv$$

As in the example of the bullet striking the steel wall, the average force on the molecule is equal to the rate of change of momentum (i.e. change of momentum per unit time).

$$F = \frac{\Delta(mv)}{\Delta t}$$
$$= \frac{-2mv}{2l/v}$$
$$= \frac{-mv^2}{l}$$

By Newton's Third Law the force acting on the wall is equal and opposite; that is it is $+mv^2/l$. It acts in the same direction as the molecule approaching the red wall.

Now let us imagine that there are N molecules in the box and that each has the same speed v. On the average $\frac{1}{3}N$ molecules would be moving up and down, $\frac{1}{3}N$ in and out, and $\frac{1}{3}N$ to and fro between the shaded walls. The *total* force exerted on the red wall would therefore be

$$F = \tfrac{1}{3}N\frac{mv^2}{l}$$

and the average pressure p would be given by

$$p = \frac{\text{force}}{\text{area}}$$
$$= \frac{\frac{1}{3}N(mv^2/l)}{b \times h}$$
$$= \frac{1}{3}\frac{Nmv^2}{lbh}$$

but $$lbh = \text{the volume } V \text{ of our box}$$

so $$p = \frac{1}{3}\frac{Nmv^2}{V}$$

or $$pV = \tfrac{1}{3}Nmv^2$$

The gas molecules are flying hither and thither at random. As they are colliding with each other and the walls millions of times a second, their speeds will be varying all the time. Our assumption, then, that all their speeds were the same is hardly likely to be correct. If, however, we take the average value of all the individual values of v^2, we should get the same result. We will write this average value as $\overline{v^2}$. It is called 'v^2 bar'. Our equation can then be written as

$$pV = \tfrac{1}{3}Nm\overline{v^2}$$

As we know that the kinetic energy of a moving body is $\tfrac{1}{2}mv^2$, we could rewrite this equation as

$$pV = \tfrac{2}{3}N(\tfrac{1}{2}m\overline{v^2})$$

where the expression in the bracket $(\tfrac{1}{2}m\overline{v^2})$ represents the average kinetic energy of a molecule. This assumes that all the kinetic energy is due to the molecules moving from place to place and ignores any kinetic energy due to their tumbling head over heels or vibrating as they go. For a given mass of gas, that is, one containing a constant number of molecules N, we have

$$pV = \tfrac{2}{3}N \text{ (average K.E. per molecule)}$$

$$pV \propto \text{average K.E. per molecule.}$$

THEORY AND PRACTICE

Our simplified theory was based on an ideal gas in which all the kinetic energy was translational, i.e. that of the molecules moving from place to place. Gases such as helium with one-atom molecules are very near to this ideal. Let us, however, compare the theory with the result of our experiments on a constant mass of air. We will *assume* that the average kinetic energy of the molecules is proportional to the absolute temperature.

Boyle's Law

$$p \propto \frac{1}{V} \text{ (or } pV = \text{constant)}$$

when the temperature is constant.
From our kinetic theory

$$pV \propto \text{average K.E. per molecule}$$

that is

$$pV = \text{constant}$$

when the temperature and thus the average kinetic energy per molecule is constant. Thus the theory agrees with experiment, bearing in mind that our experimental results were not particularly precise!

Charles' Law

$$V \propto T \text{ (or } \frac{V}{T} = \text{a constant)}$$

when the pressure is constant.
From our kinetic theory

$$pV \propto \text{average K.E. per molecule}$$

thus, when p is constant

$$V \propto \text{average K.E. per molecule.}$$

Again we have agreement between theory ar
experiment.

General Gas Law

$$\frac{pV}{T} = \text{constant}$$

or $\quad pV \propto T$

For a given mass of gas, kinetic theory predicts

$$pV \propto \text{average K.E. per molecule}$$

This is consistent with our assumption that
average kinetic energy of the molecules is prop
tional to the absolute temperature.

Avogadro's Law

Dalton had considered the possibility that the pa
ticles in all gases were the same distance apart
the temperatures and pressures were the sar
Unfortunately, experimental results seemed to
dicate that this was not so. For example, c
volume of hydrogen atoms and one of chlor
atoms produced two volumes of hydrogen chlori

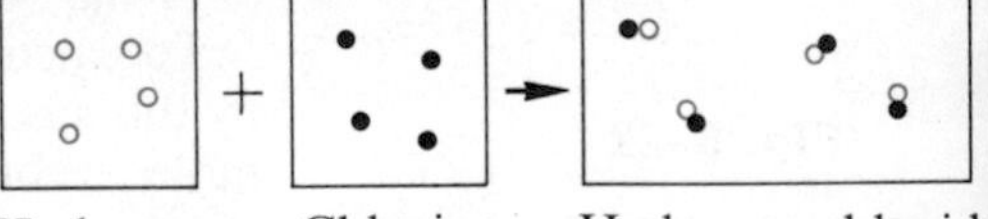

Hydrogen Chlorine Hydrogen chloride

Avogadro resolved the difficulty by assum
that in each gas the particles (molecules) consis
of two atoms.

Hydrogen Chlorine Hydrogen chloride

It was now possible to think of the molecules as being, on average, the same distance apart. Avogadro then put forward his hypothesis or law, which states that equal volumes of all gases at the same temperature and pressure contain equal numbers of molecules.

$$pV = \tfrac{1}{3}N_1 m_1 \overline{v_1^2} \quad = \tfrac{1}{3}N_2 m_2 \overline{v_2^2}$$
$$= \tfrac{2}{3}N_1(\tfrac{1}{2}m_1\overline{v_1^2}) = \tfrac{2}{3}N_2(\tfrac{1}{2}m_2\overline{v_2^2})$$

Thus $N_1 = N_2$ if the average kinetic energy per molecule is the same in each case. Kinetic theory agrees with Avogadro's hypothesis if we *assume* that the temperature is constant when the average kinetic energy per molecule is constant.

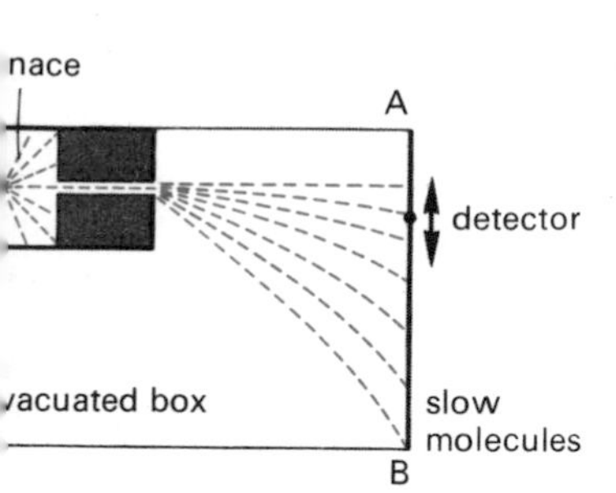

Fig. 5–23

MOLECULAR SPEEDS

To investigate the different speeds at which atoms travel, a metal can be vaporized in a small furnace (Fig. 5–23). Atoms moving in a horizontal direction pass through a narrow slit and proceed towards a target AB. Because of the gravitational pull, the slower the atoms are moving the nearer to B they will strike the target. We can discover the distribution of speed if we can measure the rate at which atoms strike different parts of the target. This can be done with a sensitive detector in which the atoms are ionized and the ion current then measured. The detector is slowly moved from A to B and readings are taken at various positions. The results show that the speeds of the atoms vary as shown in Fig. 5–24.

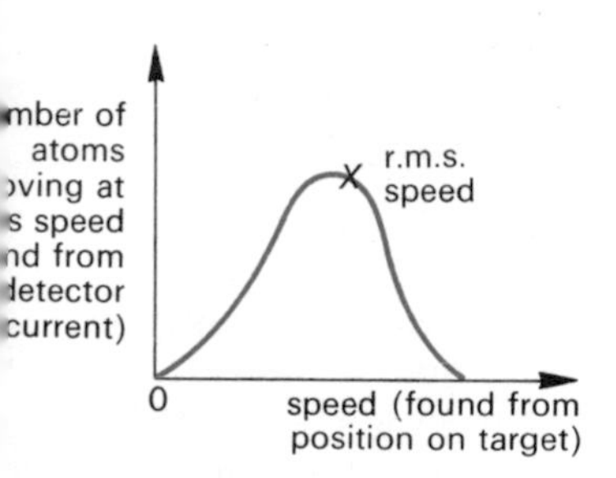

Fig. 5–24

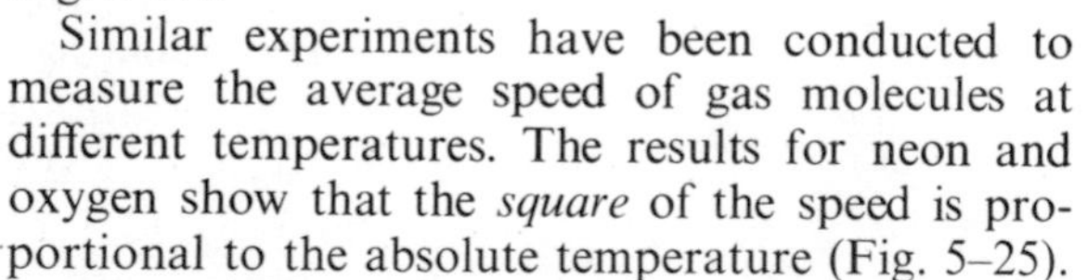

Similar experiments have been conducted to measure the average speed of gas molecules at different temperatures. The results for neon and oxygen show that the *square* of the speed is proportional to the absolute temperature (Fig. 5–25).

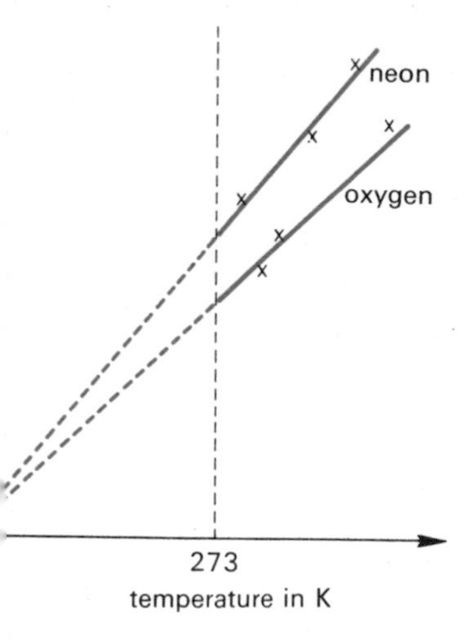

Fig. 5–25

This supports our assumption that the average kinetic energy of the molecules is proportional to the absolute temperature.

We can now use the kinetic theory to find molecular speeds. If each of N molecules in a gas has a mass m, the total mass of the gas will be Nm. The density ρ is then given by Nm/V. It follows that

$$pV = \tfrac{1}{3}Nm\overline{v^2}$$

may be written $\quad p = \tfrac{1}{3}\rho\overline{v^2}$

To find the 'average' speed of the molecules we take the square root of $\overline{v^2}$. This is the root of the mean of the squared speeds, i.e. the r.m.s. speed. It is represented on Fig. 5–24 by a point slightly to the right of the peak. From the above relation we have

$$\text{r.m.s. speed} = \sqrt{\frac{3p}{\rho}}$$

We can use this relationship to calculate the speed of nitrogen molecules. As the density of nitrogen is 1.25 kg/m^3 at s.t.p. (273 K and 10^5 Pa very nearly), we have

$$\text{r.m.s. speed} = \sqrt{\frac{3 \times 10^5}{1.25}}$$
$$= 490 \text{ m/s}$$

which is about the speed of a rifle bullet.

It is possible to measure molecular speeds directly, although not with school apparatus. The results obtained show good agreement with the calculated values, thus providing further support for the kinetic theory.

As sound is carried by longitudinal waves in air—which is mainly nitrogen—compressions and rarefactions cannot be transmitted more quickly than the speed of the gas molecules. Similarly a longitudinal wave in a slinky cannot travel more quickly than the speed of the individual turns (Fig. 5–26). Nevertheless we might expect the speed of sound in air to be of the same order of magnitude as the molecular speed calculated above. Your earlier experiments which showed that sound travels at about 330 m/s confirm this prediction.

Fig. 5–26

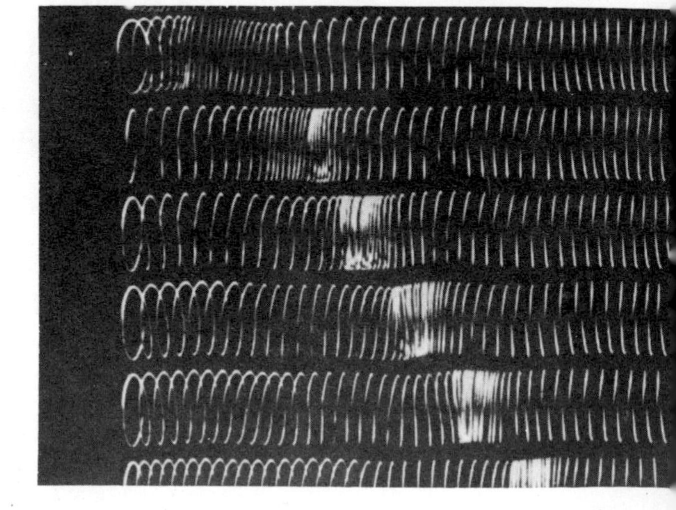

Here are some other values for the speed of sound and molecular speeds at s.t.p.

gas	speed of sound (m/s)	r.m.s. molecular speed (m/s)	relative molecular mass (molecular wt.)
hydrogen	1286	1694	2
helium	972	1202	4
neon	435	535	20
oxygen	315	425	32
mercury vapour	—	170	200

How could you show that the figures given in the above table are consistent with our earlier assumption that the average kinetic energy per molecule is the same for different gases at the same temperature? (5.1)

The wavelength of the note produced by a whistle is determined by the physical length of the whistle. *Remembering that* $v = f\lambda$ *how would you expect the frequency of the note to vary with the speed of sound? (5.2) If hydrogen or helium were used to blow the whistle, what would you expect to happen? (5.3)*

A more amusing effect can be produced by someone inhaling helium before speaking! (Fig. 5–27). *Can you explain this? (5.4)*

Fig. 5–27

MOLECULAR ARITHMETIC

In nature all the atoms of one particular element- that is, those with the same *atomic number*—a not necessarily identical even though they exhib the same chemical properties. The nucleus of eac atom contains the same number of protons but nc necessarily the same number of neutrons. As th *mass number* is the number of protons + neutron atoms of a given element can have different ma numbers. Atoms with the same atomic numb (proton number) and different mass numb (nucleon number) are called *isotopes of the elemen*

Instead of defining the mass of the hydroge atom as one unit or the oxygen atom as 16 unit it was decided in 1961 to take the most commo isotope of carbon, carbon-12, as the standard an to define its mass as exactly 12 unified atomic ma units (u). We can express the masses of the atom of other elements in these units. The masses a then referred to as *relative masses*.

As nearly all the elements found in nature are mixture of several isotopes, it is an *average* val which is found for such elements. Some of the average values are given below.

hydrogen	=	1.007 97	u
helium	=	4.0026	u
carbon	=	12.011	u
oxygen	=	15.9994	u

These relative mass values are sometimes st called 'atomic weights' and, for our purposes, w will normally use the nearest whole number.

Suppose there are N_A atomic mass units in gram; then 1 atomic mass unit = $1/N_A$th of a gran The actual mass of an atom would be as follow

hydrogen atom	$\frac{1}{N_A}$ gram
helium atom	$\frac{4}{N_A}$ gram
carbon atom	$\frac{12}{N_A}$ gram
oxygen atom	$\frac{16}{N_A}$ gram

It follows that

1 gram of hydrogen contains N_A atoms
4 grams of helium contain N_A atoms
12 grams of carbon contain N_A atoms
16 grams of oxygen contain N_A atoms

The term *mole* (mol or gram molecule) is used describe the quantity of a substance containing *N* particles. As in the case of the 'atomic masses', carbon-12 is used to define the mole. That is

> ***1 mole of a substance contains the same number of particles as there are atoms in 12 grams of carbon-12.***

This number N_A has been found to be about 6×10^2E. It is called the Avogadro Constant or Avogadro's Number. Twelve grams of carbon-12 (1 mole of carbon-12) contain 6×10^{23} atoms. Similarly, 1 g hydrogen, 4 g helium, and 16 g oxygen each contain 6×10^{23} atoms.

In the above examples we have taken the mole as the gram atom. It is also used for the gram molecule. Hence

2 grams of hydrogen gas contain 6×10^{23} molecules
32 grams of oxygen gas contain 6×10^{23} molecules

Sir James Jeans once calculated that we inhale 10^{22} molecules with each breath and that there are about 10^{44} molecules in the whole of the Earth's atmosphere. From these figures he concluded that if George Washington's last breath has now been scattered throughout the atmosphere, the chances are that we inhale one of these molecules with each breath we take!

Avogadro's Number, N_A can be expressed as
6×10^{23} atoms/gram atom
6×10^{23} molecules/gram molecule
or more simply and correctly as
6×10^{23}/mol

Avogadro's Constant has been found experimentally in many ways. Perhaps the simplest method is to find, by the electrolysis of water, the charge required to liberate 1 gram of hydrogen. This is found to be 96 500 coulombs and represents the charge involved in changing N_A hydrogen ions into N_A atoms of hydrogen gas. As each ion requires 1 electron of charge e

$$96\,500 \text{ coulombs} = N_A \times e$$

In 1909 Robert Millikan was able to measure the electronic charge e. He found it was approximately 1.6×10^{-19} coulombs. From these results we have

$$N_A = \frac{96\,500}{1.6 \times 10^{-19}}$$

$$= 6 \times 10^{23}$$

Experiment shows that 1 mole of all gases at s.t.p. (273 K and 10^5 Pa) occupies 2.24×10^{-2} m^3 (22.4 litres or dm^3). If carbon monoxide (CO) and oxygen (O_2) combine to form carbon dioxide (CO_2), the total number of molecules will be reduced and so the volume must be reduced.

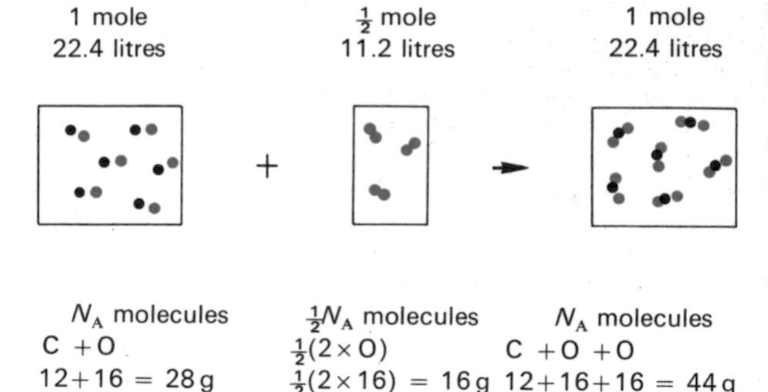

Molar Volume

If we use the experimental information that 1 mole of gas at s.t.p. occupies 2.24×10^{-2} m^3, we can obtain the r.m.s. speed of the molecules $(\overline{v^2})^{\frac{1}{2}}$ from

$$pV_m = \tfrac{1}{3}N_A m\overline{v^2}$$

where

p = standard pressure = 10^5 Pa
V_m = molar volume = 2.24×10^{-2} m^3
N_A = Avogadro's Constant = 6×10^{23}/mol
m = mass of one molecule expressed in kilograms
$$= \frac{\text{mass of 1 mole in grams}}{N_A \times 10^3}$$
i.e. $N_A m$ = the mass of 1 mole expressed in kilograms.

If these quantities, and the standard temperature ($T = 273$ K), are used in the gas equation

$$\frac{pV_m}{T} = R$$

we find the universal molar gas constant R to be 8.31 J/mol K.

THE FORCES BETWEEN REAL ATOMS AND REAL MOLECULES

We have assumed—with some success in the simple kinetic theory—that there are no forces between atoms and molecules. We cannot leave this section without mentioning the inter-particle forces that do exist in real gases, liquids, and solids.

Fig. 5–28

When atoms or molecules are fairly close to each other, forces of attraction exist between them and they move closer together. However, at very close range, strong forces of repulsion come into play and they are pushed apart (cf. Fig. 5–28). There is therefore a mean position where they exert no force on each other. The particles can vibrate to and fro about this mean position (Fig. 5–29). Whether or not the molecules or atoms are held tightly together as in the case of a solid, loosely together as in a liquid or not at all as in a gas depends on the kinetic energy each particle possesses.

Attractive Forces

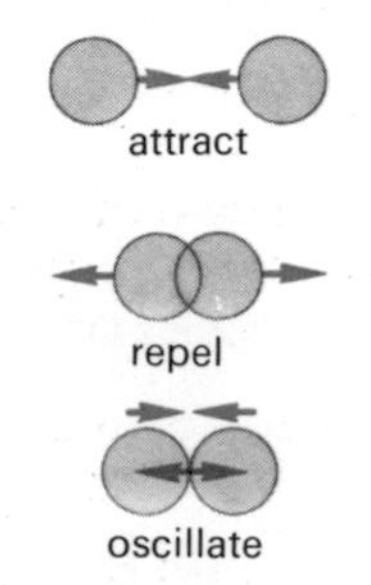

Fig. 5–29

1. *Ionic bond* The forces of attraction and repulsion between atoms and molecules are electric in nature. The first arises as a result of electron transfer. The outermost electron of one atom may transfer to another atom which requires one electron to complete its outer shell. The positive and negative ions thus formed then attract each other.
2. *Covalent bond* Covalent bonds are produced when atoms share their electrons. Two hydrogen atoms, for example, come together so that each molecule has, in effect, two electrons which provide a complete stable shell (Fig. 5–30). The atoms are bound together to form a hydrogen molecule.
3. *Metallic bond* In solid metals the atoms are ionized and exist as positive ions in a 'sea' of free electrons. The particles are held together by strong attractive forces between the ions and the electrons.
4. *Van der Waals forces* These forces exist between all molecules and atoms. They are very weak compared to the others but in gases they are the *only* attractive inter-particle forces. They arise because of some instantaneous fluctuations of the electrons near the nucleus. For an instant the electrons may be slightly displaced with respect to the positive nucleus, thus forming a *dipole*. This dipole produces an electric field which in turn may polarize another atom nearby (Fig. 5–31). The original dipole and the induced dipole then attract each other. Because of the Van der Waals forces between molecules of a gas, the *measured* pressure exerted at the walls of a container will be *less* than that due to an *ideal* gas. If we call the difference Δp, then

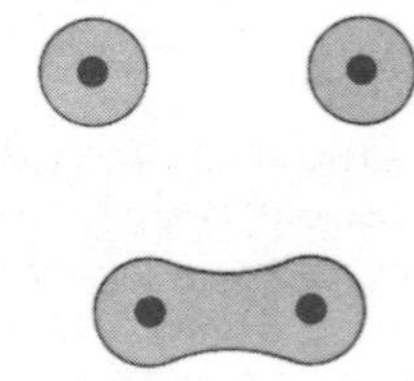

Fig. 5–30

$$p_{\text{measured}} = p_{\text{ideal}} - \Delta p$$

or

$$p_{\text{ideal}} = p_{\text{measured}} + \Delta p$$

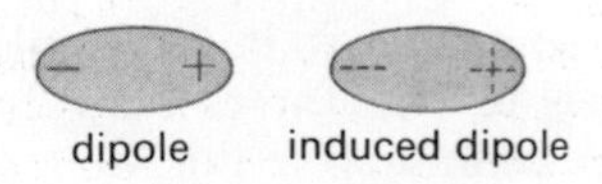

Fig. 5–31

Moreover, the *measured* volume of a real gas will include the volume of the molecules and will therefore be *greater* than the *ideal* volume of empt space.

$$V_{\text{measured}} = V_{\text{ideal}} + \Delta V$$

or

$$V_{\text{ideal}} = V_{\text{measured}} - \Delta V$$

Thus, for a real gas we have

$$(p+\Delta p)(V-\Delta V) = \text{constant} \times T$$

In 1873 Johannes Van der Waals produced a nov famous equation to describe behaviour of a rea gas. In it a, b, and R are constants, p is the measure pressure, and V_m the measured volume of 1 mol of gas

$$\left(p+\frac{a}{V_m^2}\right)(V_m - b) = RT$$

Repulsive Forces

When atoms come very close together, some kin of repulsive force must operate, as otherwise a atoms would coalesce. You can imagine the atom behaving rather like elastic spheres which repe each other when they are squeezed together. Whe the electron shell of one atom penetrates the she of another, the nuclei are no longer complete screened. The two positively charged nuclei the repel one another.

The above description of attractive and repulsiv forces is far from satisfactory. The whole story i extremely involved and cannot be told in simpl terms. However, we can summarize the resultan force between two particles in the graphs shown i Fig. 5–32.

The black lines indicate the forces of attractio and repulsion. The red line shows the combine effect.

The graphs show that the attractive forces ar very small when the molecules are more than a fev diameters apart, as in the case of a gas, in whic the molecules are separated by ten or more dia meters. In solids and liquids the nuclei tend t vibrate about an average separation distance o one diameter.

The kinetic theory is one of the greatest achieve ments of the 'mechanical view' of nature whic was popular during the eighteenth and nineteent centuries. It claimed that it was possible to describ *all* phenomena in terms of simple forces betwee unalterable objects. The 'mechanical view' wa perhaps most clearly stated by Hermann vo Helmholtz in the nineteenth century. Speaking o science he claimed that

active
force

0 separation of nuclei

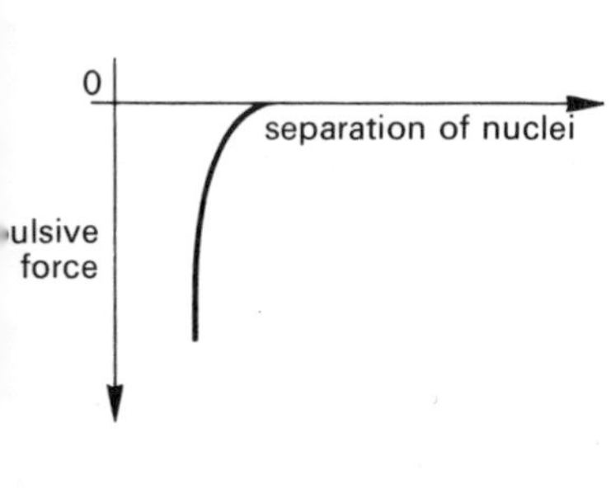

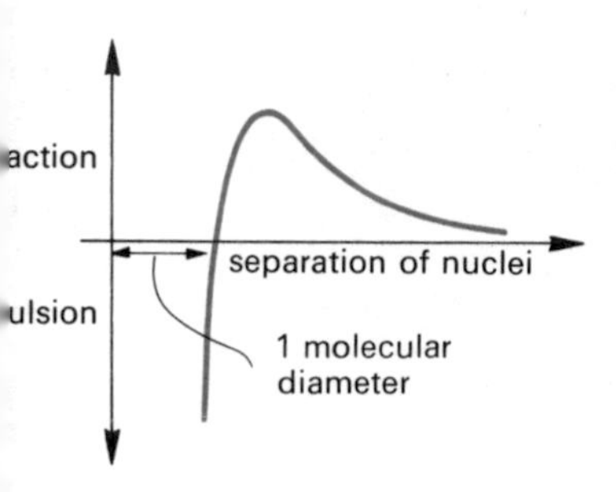

Fig. 5–32

'its vocation will be ended as soon as the reduction of natural phenomena to simple forces is complete and the proof given that this is the only reduction of which the phenomena are capable'.

Although the physics of the last century has shattered any such complacency, simple kinetic theory still stands as a fascinating and fruitful application of Newtonian mechanics.

SUMMARY

Gas Laws $pV \propto T$

or

$$\frac{p_1V_1}{T_1} = \frac{p_2V_2}{T_2}$$

If p = standard pressure = 10^5 Pa

V_m = molar volume at s.t.p. = 2.24×10^{-2} m^3

T = standard temperature = 273 K

the constant of proportionality is called the universal gas constant R = 8.31 J/mol K

$$pV_m = RT$$

From kinetic theory

$$pV = \tfrac{1}{3}Nm\overline{v^2} \quad \text{or} \quad \tfrac{1}{3}Nmc^2$$

where $c = (\overline{v^2})^{\frac{1}{2}}$ i.e. r.m.s. speed

$$p = \frac{1}{3\rho}\overline{v^2}$$

For the molar volume at s.t.p.

$$pV_m = \tfrac{1}{3}N_A m\overline{v^2}$$

where $N_A m$ = mass of 1 mole expressed in kilograms.

PROBLEMS

5.5 If the mass of 1 mole of nitrogen molecules is 28×10^{-3} kg, find the number of molecules in a cubic metre of nitrogen at s.t.p. and the mass of one nitrogen molecule.

5.6 Find the r.m.s. speed of nitrogen molecules
(a) at s.t.p. (density = 1.25 kg m^{-3}) and
(b) in a 'vacuum' at 273 K when the pressure is 10^{-2} Pa.

5.7 A mole of oxygen molecules has a mass of 32×10^{-3} kg. Find:
(a) the r.m.s. speed of the molecules at s.t.p.,
(b) the average kinetic energy of a molecule at s.t.p.,
(c) the average kinetic energy of a molecule at 546 K and a pressure of 10^4 Pa, and
(d) the pressure of the oxygen at 273 K if the density is increased to 14.3 kg m^{-3}.

5.8 In a mixture of hydrogen and oxygen gases at the same temperature which molecules would move faster? Why?

5.9 Will the pressure in a car tyre be greater or less *after* a long fast journey? Why?

5.10 Discuss each of the following in terms of the kinetic or molecular theory of matter.
(a) The three states of matter.
(b) The pressure of a gas.
(c) Expansion of a gas with rise of temperature.
(d) Latent heat.

5.11 If your gas meter measures the volume of gas flowing per second do you or the Gas Board profit when the gas pressure is low? Explain your answer.

5.12 The density of air at s.t.p. is about 1.2 kg m^{-3} and of liquid air 900 kg m^{-3}. What do these figures tell you about the spacing of the molecules in the two states?

5.13 An oil film experiment gives an estimate of 1.2 nm for the molecular length. Taking the density of olive oil as 900 kg m^{-3} and the mass of 1 mole as 0.855 kg, estimate Avogadro's Number on the assumption that the molecules are cubical and are in contact.

5.14 The r.m.s. speed of gas molecules is 400 m s^{-1} when the pressure is 7×10^4 Pa. Find the density of the gas.

5.15 The pressure in a diode valve is 10^{-9} atmospheres. How many molecules would there be in one cubic millimetre of this good 'vacuum'?

5.16 A circular patch of oil 0.4 m in diameter is formed when an oil drop 0.5 mm in diameter is placed on a water surface. Estimate the length of an oil molecule. Is this an estimate of its maximum or minimum length? What assumptions have you made?

5.17 A mercury barometer reads 760 mm at sea level and 630 mm at an altitude of 1.5 km. Compare the density of the air at these levels, assuming the temperature is the same.

5.18 A piston in a cylinder containing gas is pushed in by a force F acting through a small distance Δx. If the small volume change is ΔV, show that the work done $F\Delta x$ is equal to $p\Delta V$. Ignore any change in pressure. What happens to the energy transferred? Explain your answer in terms of molecular behaviour.

5.19 Explain the $\frac{1}{3}$ in the expression $pV = \frac{1}{3}Nm\overline{v^2}$.

5.20 Two jars contain equal volumes of two different gases A and B. They each contain the same number of molecules moving with the same r.m.s. speed but each molecule in B has twice the mass of each molecule in A. Discuss the temperature and pressure in each jar and explain your answer in terms of molecular behaviour.

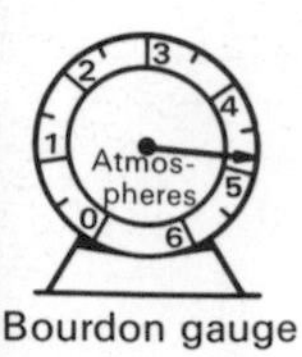

Bourdon gauge
Fig. 5–33

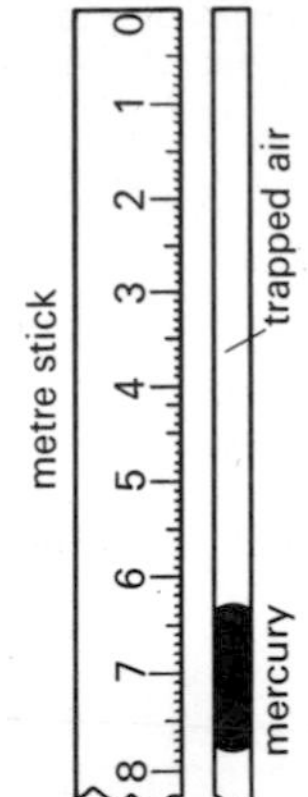

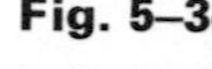
Fig. 5–34

5.21 In an experiment to investigate the relationship between the pressure and the volume of a gas, it is necessary to read the pressure of the gas on a Bourdon gauge and the length of a column of trapped air by a metre stick. The Bourdon gauge shown in Fig. 5–33 measures the pressure above normal air pressure and *not* absolute pressure.

From the sketches of the Bourdon gauge in Fig. 5–33 and the column of trapped gas and metre stick in Fig. 5–34 estimate the values of pressure and length of gas column and the degree of uncertainty in each value. Estimate also the upper and lower limits of the product of pressure multiplied by length of gas column. [S.C.E. (H) 1970]

5.22 (a) It can be shown from theoretical considerations that where p is the pressure of a sample of gas, V is its volume, N is the number of molecules present, m is the mass of a molecule, and $\overline{v^2}$ is the average of the squares of the velocities of the molecules, then

$$pV = \tfrac{1}{3}Nm\overline{v^2} = \tfrac{2}{3}N(\tfrac{1}{2}m\overline{v^2})$$

Write down the results which are obtained from three experiments, each of which is concerned with a different relation involving pressure, volume and temperature of a fixed mass of gas. Show that each of these relations is consistent with the above equation.

(b) Estimate the number of molecules of hydrogen contained in a cylinder of capacity 500 litres in which the gas is under a pressure of 10 atmospheres and is at a temperature of 15 °C. (Take Avogadro's Number as 6×10^{23} molecules per gram mole.)

(c) If the temperature of the gas in the cylinder in (b) is raised, what effect will this have on
(i) the number of molecules in the cylinder,
(ii) the average velocity of the molecules,
(iii) the average distance between molecules, and
(iv) the number of collisions per second between molecules? [S.C.E. (H) 1970]

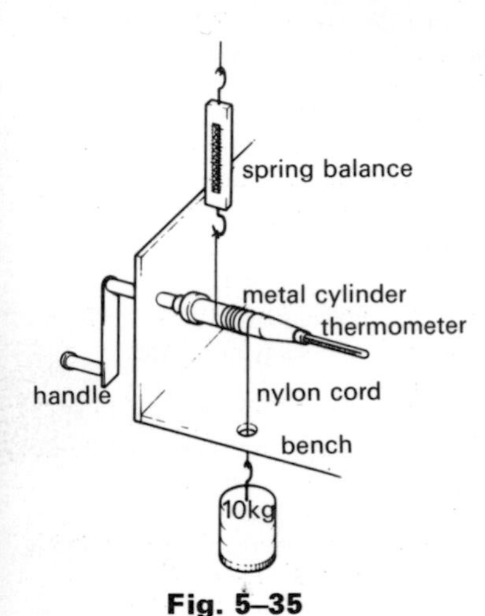

Fig. 5–35

5.23 1 mg of polonium is known to contain approximately 3×10^{18} atoms. Estimate
(a) the mass of a polonium atom,
(b) the 'volume' occupied by a polonium atom if the density of polonium is $9.5 \times 10^3\ \text{kg m}^{-3}$. [S.C.E. (H) 1971]

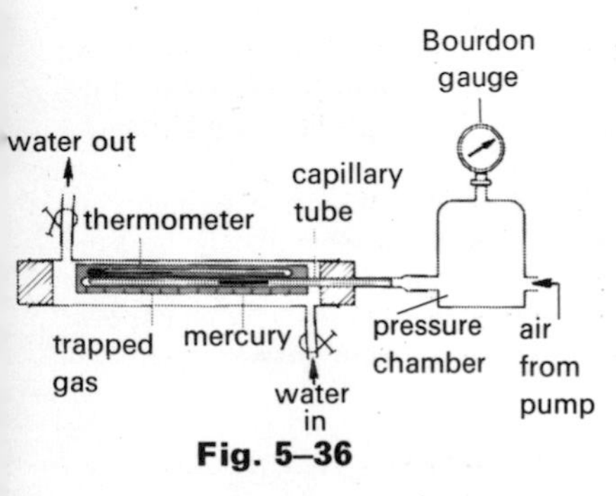

Fig. 5–36

5.24 The pressure p of an ideal gas may be written in the form $pV = \frac{1}{3}Nmv^2$. Explain why the pressure can also be written as $p = \frac{1}{3}\rho v^2$ where ρ is the density of the gas. Hence estimate the root mean square speed of air molecules at standard pressure (10^5 Pa) and temperature, when the density of air is $23\ \text{kg m}^{-3}$. [S.C.E. (H) 1971]

5.25 (a) With reference to the Kinetic Theory of Matter, distinguish between the specific heat capacity and the specific latent heat of a substance.

(b) In the apparatus shown in Fig. 5–35, the number of turns of the cord round the metal cylinder and the rate at which the cylinder is rotated are adjusted until the reading on the spring balance becomes negligible and the 10 kg mass hangs at a constant height above the floor.

The metal cylinder has a mass of 0.25 kg and a circumference of 0.025 m. A temperature rise of 5 °C is produced by turning the handle 150 times. Calculate
(i) the total work done and (ii) the specific heat capacity of the metal.

Discuss whether your answer to (ii) is an overestimate or an underestimate.

(c) 50 g of water at 20 °C in a copper vessel of mass 40 g are raised to a temperature of 70 °C in 6 minutes by an immersion heater. When the experiment is repeated with the vessel lagged the time taken is only 4 minutes. Calculate the average rate of loss of heat per minute from the unlagged vessel. State any assumptions you have made. [S.C.E. (H) 1971]

5.26 It is possible, using the apparatus shown in Fig. 5–36 to change the volume, temperature and pressure of a trapped mass of gas.

(a) The temperature of the trapped gas is kept constant and the pressure increased. Sketch the shape of graph you would expect to obtain if the pressure of the gas were plotted against the length of the trapped gas column.

(b) Water at 27 °C surrounds the capillary tube and the gauge reads 1.00×10^5 Pa (atmospheric pressure).

Air is pumped into the chamber until the gauge reads a steady value of 1.20×10^5 Pa. What is the new length l of the column of trapped gas if it was 26 cm originally?

(c) The water temperature is now changed to 57 °C.
(i) Should air be released from or pumped into the chamber to maintain the length of the column at the value l?
(ii) What will then be the reading on the Bourdon gauge?

(d) The tube containing the trapped gas is now mounted vertically with end E uppermost. Explain whether the pressure of this gas as read from the gauge would be too high or too low.

(e) With the aid of a diagram explain briefly how a Bourdon Gauge works and how it could be calibrated to read gas pressures in Pa. [S.C.E. (H) 1971]

5.27 Give details of how you could obtain experimentally the information necessary for:
(a) estimating maximum molecular diameter;
(b) determining the density of air.
Show in *both* cases how you would use the information in the calculation or calibration. [S.C.E (H) 1972]

5.28 Discuss the practical difficulties involved in finding the absolute zero of temperature in degrees Celsius from a study of the change of pressure of air at constant volume. [S.C.E (H) 1972]

UNIT TWO

Charges in Motion

The Electric Model

6 Charge Carriers

Fig. 6–1

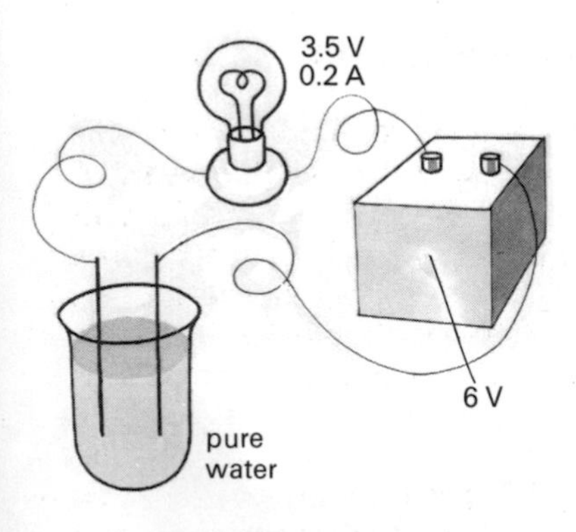

Fig. 6–2

When you slide off the seat of a car and then touch the door, you may discover that you can be a charge carrier! The label is, however, usually used for something a bit smaller, namely one of those particles in the atomic or sub-atomic world which carry electric charge.

You have no doubt seen that pure water will not allow electric current to flow, so that in the circuit of Fig. 6–2 the bulb will not light up.

When salt is added to the water the bulb starts to glow, and we explain this by assuming that charged particles (ions) move in the salt solution. In Fig. 6–3 the positive *ion* is represented as an atom which has lost some negative charge and is therefore positively charged. In semiconductors the positive charge carriers are called *holes*.

The negative ion is shown in Fig. 6–3 as an atom with an extra negative charge stuck to it. In most solid conductors and in thermionic tubes the negative charge carriers are *electrons*. But we are going too fast! The evolution of the idea of the electron was a long slow process. Let us look briefly at some of the landmarks in its 'discovery'.

THE ELECTRON

In the sixth century before Christ, the Greek philosopher Thales found that a piece of amber rubbed with fur would attract tiny fragments of straw. Our modern term *electricity* comes from the Greek word for amber, which is ἠλεχτρου. During the next 2000 years little, if any, progress was made towards finding a model which would make sense of this phenomenon. At the beginning of the eighteenth century Isaac Newton described electricity as a 'weightless fluid', a description which was generally accepted at the time. Benjamin Franklin, about 1750, seems to have been the first person to speak of 'electrical particles', but he produced no evidence for their existence.

In 1833 Michael Faraday discovered that the mass of a substance liberated during electrolysis was directly proportional to the product of current

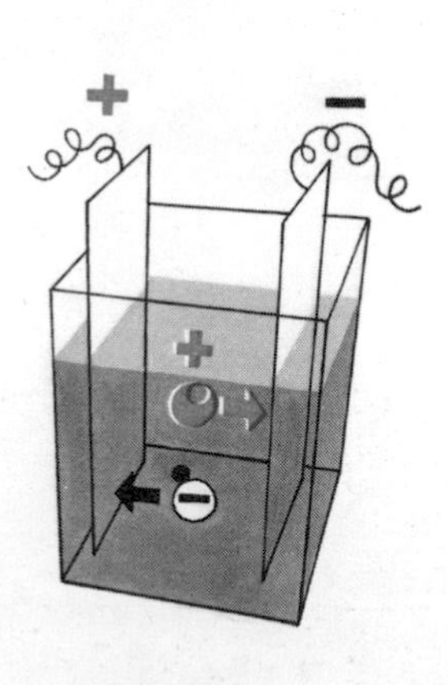

Fig. 6–3

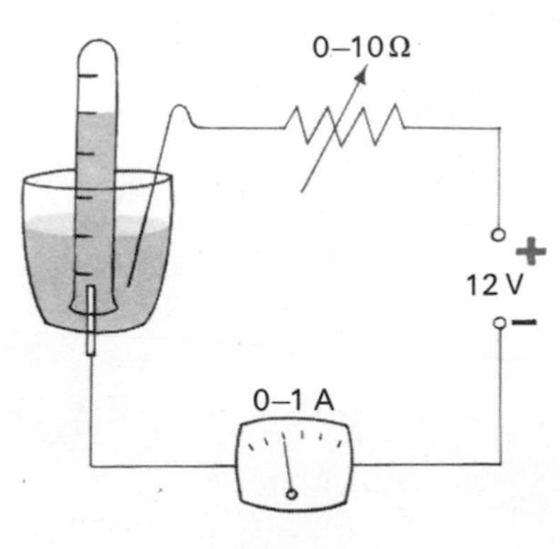

Fig. 6–4

and time, i.e. charge. By measuring the charge and the mass M liberated during the electrolysis water (Fig. 6–4), Faraday found the charge per u mass for the hydrogen ion.

$$\frac{Q}{M} \text{ for hydrogen ion } = 9.6 \times 10^7 \text{ C/kg}$$

Faraday's results led him to say,

> 'Although we know nothing of what an atom is, and though we are in equal if not greater ignorance of electricity, yet there is an immensity of *facts* which justify us in *believing* that the atoms of matter are in some way endowed with electrical powers.'

In 1871 Wilhelm Weber spoke of 'charged pa ticles' to describe his theory of electromagnetis and in 1881 Hermann von Helmholtz quot Faraday's evidence from electrolysis to support h own belief that electricity 'positive as well as neg tive, is divided into definite elementary portio which behave like atoms of electricity'.

It seems, then, that during the nineteenth centu the idea of charged particles gained popularity, a the results of many experiments were explained terms of such particles. Yet it was equally possil to explain these results in terms of a kind of co tinuous electric juice. There was still no dir evidence for the belief in a charged particle.

Fig. 6–5

Fig. 6–8

Cathode Rays

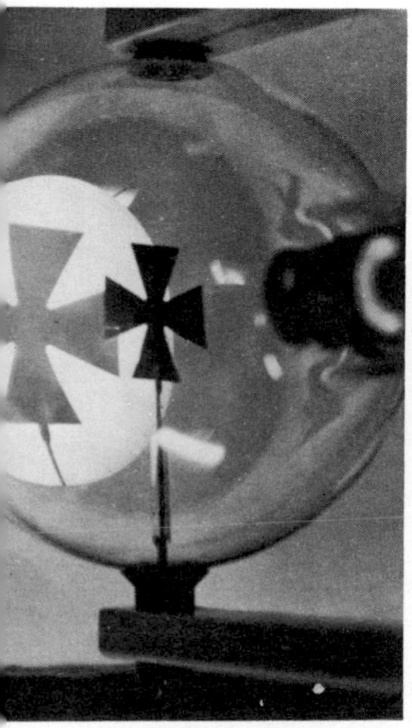

The invention of an effective vacuum pump by Henry Geissler in 1855, together with the development of good glass-to-metal seals, made possible the study of currents through gases at very low pressure. (Fig. 6–5 shows a modern version.) As the pressure was reduced the glow of the gas in the tube disappeared and the glass itself started to glow. This was particularly noticeable at the end of the tube opposite the cathode. Could it be that some kind of radiation was being emitted from the cathode? This was the question that puzzled scientists. Some German scientists favoured the idea of a wave or ray—a kind of electromagnetic radiation—and Eugen Goldstein coined the phrase *cathode rays*. Johann Hittorf and others showed that the rays travelled in straight lines and could be bent by a magnetic field. In England, William Crookes (1879) and others devised a variety of tubes to investigate the properties of cathode rays. They found that the rays could heat a thin foil to red heat. *What does this tell you about cathode rays? (6.1)*

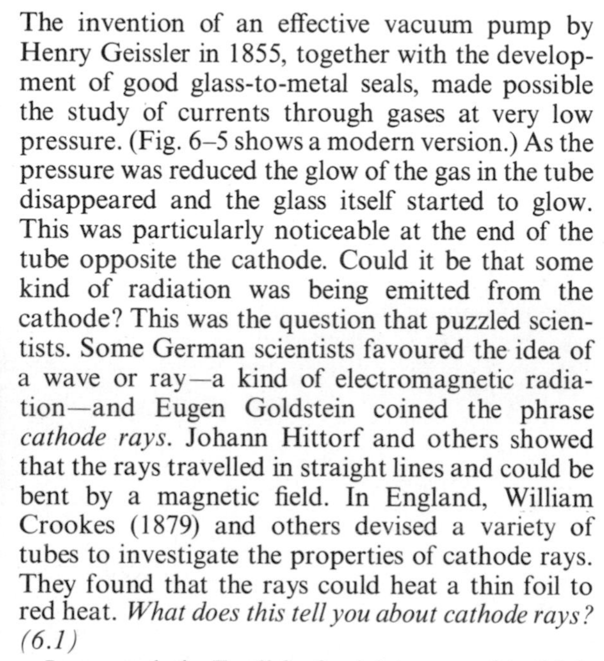

Fig. 6–6

In general, the English physicists seemed to think that cathode rays behaved more like a stream of particles (Fig. 6–6) than waves or rays. It has been suggested that the Englishman's passion for cricket may have influenced this belief! The deflection of the cathode rays by a magnetic field could not be easily explained by the wave model and, when in 1895 the Frenchman Jean Perrin showed that cathode rays definitely carried negative charges (Fig. 6–7), the case for the particle model was strengthened still further.

J. J. THOMSON (1856–1940)

In 1897 J. J. Thomson (Fig. 6–8) 'discovered the electron'. What exactly does this statement mean? Certainly not that the idea was new. The Irishman G. Johnstone Stoney had given the name *electron* to a negatively charged particle many years earlier. The fact was, however, that no-one had demonstrated conclusively that cathode rays behaved as particles. This was Thomson's contribution to the story. He showed that cathode rays not only had charge but also mass. What is more important, he was able to measure the actual charge per unit mass, that is, the ratio e/m, where e is the charge on the electron and m its mass. Moreover, he found that this ratio remained the same with different gases in the tube and using different metals for the electrodes. J. J. Thomson was awarded the Nobel prize in physics in 1906.

As with so many other scientific discoveries, the electron was not immediately seen to have great practical significance. In fact one of the toasts at the annual dinner of the Cavendish Laboratory was 'The electron: may it never be of any use to anybody!'

With the apparatus shown in Fig. 6–9, Thomson

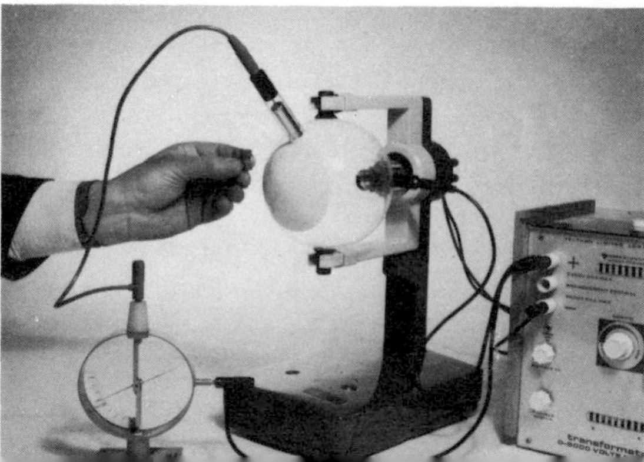

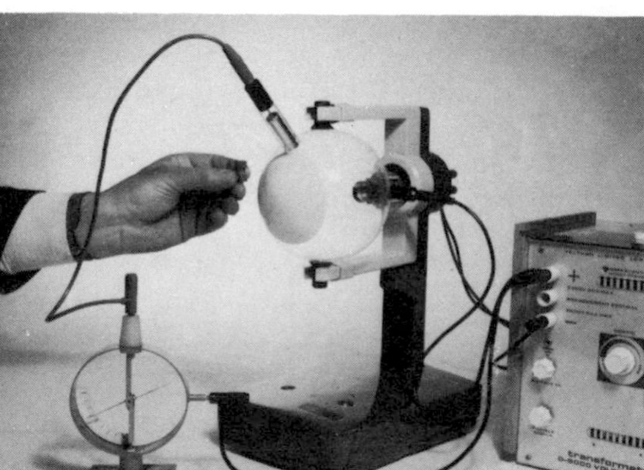

Fig. 6–7

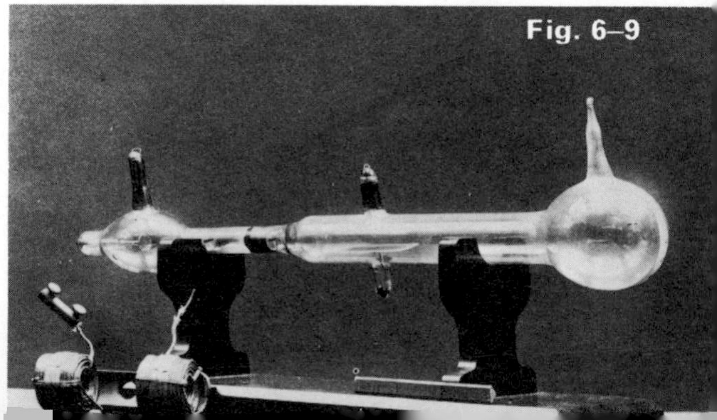

Fig. 6–9

was able to deflect a beam of cathode rays by an electric field and by a magnetic field. He then assumed that, if the rays consisted of particles of charge e and mass m, they should obey Newton's Laws of motion. By applying these Laws he found the ratio e/m to be constant.

The importance of Thomson's experiment lay in showing that not only was e/m the same under a variety of different conditions, but also the particles carried a charge per unit mass which was nearly two thousand times greater than that carried by the lightest known ion. As we have seen, Faraday found Q/M to be 9.6×10^7 C/kg for the hydrogen ion. Thomson found the charge/mass ratio of the electron (e/m) to be 1.76×10^{11} C/kg. Thus

$$\frac{e/m}{Q/M} = \frac{1.76 \times 10^{11}}{9.6 \times 10^7} \approx 2000$$

Either the charge on the electron was very much bigger than the charge on the hydrogen ion, or the mass of the electron was very much smaller than that of the hydrogen ion. Thomson believed that the latter was the case, but, until someone could measure the actual charge e or the mass m, the problem could not be solved.

R. A. MILLIKAN (1868–1953)

The task of accurately measuring the charge on the electron was eventually accomplished by R. A. Millikan in the early years of the twentieth century. He constructed a device so sensitive that with it he could measure the small forces exerted on a tiny charged oil drop as it fell through the air. The air was then ionized so that the drop collected or lost some charge as it fell. By measuring the new force acting on the drop, he discovered that it had gained or lost charge only in multiples of a basic unit of charge which he was able to calculate. To appreciate how he did this we must first revise some basic principles.

Between two charged parallel plates the electric field is essentially uniform (Fig. 6–10). If the positively charged particle Q is moved from plate A to plate B, work has to be done against the field. If the potential difference (p.d.) between the plates is kept constant at 1 volt, 1 joule of work will be done on every coulomb of charge transferred.

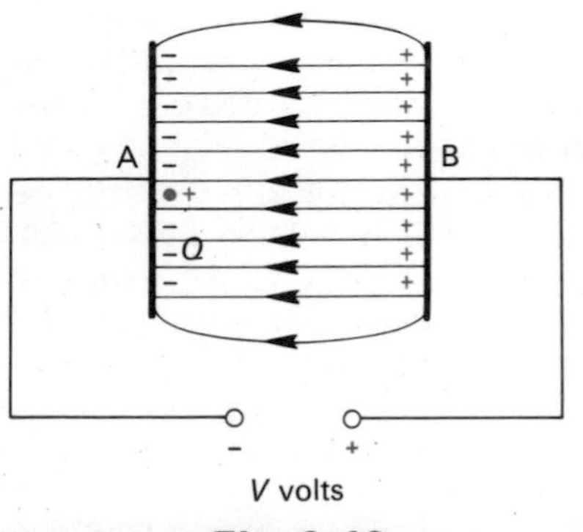

Fig. 6–10

When 1 coulomb moves across a p.d. of 1 volt, the work done is 1 joule.

When Q coulombs move across a p.d. of 1 volt, the work done is Q joules

When Q coulombs move across a p.d. of V volts, the work done is QV jou

$$\text{Work done} = QV$$

As the field is uniform, it will exert a constant for F on the charge Q as it moves from one plate the other. If the plates are d metres apart, the wo done in moving Q coulombs from A to B is giv by the product $F \times d$.

$$\text{Work done} = Fd$$

As equations (1) and (2) both represent the sa amount of work, we have $Fd = QV$. This gives t force acting on the charge as

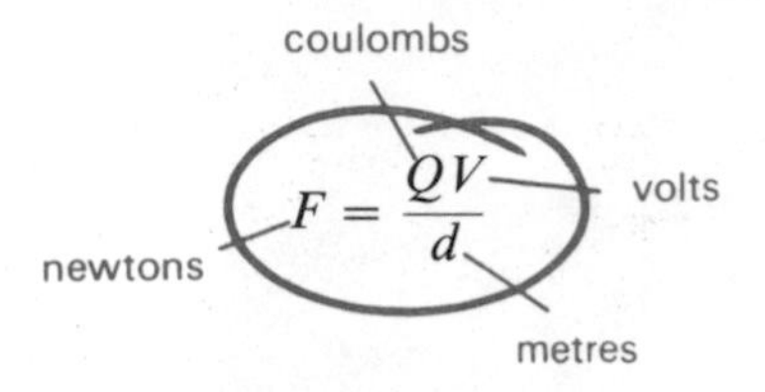

In the above example, we considered the force o *positive* charge in an electric field. A force of same size but acting in the opposite direction wo be exerted on a *negative* charge of Q coulom Consider, for example, the force on a small ne tively charged metal cylinder placed between t parallel plates. The plates could be made fr aluminium foil glued to the lid and the base o plastic food box and the cylinder made from li aluminium foil and supported by a nylon thre When the plates are connected to a Van de Gra generator as shown in Fig. 6–11, there will be upward force exerted on the cylinder. Gravity w exert a downward force. By carefully adjusting potential difference across the plates, it is possi to get the cylinder to 'hover' in the middle of box (Fig. 6–12). The two forces will then be equ that is,

$$\frac{QV}{d} = mg$$

$$\Leftrightarrow \quad Q = \frac{mgd}{V}$$

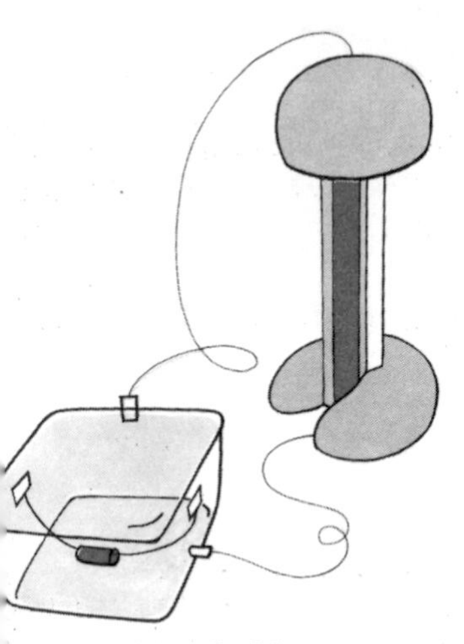

Fig. 6–11

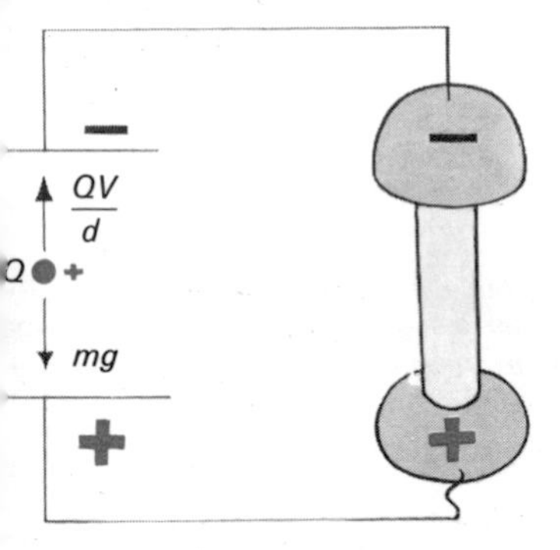

Fig. 6–12

The charge Q (coulombs) on the cylinder could thus be found by measuring m (kilograms), g (10 N/kg), d (metres), and V (volts).

This analogue illustrates the *principle* which Millikan used to investigate electric charges.

Millikan's Experiment

An aluminium cylinder would have been far too big for Millikan's purpose. He looked for the smallest thing which he could charge and then observe. He decided to use minute liquid drops produced by a spray and charged electrically by friction as they left the nozzle (Fig. 6–13). Obviously the drops must not evaporate during the experiment; so Millikan chose the special kind of oil used in vacuum pumps.

When the oil drop is balanced, the charge on it is found from the equation

$$Q_1 = \frac{mgd}{V_1}$$

The charge on the drop can now be altered by holding a radioactive source nearby, or by sending a beam of X-rays between the plates. In either case the air is ionized and, when an ion collides with the oil drop, the charge on the drop changes. The drop then starts to move, and the p.d. across the plates has to be altered in order to stop it once more. If the new p.d. is V_2, the new charge on the drop will be

$$Q_2 = \frac{mgd}{V_2}$$

This process can be repeated several times to give Q_3, Q_4, Q_5, etc. as the values of the charge on a particular drop. As the mass of the drop m will be constant, the product mgd will be constant throughout the experiment. The reciprocal of the voltage will therefore be proportional to the charge and we can use this as a measure of charge. When the results of such an experiment are arranged on a chart as shown in Fig. 6–14, they tend to lie in

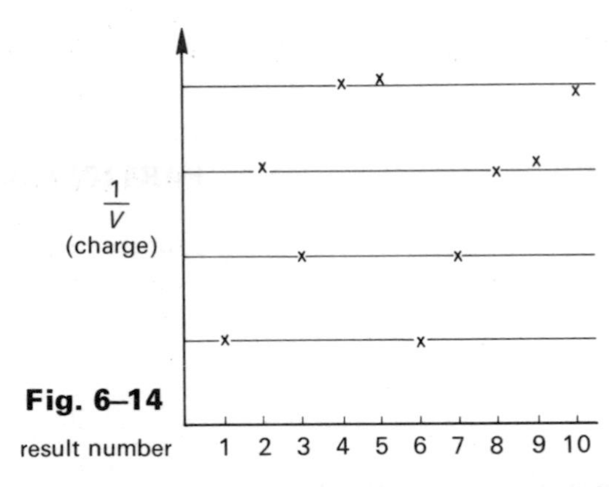

Fig. 6–14

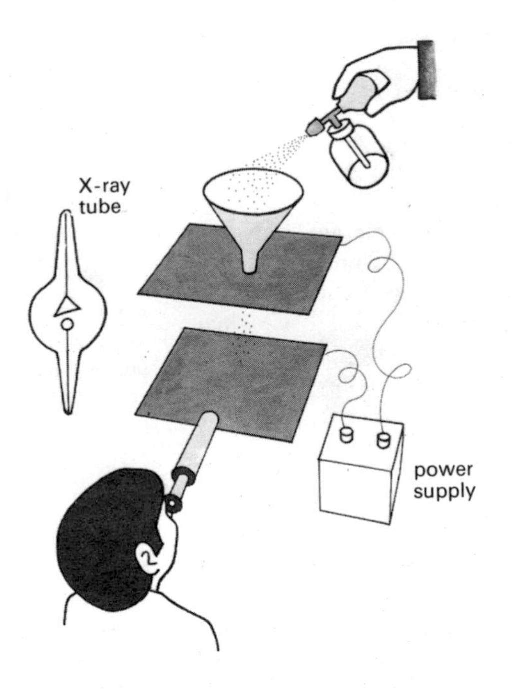

Fig. 6–13

bands which are equally spaced. This strongly suggests that charge exists in 'units' which cannot normally be split up further. To find the actual value of this minimum unit of charge, the mass of the oil drop must be found and substituted in each of the equations for Q_1, Q_2, etc. Finding the mass is, however, a difficult process beyond the scope of this course.

From thousands of measurements Millikan discovered that the value of Q_1, Q_2, Q_3, etc. always produced the same kind of pattern, which showed that that the minimum quantity of charge was 1.6×10^{-19} C. This must therefore be the charge on an electron. For his work on the measurement of the electronic charge Millikan received the Nobel prize for physics in 1923.

Millikan's work made it possible to explain Faraday's electrolysis experiment, in terms of electronic charges. Faraday had found that the mass of hydrogen liberated per coulomb did not depend on the strength of the current nor on the electrolyte. This makes sense if we assume that

(a) a quantity of charge corresponds to a definite number of electrons,
(b) hydrogen ions (hydrogen atoms minus one electron) are present in the electrolyte, and
(c) when hydrogen is liberated, the hydrogen ion gains one electron from the cathode and becomes electrically neutral.

One hydrogen atom is thus liberated by the passage of one electron, and so the number of atoms of hydrogen (mass) is proportional to the number of electrons passing (charge).

If the hydrogen ion is a neutral atom less an electron, i.e. a proton, then the electron and the proton must carry the same size of charge. Faraday's value of Q/M, Thomson's value of e/m, and Millikan's value of electronic charge can be used to calculate the mass of the proton and the mass of the electron.

PARTICLE ACCELERATORS

From the definition of the volt (1 joule per coulomb), we see that the

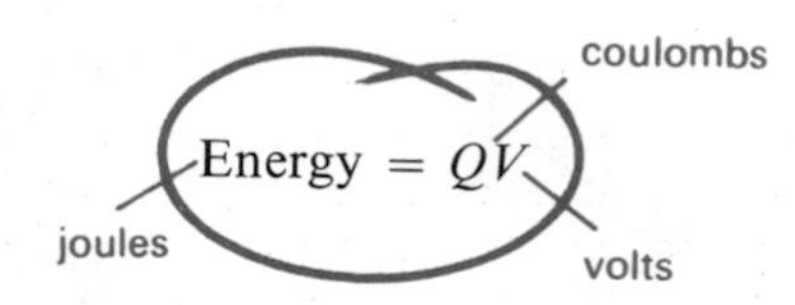

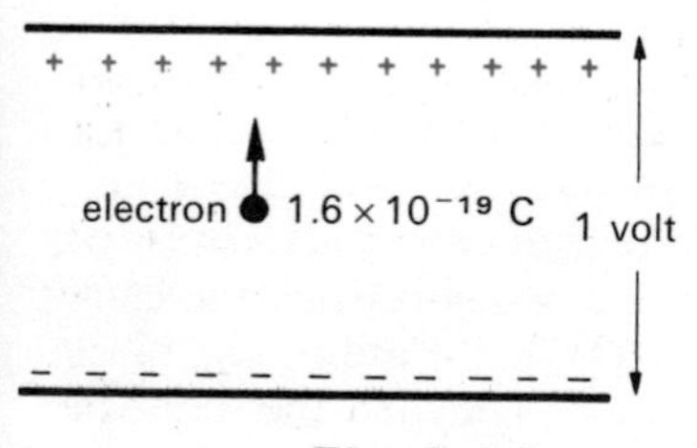

Fig. 6–15

If a charge of 1 coulomb is accelerated through a p.d. of 1 volt, it gains 1 joule of kinetic energy. For some purposes a smaller unit is used. It is the energy gained by one electron (1.6×10^{-19} C) when accelerated through a p.d. of 1 volt (Fig. 6–15). This amount of energy is called the electron volt.

$$1 \text{ electron volt (eV)} = 1.6 \times 10^{-19} \text{ joules}$$

In particle physics it is often necessary to accelerate particles, for example electrons, to speeds much greater than those which can be achieved with even the highest voltages available (i.e. several millions of volts). An electron accelerated by a 10 million volt Van de Graaff generator would attain a kinetic energy of ten million electron volts. Modern particle accelerators can produce thousands of millions of electron volts.

SUMMARY

1897 J. J. Thomson measured e/m
1917 R. A. Millikan measured the charge e an electron.

Force on charged oil drop $= \dfrac{QV}{d} = mg$

thus $Q = \dfrac{mgd}{V}$

Q/M for hydrogen ion (proton) $= 9.58 \times 10^{7}$ C/k
e/m for electron $= 1.76 \times 10^{11}$ C/
Electronic charge e $= 1.60 \times 10^{-19}$ C
i.e. 1 coulomb $= 6.24 \times 10^{18}$ e
Rest mass of electron m $= 9.11 \times 10^{-31}$ k
Rest mass of proton $= 1.67 \times 10^{-27}$ k
1 electron volt $= 1.60 \times 10^{-19}$ J
i.e. 1 joule $= 6.24 \times 10^{18}$ eV

PROBLEMS

6.2 To understand how Millikan's results can be int preted, imagine that you are given twenty-one sealed ba which you are asked to weigh. The results in grams are follows:

78.0	39.1	38.9	65.2	52.1	78.1	26.0
39.2	64.8	78.2	25.8	51.9	65.9	51.8
26.1	39.0	77.1	26.0	51.7	65.0	38.9

Record these results on a sheet of graph paper. Indicate t mass on the y axis, and mark each result with a cross (say) 1 cm intervals in the x direction. Can you disco anything about the contents of the bags? Are they m likely to be filled with marbles or marmalade?

6.3 If the results of Millikan's experiment produced following results, what would you deduce to be the small unit of electric charge?

Charges on the oil drop ($\times 10^{-19}$ C).

6.4	4.8	12.8	14.4	3.2	8.0	12.8
9.6	8.0	16.0	4.8	11.2	6.4	12.8
3.2	4.8	6.4	8.0	12.8	3.2	9.6
8.0	3.2	9.6	6.4	16.0	11.2	12.8

6.4 If the ratio of charge/mass for an electron 1.76×10^{11} C kg^{-1} and the electron charge is 1.6×10^{-19} what is the mass of the electron? Compare the mass of t electron with the mass of a hydrogen nucleus (protor which is 1.67×10^{-27} kg. Was J. J. Thomson's hunch (s page 56) correct?

6.5 Using the values stated in the previous question, find the increase in the speed of an electron accelerated by a p.d. of 500 V. (Hint: find the gain in the kinetic energy of the electron.)

6.6 If the electronic charge is 1.6×10^{-19} C, how many electrons are equivalent to a coulomb? How many of these charges pass a given point in a wire per second when a current of 1 A flows?

6.7 If a small sphere carries a negative charge of 0.1 μC, how many excess electrons does it have?

6.8 Two parallel metal plates are placed 5 cm apart in a vacuum with a p.d. of 1000 V between them. A small plastic sphere of mass 3×10^{-9} kg carries a negative charge of −3 pC. It is held on the lower metal plate (Fig. 6–16).

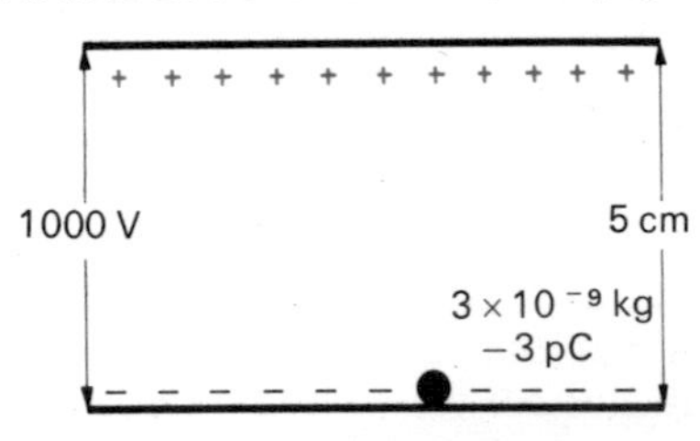

Fig. 6–16

(i) If the sphere is released what electric force is acting on the sphere as it rises? Compare this with the weight of the sphere.
(ii) At what speed will it be travelling when it reaches the top plate?
(iii) Show that energy has been conserved.

6.9 In a Millikan experiment an oil drop has a mass of 1.6×10^{-15} kg. If, in order to hold the drop stationary, a p.d. of 200 V has to be applied between the plates, which are 1 cm apart, what charge does the drop carry? How many electronic charges does this represent?

6.10 Find the energy, in electron volts, of an electron travelling at 10^6 m s^{-1} if the mass of the electron is 9.1×10^{-31} kg. Compare this with the average kinetic energy of an air molecule at room temperature which is 0.03 eV.

6.11 What is the acceleration of an electron (1.6×10^{-19} C) of mass 9.1×10^{-31} kg in a field of 400 N C^{-1}? What would be its acceleration due to a gravitational field of 10 N kg^{-1}?

6.12 A tiny charged droplet carries one surplus electron (1.6×10^{-19} C). If this droplet is balanced between two parallel plates in a field of 10^5 N C^{-1}, find the mass of the droplet.

6.13 The diagram (Fig. 6–17) shows a pair of parallel plates which may be charged by connection to a high voltage supply. Tiny dust particles, which become charged on passing through the plastic funnel, are allowed to fall into the space between the plates.

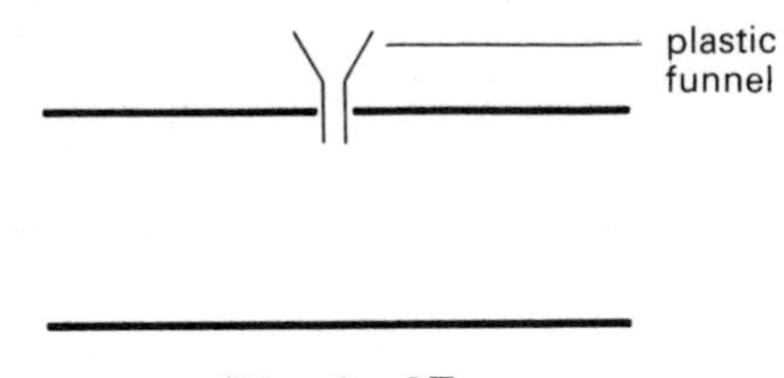

Fig. 6–17

Discuss:

(i) the forces acting on a particle on passing into the air space between the uncharged plates,
(ii) the type of motion that a particle will undergo in the region between the uncharged plates, and
(iii) how a particle may become charged on passing through the funnel.

A potential difference of 200 V is applied between the plates, which are separated by a distance of 1 cm. If the upper plate is charged positively, one of the particles of mass 1.6×10^{-15} kg is seen to become stationary.

(iv) Explain the equilibrium of the particle.
(v) Estimate the charge on the particle.
(vi) A source of β radiation is now directed into the space between the plates and the particle is seen to fall. Suggest a possible explanation.
(vii) It is found that the motion of certain particles is unaffected by charging the plates, while other particles are seen to rise. Give possible explanations.

[S.C.E. (H) 1968]

7 Particles in a Field

ELECTRONS

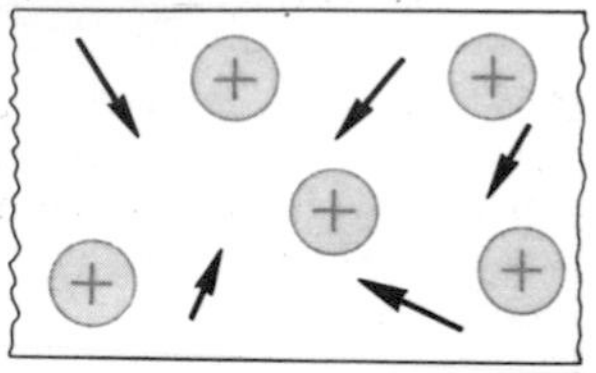

Fig. 7–1

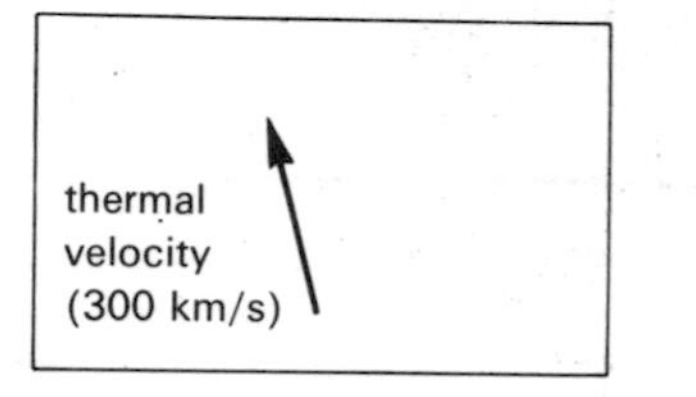

Fig. 7–2

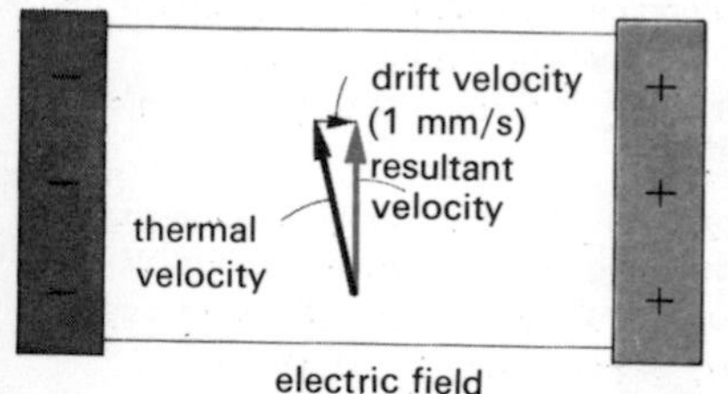

Fig. 7–3

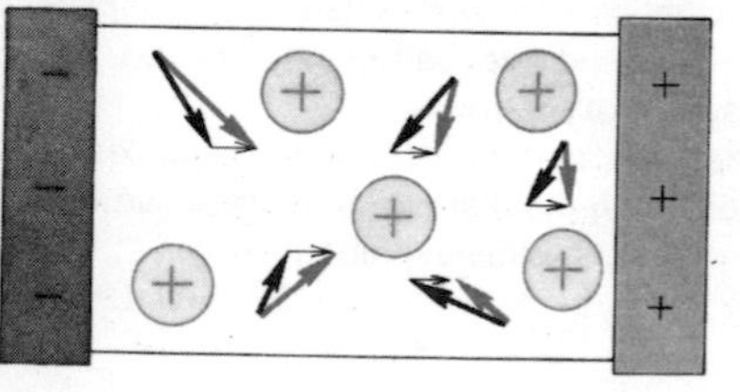

Fig. 7–4

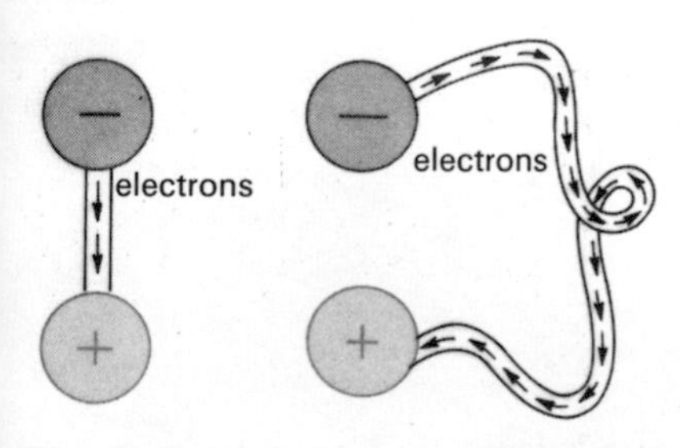

Fig. 7–5 Fig. 7–6

You have never seen and never will see an electron. Yet you no doubt imagine it as a wee black or blue ball. It may even have a minus sign painted on the side! Fair enough. Most people have their favourite analogues, but a model which 'makes sense' to one person may be of little help to another. It is wise to accept whatever you find valuable in an analogue while retaining a healthy suspicion of it.

The work of Thomson, Millikan, and others has certainly encouraged us to think of an electron as an incredibly light particle carrying a fixed amount of electric charge. It has a certain mass and can be accelerated by a force exerted on it in a gravitational, electric or magnetic field. It is, however, so light and usually travels so quickly that the effect of gravitational forces can be neglected. We will be dealing with magnetic fields later, and so for the moment our main concern will be with the motion of electrons in an electric field, i.e. an electric current. As with the kinetic theory, we are going to use a simplified model. It ignores, for example, the dual nature of an electron, i.e. that it behaves like a particle *and* as a wave.

In our model a metallic conductor consists of millions of positive ions surrounded by a sea of free electrons (Fig. 7–1). Normally the electrons move around in a random fashion with an average thermal speed of the order of 300 km/s (Fig. 7–2). If, however, an electric field is applied to the conductor, each of these electrons will be accelerated. The resultant velocity of each electron at any instant will then be the vector sum of the original (thermal) velocity plus the (drift) velocity gained as a result of the field (Fig. 7–3). This diagram is not drawn to scale, as the average drift velocity may be of the order of a millimetre per second!

If we add a small drift velocity to all the thermal velocity vectors in Fig. 7–1, the result looks something like Fig. 7–4. It is this small systematic drift superimposed on the random motion of the electrons that constitutes an electric current in a conductor.

If the electrons were in a vacuum, they would continue to accelerate in the electric field until they reached a very high velocity in that direction. This they do in a thermionic valve. However, they cannot move far in a conductor without encountering a positive ion. At each such encounter, momentu and kinetic energy are transferred. Consequen the electrons are, on average, continually gaini kinetic energy as they accelerate in the field a then almost immediately losing at least some of in collisions. The *average* drift velocity of the ele tron is therefore small and the positive ions in t conductor are continually gaining energy of vibr tion; that is, the temperature of the conduct rises.

When two spheres are charged as shown in Fi 7–5, there will be an electric field between them. now a conductor is connected between the sphere a weak field is produced in it. This field exerts a accelerating force on the free electrons in the dire tion shown. The conductor does not of course ha to lie physically between the spheres, as the fie will follow the line of any wire (Fig. 7–6), a electrons will drift in the direction indicated un the field disappears.

E.M.F.

To maintain a steady flow of charge, we requi some source of energy capable of maintaining th electric field that drives the electrons round th circuit. One way of doing this is to carry the ele trons up on a non-conducting belt as in the Van d Graaff generator. In this a series of sharp poin 'spray' the electrons on to the belt. Once there the must be pushed against the field; that is, wor must be done. The energy necessary to do this usually supplied by hand or by an electric moto (Fig. 7–7). This energy is the source of the *electr motive force* which maintains the current in th circuit.

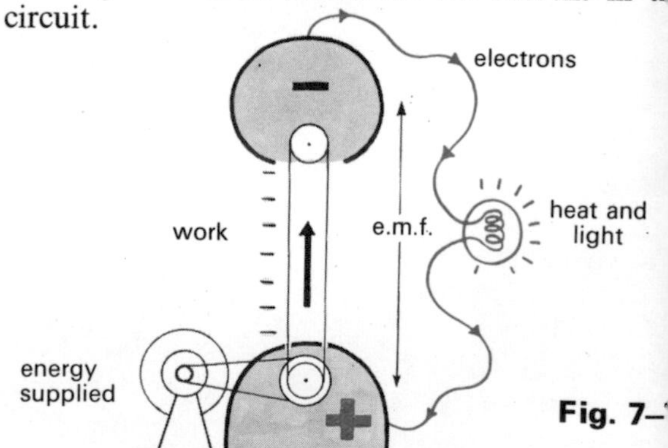

Fig. 7–

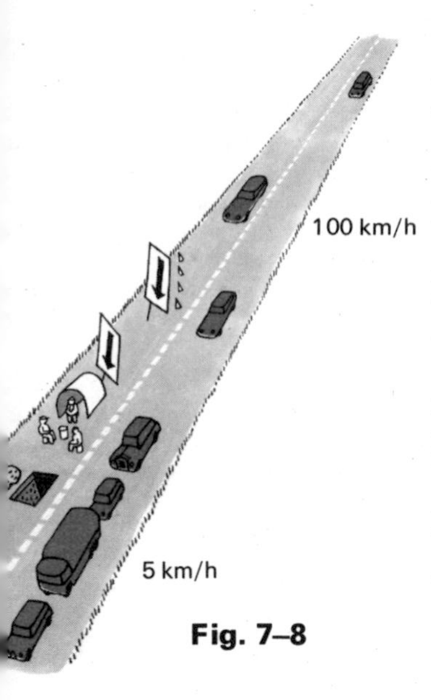

Fig. 7–8

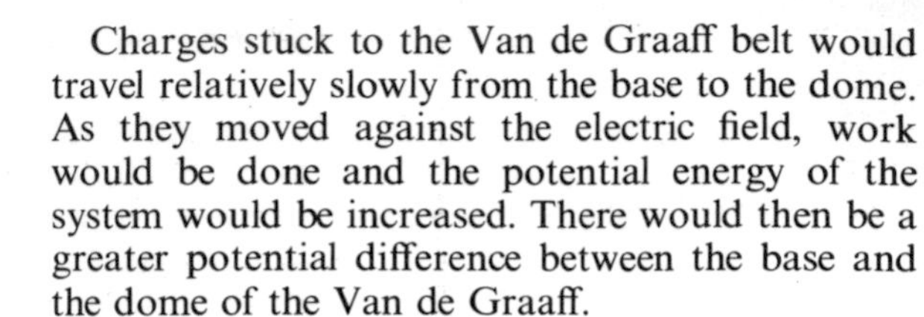

There are, of course, many other ways of producing an e.m.f. and maintaining an electric field in a conductor. Dynamos and batteries are perhaps the most common, but thermocouples, photovoltaic cells, and piezo-electric crystals will also produce e.m.f.

In each case one form of energy is changed to another to produce the electric field which drives the electrons through the circuit. This flow of charge, in turn, produces *work* or *heat*.

Although the electrons in the conductor do not necessarily move very quickly, the *electric field* is propagated at very nearly the speed of light. If a transatlantic cable 3000 km long were used in a lighting circuit, the electrons in a bulb at one end would be accelerated about one hundredth of a second after a switch at the other end was closed.

CURRENT

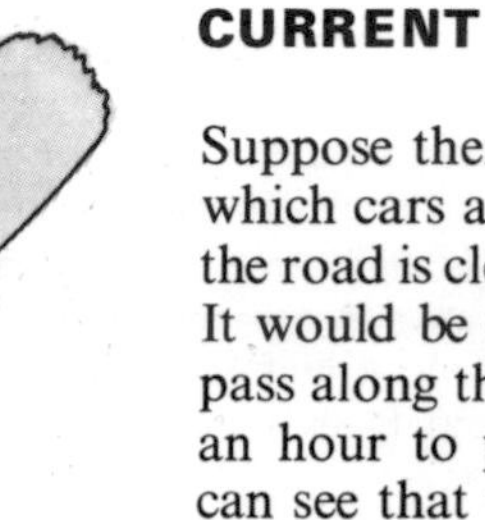

Fig. 7–9

Suppose there is an obstruction on a motorway at which cars are slowed down to 5 km/h. The rest of the road is clear and cars travel along it at 100 km/h. It would be quite possible for 200 cars an hour to pass along the clear road and for the same 200 cars an hour to pass the obstruction (Fig. 7–8). You can see that the 'number of cars passing per hour' is not necessarily related to *speed*.

We call the amount of electric charge per second crossing any section of a conductor (i.e. AA′, BB′ in Fig. 7–9)—the *current*.

For practical purposes, the number of electronic charges per second is too large a number, and so we use the *coulomb* (about 6.25×10^{18} electronic charges) as our unit of charge. When the current in a conductor is 1 ampere, 1 coulomb of charge per second crosses any section of the conductor.

Do not, however, confuse this rate of flow with the speed of the electrons. The average drift velocity in a conductor may be a fraction of a millimetre per second, and the electron velocity in a thermionic valve may reach hundreds of thousands of metres per second, yet the *current* in each could be the same. We might illustrate the difference in the rather unlikely circuit of Fig. 7–10. Here electrons move round a circuit at many different speeds, yet the current (coulombs/second) is the same everywhere. The speeds indicated are for comparative purposes only. In practice they could vary considerably depending on the actual components used.

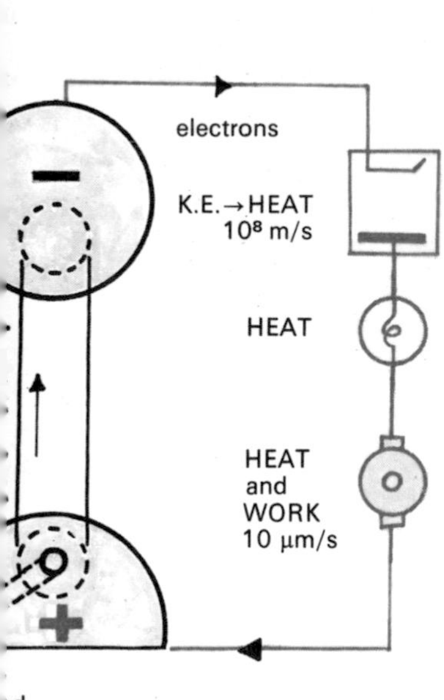

Fig. 7–10

Charges stuck to the Van de Graaff belt would travel relatively slowly from the base to the dome. As they moved against the electric field, work would be done and the potential energy of the system would be increased. There would then be a greater potential difference between the base and the dome of the Van de Graaff.

In the thermionic valve the electrons are accelerated to high speeds and their kinetic energy transformed to heat at the anode. In the bulb and wires heat is produced by collisions and mechanical work is available from the motor.

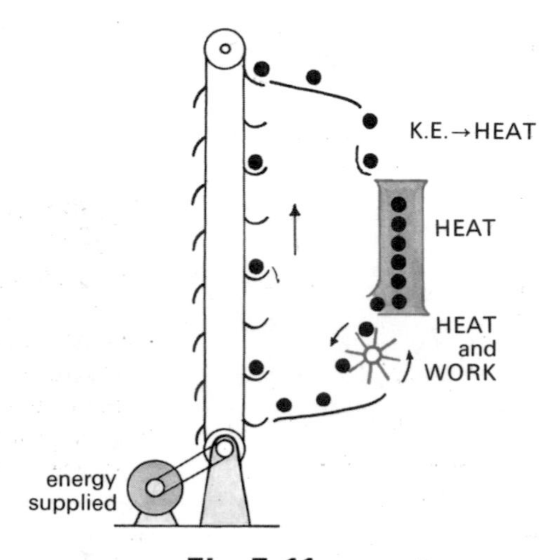

Fig. 7–11

Here is an analogue which may help, provided you don't press it too far. A conveyor belt (Fig. 7–11) is used to raise ball-bearings in the Earth's gravitational field. Work is done and the gain in potential energy could be expressed in joules per ball-bearings. The ball-bearings then fall freely in the Earth's field gaining K.E. which is transformed to heat when they strike a chute which directs them into a jar of treacle. The Earth's field now forces the ball-bearings against the viscous forces in the treacle. As they move *very slowly*, they have practically no kinetic energy, but the temperature of the treacle rises as a result of continuous energy transfer in molecular collisions. The balls now leave the jar through a magic door which doesn't allow treacle to flow out (!) and operate a paddle wheel as they fall through the last stage of their journey. Mechanical work and heat are thus produced. The energy needed to maintain this flow of ball-bearings is supplied by the motor driving the conveyor belt.

E.M.F.

If the e.m.f. maintained between the sphere and base of the Van de Graaff is very large, a large amount of work will have to be done on each charge stuck to the belt. In fact we *measure* the e.m.f. in terms of the 'work done per unit charge' as it is carried from one sphere to the other. As for practical purposes we take the unit charge as the coulomb, the unit of electromotive force, the *volt*, is the number of *joules per coulomb*.

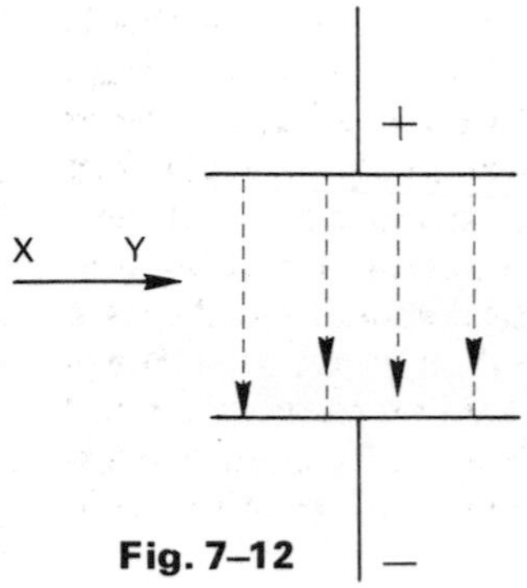

Fig. 7–12

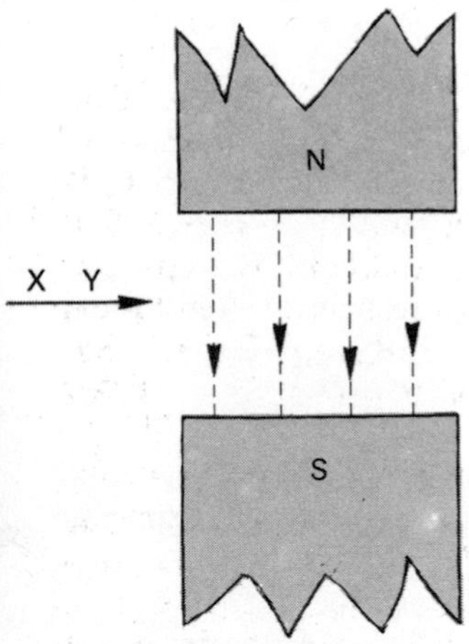

Fig. 7–13

SUMMARY

A steady electric current (i.e. a constant quantity of charge passing each second) is possible at various parts of a circuit, even though there are enormous differences in the speeds of the electrons. In a metal conductor the drift velocity could be less than a millimetre per second, yet in a thermionic valve carrying the same current electrons could be moving at many kilometres per second.

A source of e.m.f. produces an electric field in a conductor and the free electrons are accelerated along the field lines. Much of the energy they gain by accelerating in the electric field is immediately transferred by work (e.g. in an electric motor) or heat (e.g. in an electric radiator).

The source of e.m.f. supplies the energy needed to maintain the field in the conductor when the current flows.

PROBLEMS

7.1 A negatively charged particle is projected as indicat by the arrow, XY, from left to right and in the plane of t paper.

(a) What is the direction of the force on the particle in
   (i) the electric field 7–12 and
   (ii) the magnetic field (Fig. 7–13)?

(b) Give an example of the application of *one* of the effects. [S.C.E. (H) 197

7.2 Two insulated metal plates X and Y were mount parallel and close to each other. A light metallized sphere initially negatively charged, was hung between them. X w connected to the negatively charged dome of a Van Graaff generator and Y to one terminal of a microammet The other terminal of the meter was connected to ea (Fig. 7–14).

Explain why

(a) S oscillated between X and Y and

(b) the meter registered a deflection. [S.C.E. (H) 197

7.3 (a) Compare the processes by which current flo between electrodes A and B in the two circuits shown Figs. 7–15 and 7–16. Discuss the effect on the current reversing the H.T. supply in each case.

(b) In the thermionic vaccum tube illustrated in Fig. 7–1 h-h is the heater, k the cathode, and a the anode.

Between cathode and anode is a potential difference 1.14 kV. A uniform electric field is applied across the plat P_1 and P_2 by connecting them to a source of extra-hig voltage.

(i) Describe the nature of the motion of an electron
   A. between cathode and anode,
   B. between anode and left-hand edge of P_1 and P_2, a
   C. in the uniform field between P_1 and P_2.

(ii) What is the total work done on an electron by t electric field between cathode and anode?

(iii) With what kinetic energy does an electron leave t anode?

(iv) What will be its velocity?

(v) In what direction would a magnetic field require to applied in order to counteract the effect on the electr beam of the electric field between P_1 and P_2?

[S.C.E. (H) 197

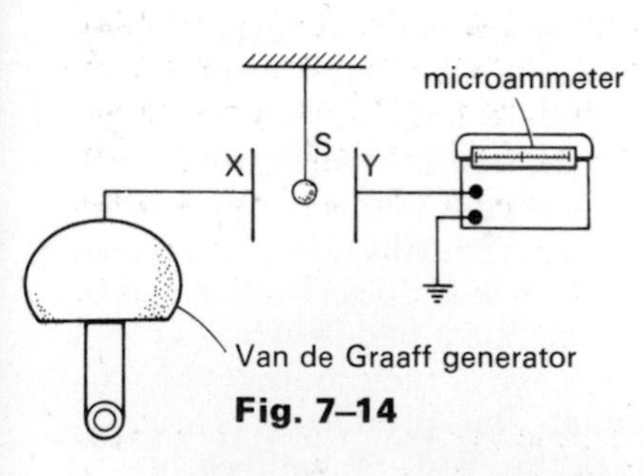

Fig. 7–14

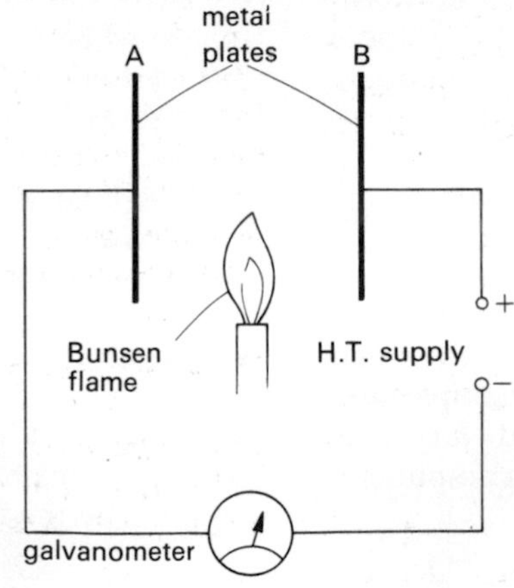

Fig. 7–15

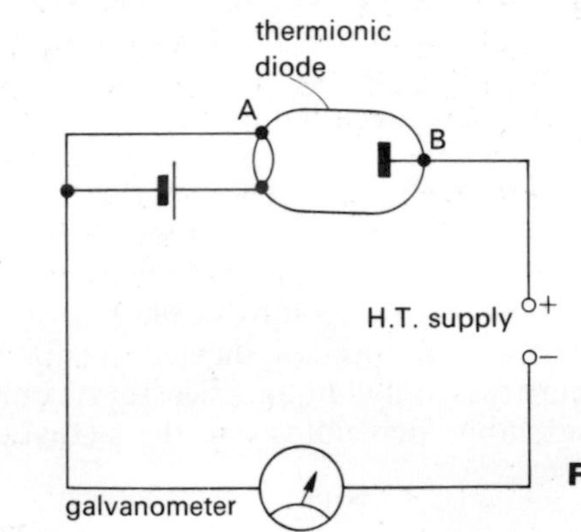

Fig. 7–16

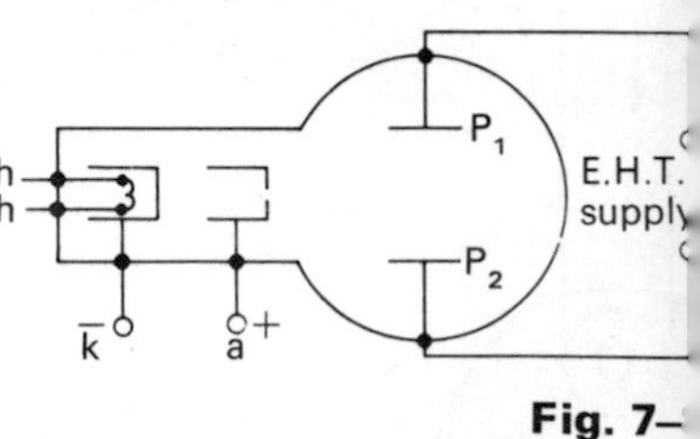

Fig. 7–

8 Measuring Direct Current

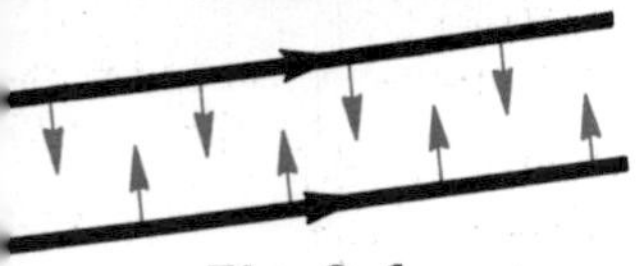

Fig. 8–1

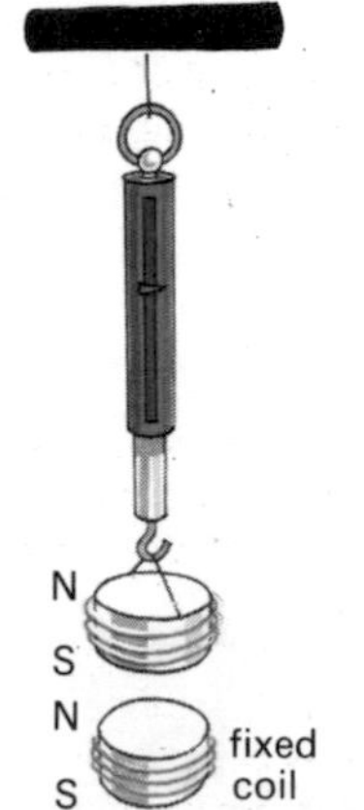

Fig. 8–2

AMMETER

Although like charges repel each other, we have seen that, when charges flow in the same direction through two parallel conductors, there is an *attractive* force between them (Fig. 8–1). This attractive force increases as the current increases and can be used to measure the strength of the current. In fact the ampere is defined in terms of this force. In principle the current balance, which is used for this purpose, consists of two coils, one fixed and one movable (Fig. 8–2). The same current flows through the coils and the force on the movable coil is then measured. In practice the apparatus is, of course, very much more complicated.

The current balance is, then, an *ammeter* and the spring balance could be calibrated in amperes. For everyday use we need something much simpler, more robust and more easily carried around. In the moving coil meter we replace one of the coils by a permanent magnet and allow the other coil to *rotate* in the radial magnetic field (Fig. 8–3). A small hair spring exerts a restoring torque which is proportional to the angle through which the coil turns; thus the deflection of the coil is proportional to the current through it.

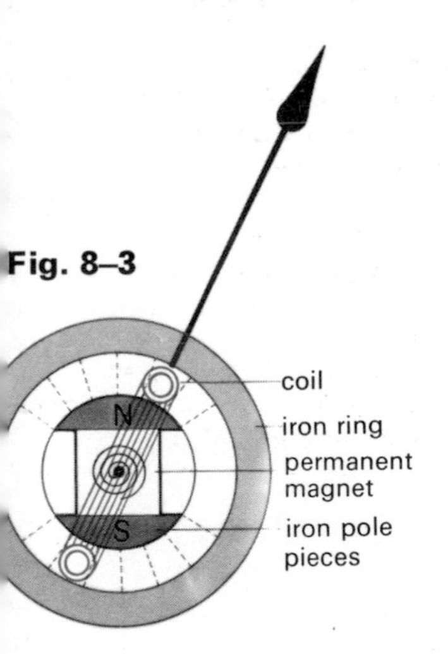

Fig. 8–3

Instruments have a nasty habit of affecting the thing they are supposed to be measuring. If you put a cold thermometer into a cup of warm tea it will cool it down slightly and indicate the reduced temperature. A tyre pressure gauge works by allowing some air to escape from the tyre thus reducing the pressure a little. Ammeters are no exception. If you wanted to measure the current in the circuit of Fig. 8–4, you might insert an ammeter as shown in Fig. 8–5. If, however, the ammeter had a resistance of 1 Ω, the current reading would be 1.67 A instead of 2 A.

Fig. 8–4

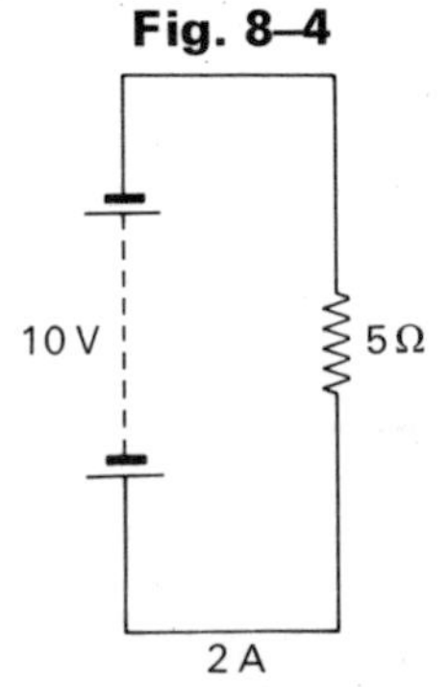

Fig. 8–5

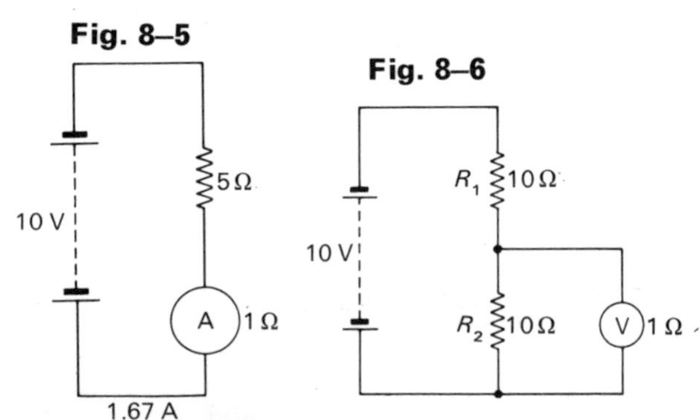

Fig. 8–6

The greater the resistance of an ammeter, the greater will be its effect on the circuit. Ideally then an ammeter should have no resistance but this is impossible. To keep its resistance to a minimum we could use heavy wire for the coil but the meter would then be clumsy. Alternatively, we could keep the resistance low by using a coil of only a very few turns; then the force, and therefore the deflection, would be small. The practical solution is always a compromise which takes into account such things as size, cost, sensitivity, and reliability. A good ammeter will, however, always have a small resistance relative to the resistance of the other components.

VOLTMETER

A good voltmeter is a very different kettle of fish! If we were to use a voltmeter with a resistance of 1 Ω to measure the p.d. across R_2 in Fig. 8–6, it would read less than 1 V instead of the correct 5 V. If it is to have no effect on the circuit, the voltmeter must have an infinitely high resistance. A gold-leaf electroscope is very near to this ideal but it has grave disadvantages as a practical voltmeter. *What are they? (8.1)*

A cathode ray oscilloscope is one of the best voltmeters we use in school, as it can have a resistance of several megohms. *What are some of its disadvantages? (8.2)*

For general use the moving coil instrument is still the most popular voltmeter. If the full-scale deflection current is small, a large series resistance can be used to measure a particular p.d. The greater the resistance of the meter for a particular full-scale deflection, the better the instrument. A good moving coil voltmeter might have a resistance of '100 000 ohms per volt'. On the 100 V range, for example, such an instrument would have a total resistance of 10 MΩ.

Potentiometer

One of the most accurate ways of measuring potential difference is to use a potentiometer. This is, in effect, a voltmeter with an infinitely high resistance.

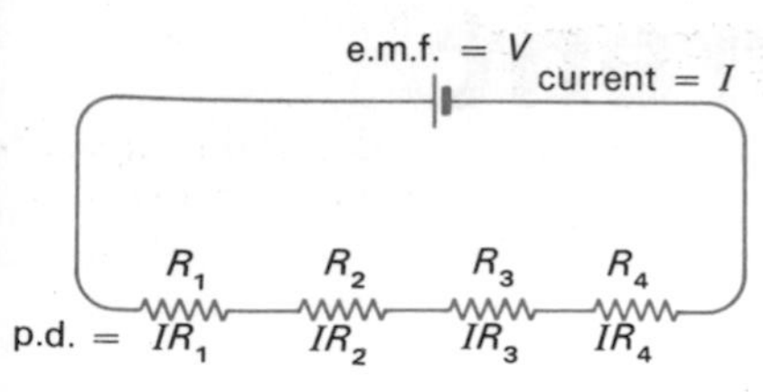

Fig. 8–7

We achieve this effect by balancing the p.d. to be measured against a known p.d. of equal magnitude. In this way no current flows *when the measurement is being taken.*

First consider what happens when four resistors are connected in series (Fig. 8–7). If we call the total resistance R_T (where $R_T = R_1 + R_2 + R_3 + R_4$) and the current through the resistors I, then we have

$$\text{p.d. across } R_1 = IR_1 = \frac{V}{R_T}R_1 = \left(\frac{R_1}{R_T}\right)V$$

$$\text{p.d. across } R_1 + R_2 = I(R_1 + R_2) = \frac{V}{R_T}(R_1 + R_2) = \left(\frac{R_1 + R_2}{R_T}\right)V$$

$$\text{p.d. across } R_1 + R_2 + R_3 = I(R_1 + R_2 + R_3) = \frac{V}{R_T}(R_1 + R_2 + R_3) = \left(\frac{R_1 + R_2 + R_3}{R_T}\right)V$$

$$\text{p.d. across } R_1 + R_2 + R_3 + R_4 = I(R_1 + R_2 + R_3 + R_4) = \frac{V}{R_T}(R_1 + R_2 + R_3 + R_4) = \left(\frac{R_1 + R_2 + R_3 + R_4}{R_T}\right)V$$

If the resistors were four 25-cm lengths of uniform resistance wire joined end to end, i.e. 1 metre of wire, the resistance would be proportional to the length. Then the

$$\text{p.d. across 25 cm wire} = \left(\frac{25}{100}\right)V$$

$$\text{p.d. across 50 cm wire} = \left(\frac{50}{100}\right)V$$

$$\text{p.d. across 75 cm wire} = \left(\frac{75}{100}\right)V$$

$$\text{p.d. across 100 cm wire} = \left(\frac{100}{100}\right)V$$

This would give the voltages indicated in Fig. 8–8 if a 2 V supply were used.

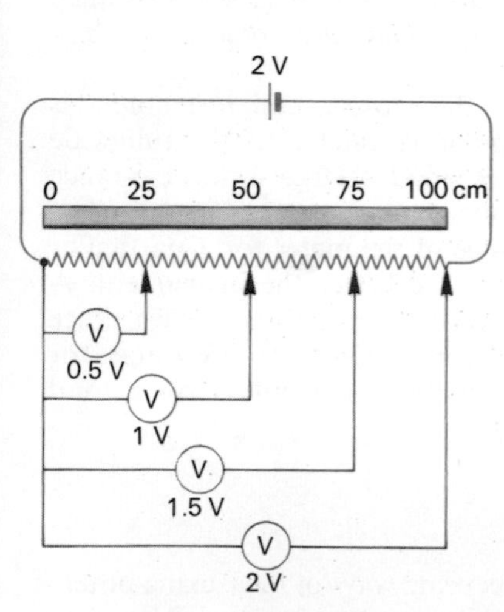

Fig. 8–8

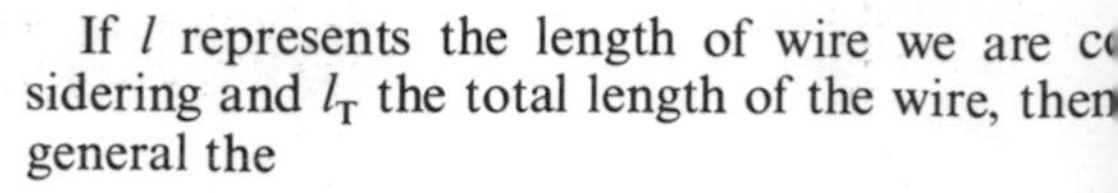

If l represents the length of wire we are co sidering and l_T the total length of the wire, then general the

$$\text{p.d. across the wire} = \left(\frac{l}{l_T}\right)V$$

The above expression may be rearranged to rea

$$\text{p.d.} = \frac{V}{l_T} \times l$$

where V/l_T is the 'change in potential per u length'. This quantity is sometimes called *potential gradient*.

We can use the above device to measure potent differences, for example, the e.m.f. of a cell, p vided we have a steady supply voltage which *greater* than the e.m.f. of the cell. Let us imag for the moment that a well charged accumula produces a p.d. of exactly 2 V across our 1-me resistance wire. To measure the e.m.f. of a cell connect one of its terminals to the end of the w

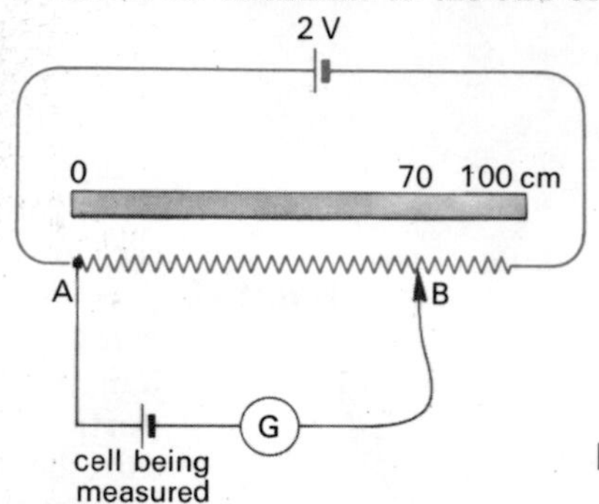

Fig. 8

as shown in Fig. 8–9. The other terminal is wired a galvanometer and then to a sliding contact the wire. If we slide the contact along the wire u *no current* flows in the galvanometer, then the e.m of the cell must be exactly equal to the p.d. acr AB. That is the

$$\text{e.m.f.} = \left(\frac{70}{100}\right)2 = 1.4 \text{ volts}$$

If the contact is slightly to the *left* of the 70 mark, current will flow in one direction throu the galvanometer and if it is slightly to the *rig* current will flow in the other direction. Howev there is no current through the galvanometer a thus no current being taken from the cell when measurement is made. This is equivalent to usi a voltmeter with an infinitely high resistance.

The above method is all very well if we kn that the accumulator is supplying exactly 2 V. practice this cannot be assumed and for rea accurate work we must calibrate the potentiome

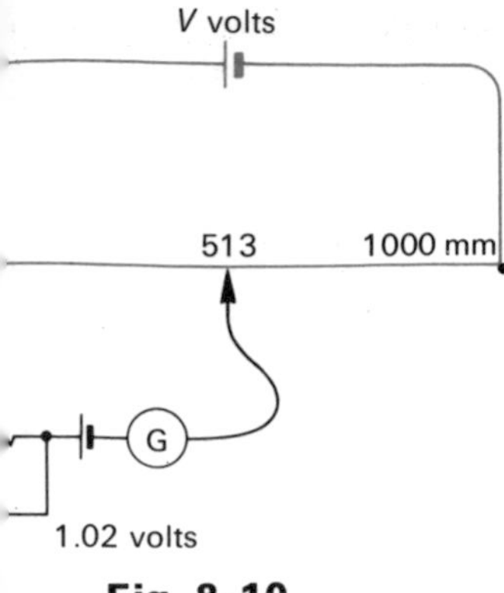

Fig. 8–10

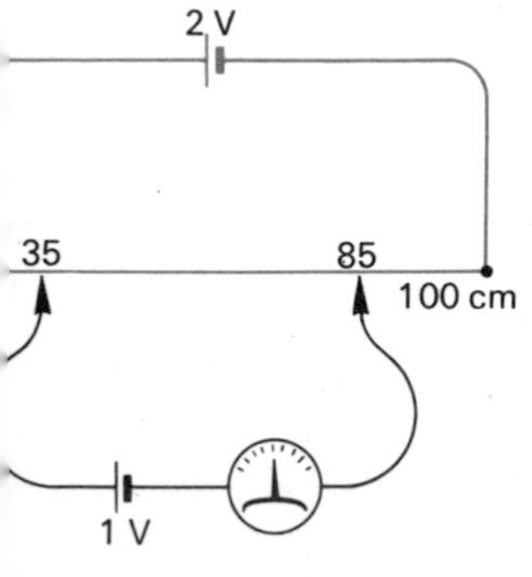

Fig. 8–11

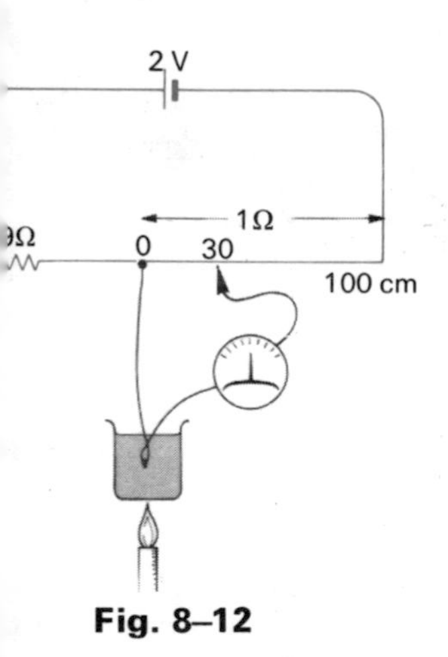

Fig. 8–12

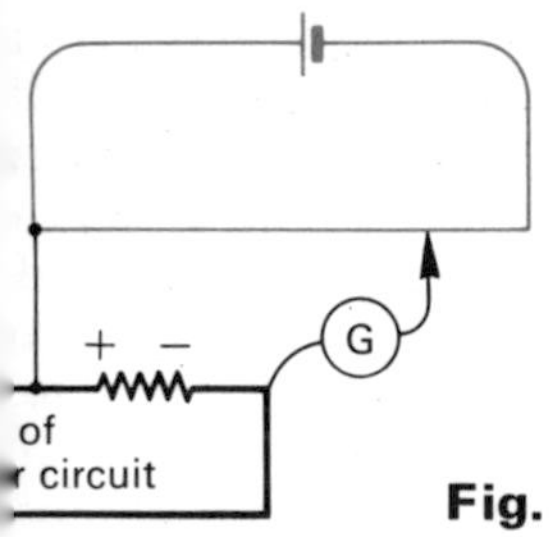

Fig. 8–13

against some known standard. The Weston Standard Cell produces an e.m.f. which is known precisely to four or more significant figures. For our example we will take its e.m.f. as 1.02 V. Such a cell can be used to calibrate the potentiometer. Suppose that a balance position is achieved at 513 mm using the Standard Cell (Fig. 8–10). Then

$$\left(\frac{513}{1000}\right)V = 1.02$$

If, when the Standard Cell is replaced by another cell of e.m.f. E volts, the balance position is at 768 mm, then

$$\left(\frac{768}{1000}\right)V = E$$

Thus $$E = \frac{768}{1000} \times 1.02 \times \frac{1000}{513}$$

$$E = 1.53 \text{ volts}$$

We have assumed that although the actual value of the supply voltage V is not known, it does remain constant throughout the experiment. A resistor R, of about 2 kΩ, limits the current through the galvanometer if the slider is originally some distance from the balance position. To make final adjustments the resistor is short-circuited by the switch.

For convenience it is usual to measure from one end of the wire the length over which balance is achieved. This is not, however, essential. With a 2 V accumulator and a 1 V test cell, balance could be obtained just as easily between, say, the 35 cm and 85 cm marks (Fig. 8–11).

To read very small e.m.fs we can extend the effective length of the metre wire by putting a resistor in series with it (Fig. 8–12). This produces the same effect as we would obtain from a 20-metre wire on which the slider moved only on the last metre. For the values shown in Fig. 8–12, the e.m.f. of the thermocouple would be 0.03 volts.

In the above examples, potentiometers have been used to compare the e.m.fs of two sources. A potentiometer can also be used to measure the p.d. across a resistor in another circuit, provided the polarity is as shown and the p.d. is less than the p.d. across the metre wire (Fig. 8–13).

RESISTANCE MEASUREMENTS

We will consider three methods of measuring resistance—the ammeter/voltmeter method, the ohmmeter, and finally the Wheatstone Bridge

Ammeter Voltmeter

The first involves measuring the current through and the p.d. across the resistor. From these readings the resistance is then obtained, using $R = V/I$.

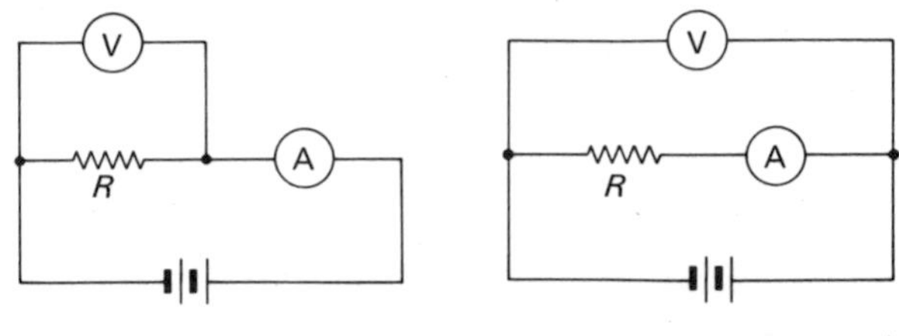

Fig. 8–14 **Fig. 8–15**

If we had ideal meters, the circuits shown above would be suitable for measuring any value of resistance. However, moving coil meters have a finite resistance and some current flows through them. We are therefore faced with two possibilities, neither of which gives the exact value of R from the ratio of the voltmeter and ammeter readings! If we use the circuit of Fig. 8–14, the ammeter reads the sum of the currents through the voltmeter and the resistor. If we use the circuit of Fig. 8–15, the voltmeter reads the sum of the p.ds across the ammeter and the resistor. You can't win—but you can get a reasonable estimate if you consider each situation carefully.

It is easy enough to say, 'Use the first circuit for low resistances and the second for high'—but what is 'low' and what is 'high'? Perhaps the first question to ask is, 'Which meter is more likely to upset the reading in the other?' Let us look at two extreme examples, using a voltmeter with a resistance of 5 kΩ, an ammeter with a resistance of 0.02 Ω, and a milliammeter with a resistance of 10 Ω.

1. ***Unknown resistance comparable to resistance of voltmeter***
   (a) voltmeter connected across resistor (Fig. 8–16).

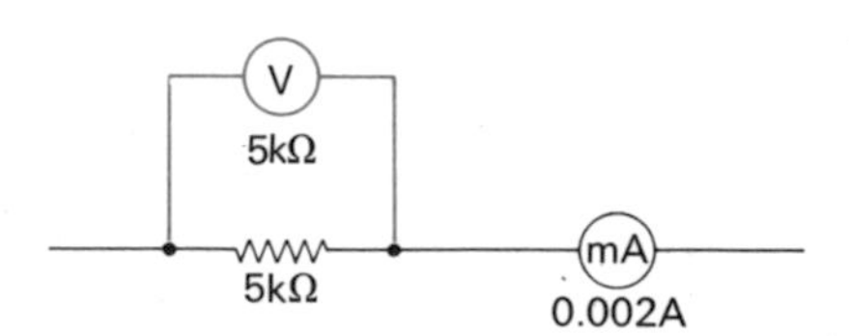

Fig. 8–16

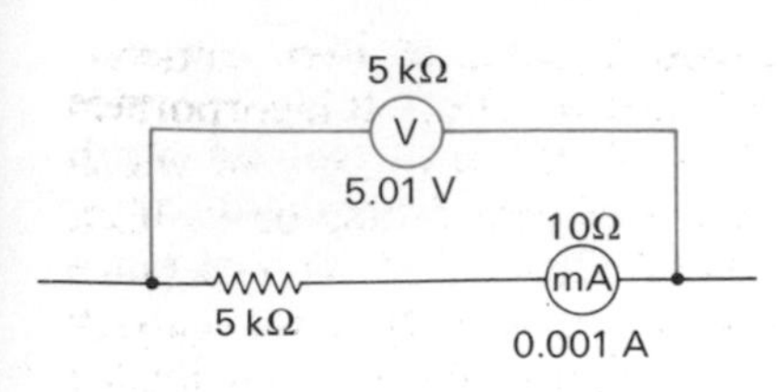

Fig. 8–17

Only *half* the current flowing through the ammeter will go through the resistor. The ratio V/I will therefore be $5/0.002 = 2.5\,\text{k}\Omega$, which is *half* the required value. Of course you could do a calculation to find the value of R if you also knew the voltmeter resistance. But this is giving yourself unnecessary work.

(b) voltmeter connected across resistor plus milliammeter in series (Fig. 8.17).

If the voltmeter now reads 5.01 V and the milliammeter 0.001 A, the unknown resistance would be $V/I = 5.01/0.001 = 5010\,\Omega$, which is for most purposes near enough the required value of $5000\,\Omega$.

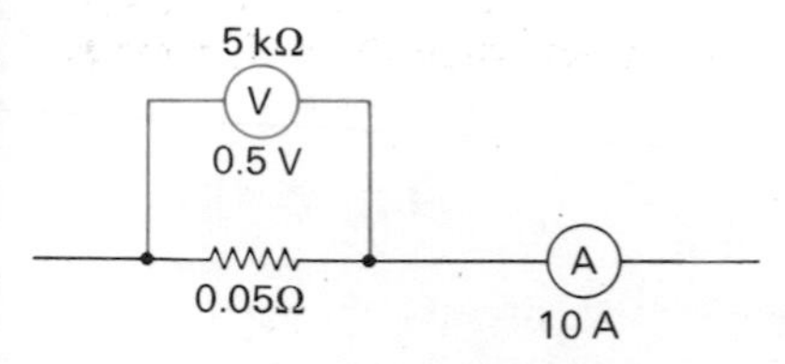

Fig. 8–18

2. ***Unknown resistance comparable to resistance of ammeter***

(a) voltmeter connected across resistor (Fig. 8–18).

In this example a very low resistance, perhaps a piece of wire, is being measured. The voltmeter makes no appreciable difference to the total resistance. Here the ratio V/I gives $0.5/10 = 0.05\,\Omega$.

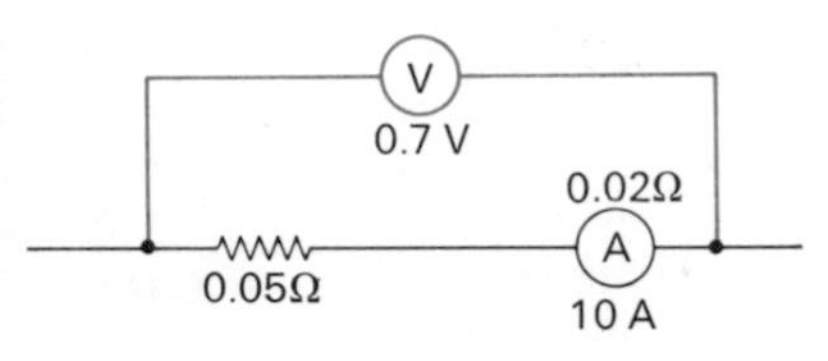

Fig. 8–19

(b) voltmeter connected across resistor plus ammeter in series (Fig. 8–19).

The ratio V/I now gives $0.7/10 = 0.07\,\Omega$, which is greater than the required value.

As we might expect, if we put a high resistance voltmeter in parallel with a high resistance, the ratio V/I gives the *total* resistance of the two in parallel (Fig. 8–16). If we put a low resistance ammeter in series with a low resistance, the V/I ratio indicates the *sum* of the two resistances (Fig. 8–19). The arrangements shown in Figs. 8–17 and 8–18 avoid these difficulties.

There still remains the border-line problem. Which circuit do we use with a resistance of, say, $1\,\Omega$? If we use the circuit of Fig. 8–19, the ratio V/I will give us 1.02Ω, and, if we use Fig. 8–18 we get $0.999\,\Omega$. As both answers are reasonably accurate, we might assume that *with these particular meters* we could take $1\,\Omega$ as a reasonable border line between a 'high' and a 'low' resistance.

In these examples we have used an ammeter (with a resistance of $0.02\,\Omega$) and a milliammeter (with a resistance of $10\,\Omega$). In another situation we might have to use a milliammeter reading 0–100 mA with a resistance of, say, $1\,\Omega$. *What V/I ratio would we obtain in each of the two methods if such a meter were used in conjunction with a $5k\Omega$ voltmeter to measure a resistance of 100Ω? (8.3)*

In general then we can, for measurement pu poses, treat a resistance of a few ohms or more a *high* resistance.

Ohmmeter

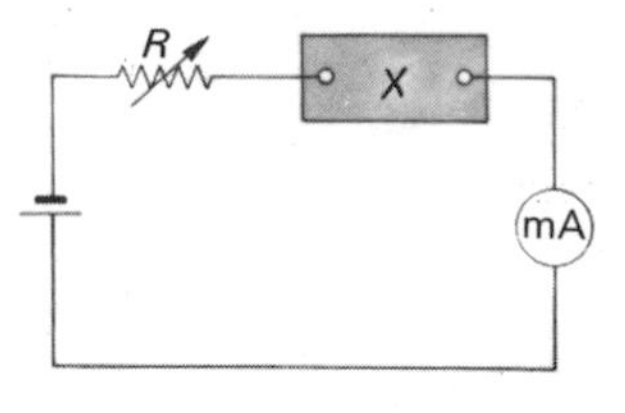

Fig. 8–20

To see how an ohmmeter works, consider the c cuit of Fig. 8–20, in which X represents an unkno resistance. First, however, assume that X is sh circuited and the resistor R adjusted until t milliammeter reads full-scale deflection. The sh circuit should then be removed and the curr reading noted. X is replaced by a calibrated re

Fig. 8

tance box such as the one illustrated in Fig. 8– The resistance of the box can be adjusted to p duce the same current as before. The unknc resistance X must then be equal to the resista in the box which can be read directly off the di

Of course it would be possible to mark the sc of the milliammeter in 'ohms' using the resista box to make the calibration. This is what is d in commercial ohmmeters. *What is the rheosta normally labelled in such a meter? (8.4) Can explain why the ohms scale is not linear on, sa multi-range meter? (8.5)*

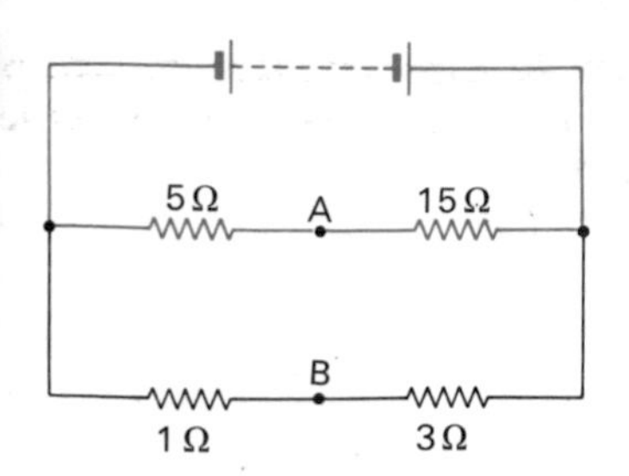

Fig. 8–22

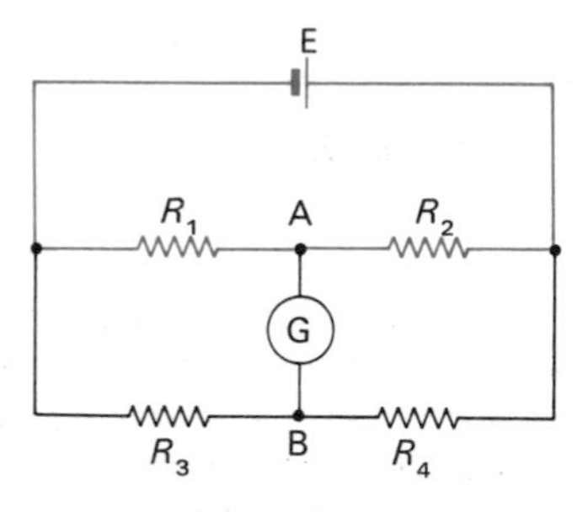

Fig. 8–23

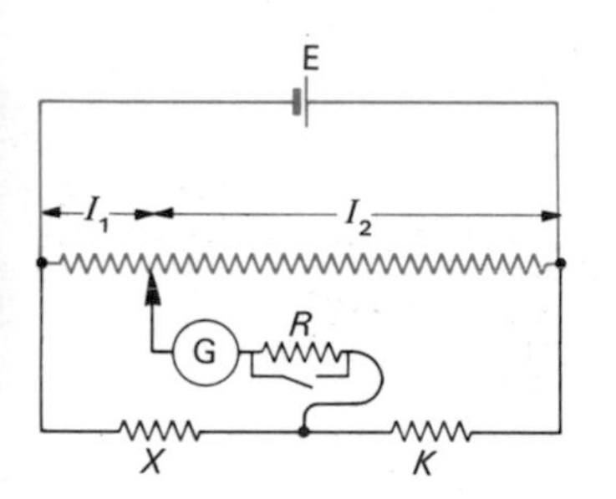

Fig. 8–24

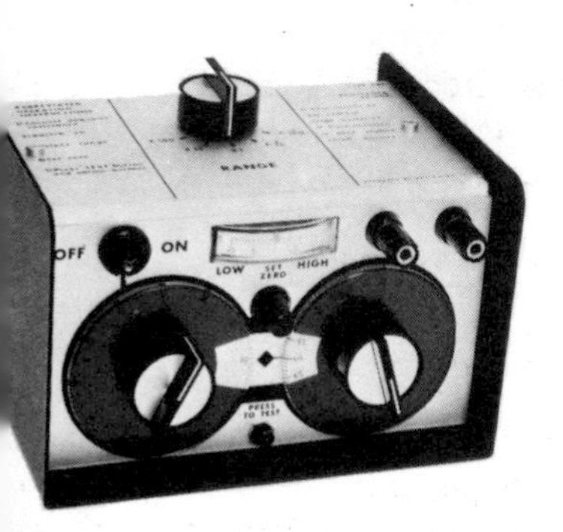

Fig. 8–25

Wheatstone Bridge

Among the many inventions of the English physicist Charles Wheatstone were the electric telegraph—which he sold for £30 000—and the concertina! Yet his name is best remembered for a device which he openly admitted he did not invent—the resistance bridge. It was, however, his use of the bridge that brought him into prominence.

If two resistors wired in series are connected to a 10 V supply as shown in black in Fig. 8–22, the same current, 0.5 A, will flow through each. The p.d. across the 5 Ω resistor will therefore be 2.5 V. Similarly, if the resistors shown in red are connected to the supply, the p.d. across the 1 Ω resistor will be 2.5 V. The p.d. across AB will therefore be *zero*. If a galvanometer is connected across AB, no current will flow through it.

In general, no current will flow in the galvanometer if $R_1/R_2 = R_3/R_4$ (Fig. 8–23). If R_1, R_2, and R_4 are known resistances (for example, they could be three resistance boxes such as the one illustrated in Fig. 8–21), we can use the circuit to find the value of R_3.

$$R_3 = \left(\frac{R_1}{R_2}\right) R_4$$

Can you deduce this relationship, assuming the current is I_A in R_1 and R_2 and I_B in R_3 and R_4? (8.6)

Instead of using three resistance boxes, we can use a length of uniform resistance wire in place of R_1 and R_2 and a fixed resistance of known value for R_4 (Fig. 8–24). The resistance of the wire is proportional to its length.

To find the balance position, a *jockey* is moved along the wire until the current in the galvo is zero. A limiting resistor R prevents large off-balance currents flowing through the galvanometer. It should, however, be short circuited to make the final adjustments. The unknown resistance X is then found as before.

$$X = \left(\frac{l_1}{l_2}\right) K$$

A modern miniature form of Wheatstone Bridge in which all the components are housed in a small box is shown in Fig. 8–25.

One great advantage of the Wheatstone Bridge circuit is that the balance point and thus the calculated resistance do not depend on the supply voltage. E does not come into the above relationship. This fact is used in some modern cameras, for example, the Praktica LTL. It incorporates a cadmium sulphide cell, the resistance of which varies with the amount of light falling on it. If an 'ohmmeter' method of measuring its resistance were used, the reading would vary if the battery voltage changed. By using a Wheatstone Bridge circuit, the balance position is always the same, regardless of the output of the battery. The aperture control is coupled to the sliding contact of a variable resistor as shown in Fig. 8–26 and 8–27. The aperture is correctly set when the meter reads zero.

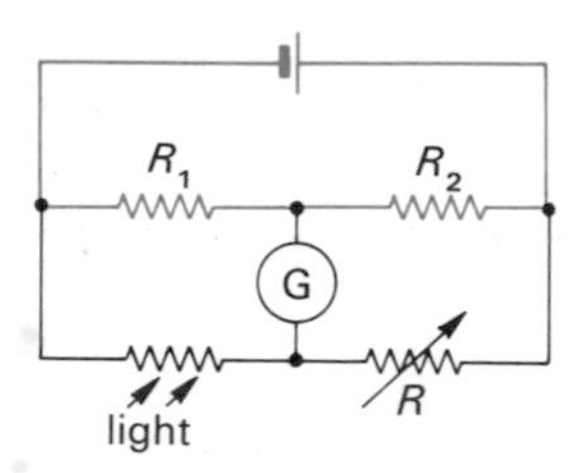

Fig. 8–26

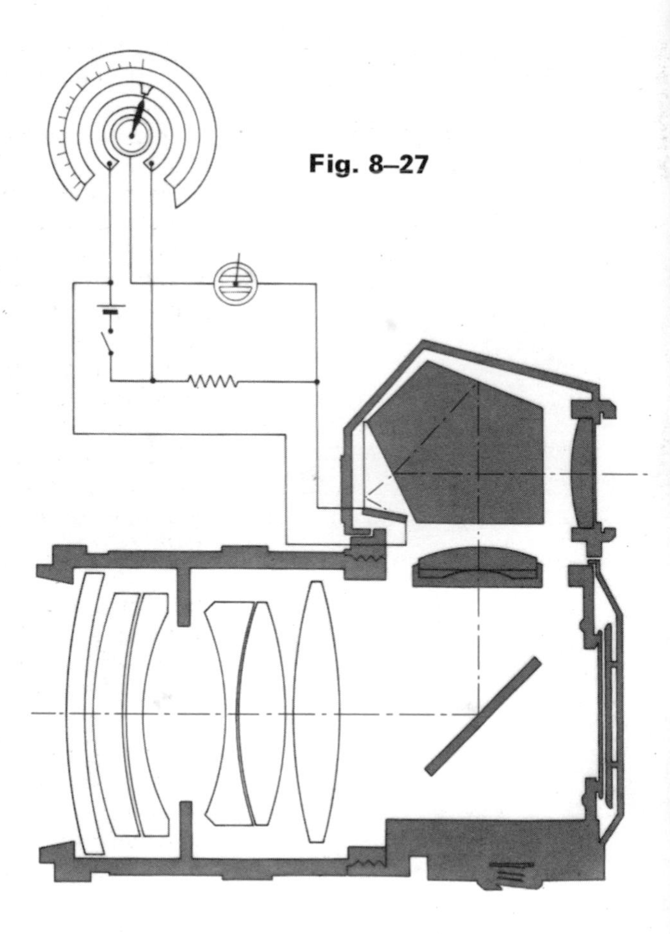
Fig. 8–27

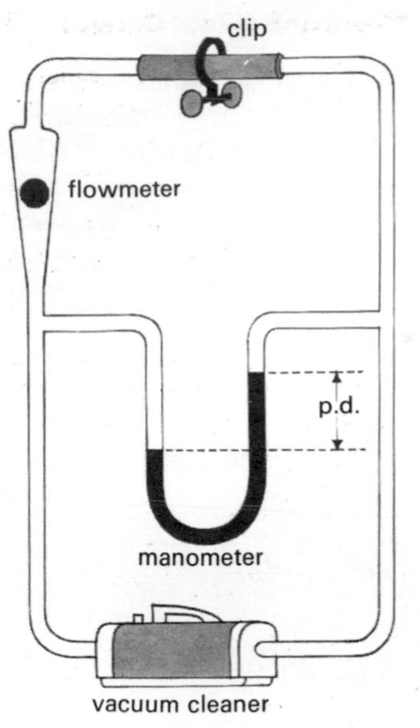

Fig. 8–28

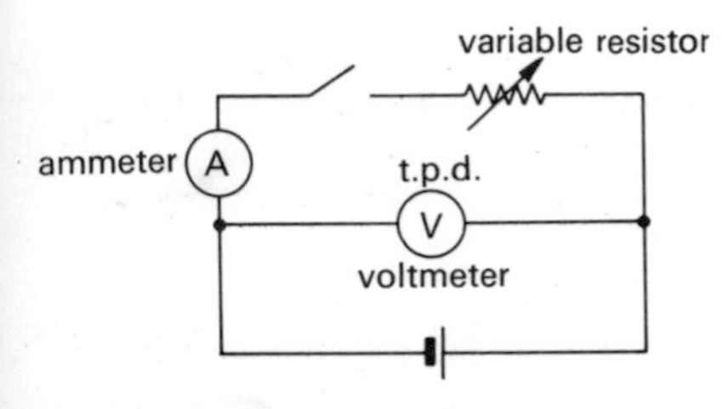

Fig. 8–29

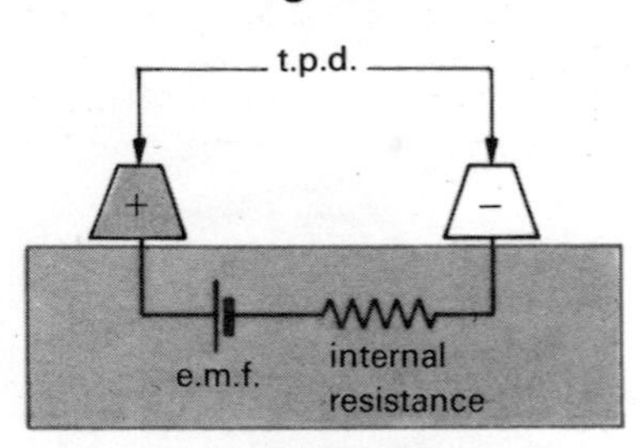

Fig. 8–30

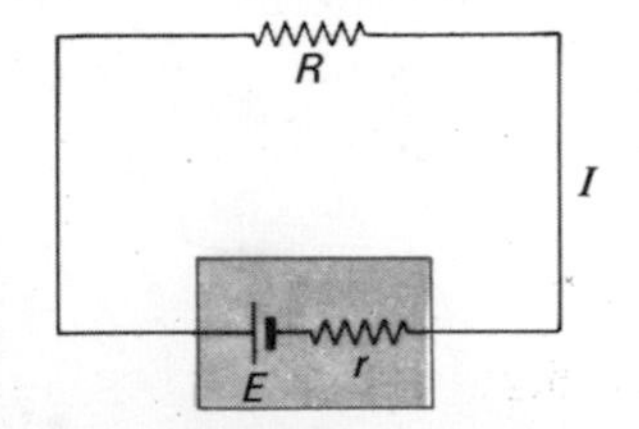

Fig. 8–31

INTERNAL RESISTANCE

If a vacuum cleaner is used to blow air round a circuit (Fig. 8–28), the pressure difference, measured on the manometer, varies if the air flow varies. If we loosen the clip the air flow *increases* and the p.d. *decreases*.

In an electric circuit (Fig. 8–29) the potential difference across the terminals of a cell *decreases* when the current taken from the cell *increases*. Watch what happens to the headlamps of a car if you leave them on and operate the starter motor. But don't make a habit of this!

When a battery drives a current around a circuit, some of the energy (from chemical reactions inside) is used to drive the current through the battery itself. The greater the current, the greater is the energy required for the 'internal operation'. For simplicity you can think of a cell as if it consisted of a constant source of e.m.f. in series with a resistance (Fig. 8–30). Although this is a fanciful picture, it helps us to see that the voltage across the terminals (sometimes called the terminal potential difference or t.p.d.) is not necessarily the same as the e.m.f. The difference between them is sometimes called the 'lost volts'. The resistance is really part of the cell itself and is called the *internal resistance*. In general, dry cells have a much greater internal resistance than accumulators. The internal resistance of a particular type of cell depends on its dimensions. Normally the larger the cell, the smaller the internal resistance.

A cell might be considered as an energy transformer in which chemical energy is changed to electrical energy. Like other machines, it is not 100% efficient and the difference between the input energy per coulomb (e.m.f.) and the output energy per coulomb (t.p.d.) is the 'lost volts'. In Fig. 8–31,

E = e.m.f., I = current,
R = external resistance,
r = internal resistance.

Input	=	*'Lost volts'*	+	*Output*
Energy supplied by the cell per coulomb		Energy used in the cell per coulomb		Energy transformed in the external load per coulomb
E	=	Ir	+	IR

Torch Battery and Bulb

Some torch batteries are marked 3 V, yet the bul which they supply are marked 2.5 V 0.3 A. T difference can be accounted for by considering t internal resistance of the battery. Suppose, f

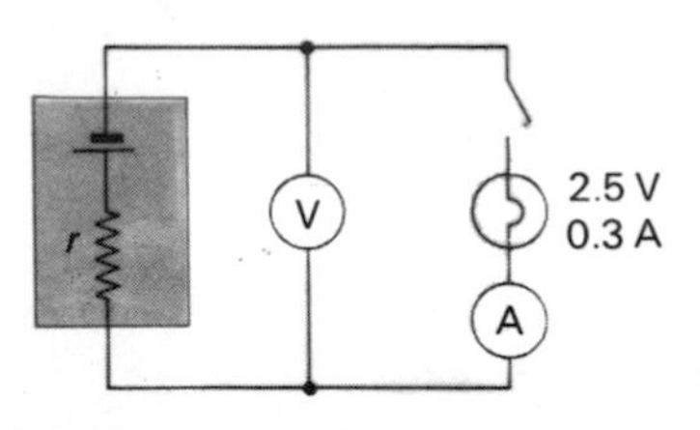

Fig. 8–32

example, that the voltmeter in Fig. 8–32 reads 3 when the switch is open and 2.5 V when it is close If the current in the bulb is then 0.3 A, and w ignore the current through the voltmeter, we c say that the 'lost volts' across the internal resistan is 0.5 V when the battery is supplying 0.3 A. T internal resistance of the battery is therefo $V/I = 0.5/0.3 = 1.67\,\Omega$.

SUMMARY

Measuring instruments

Direct current: current balance
moving coil meter
potentiometer and known resistance

Direct voltage: moving coil meter
potentiometer
C.R.O.

Resistance : ammeter and voltmeter
ohmmeter
Wheatstone Bridge

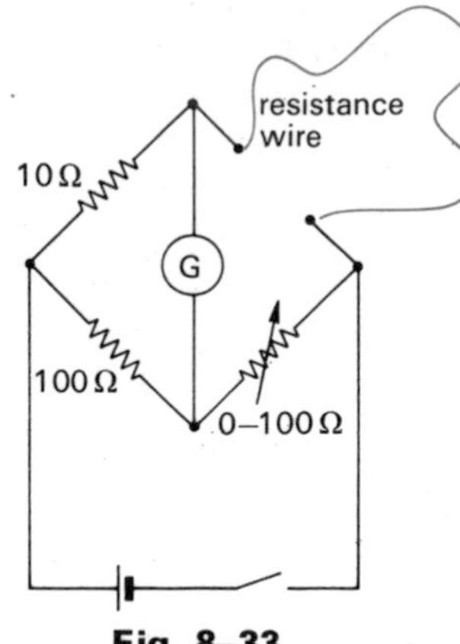

Fig. 8–33

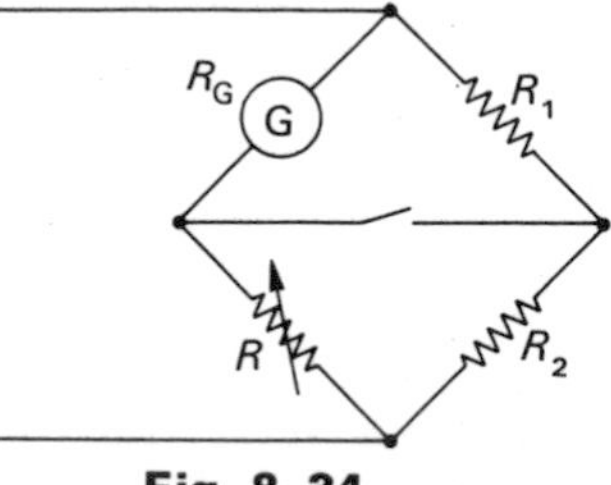

Fig. 8–34

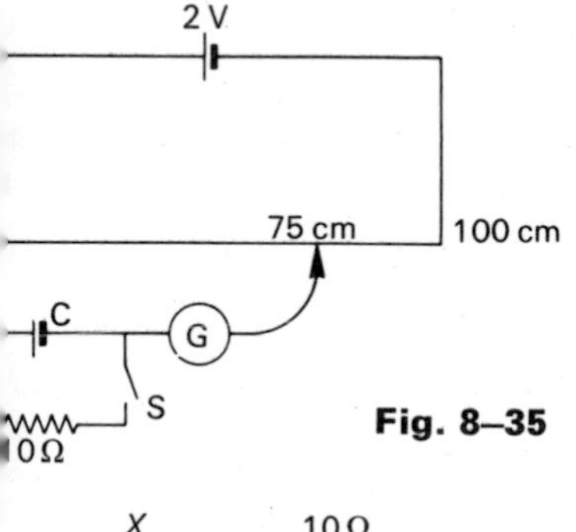

Fig. 8–35

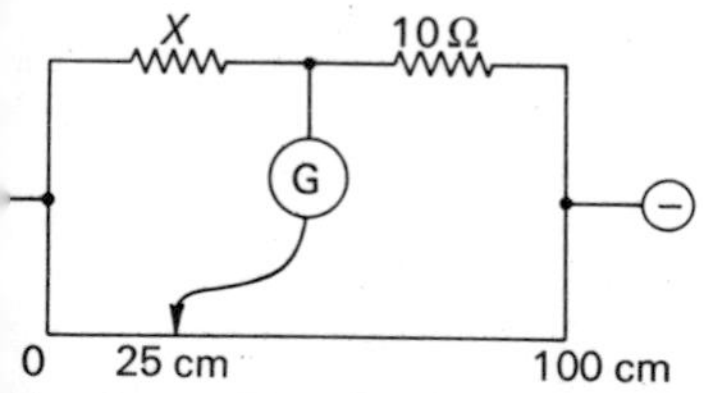

Fig. 8–36

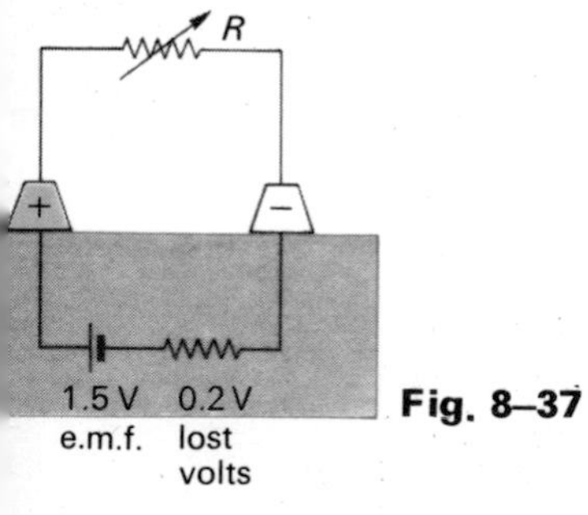

Fig. 8–37

PROBLEMS

8.7 A Wheatstone Bridge has a 15Ω resistor in one arm and resistances of 10Ω and 30Ω joined in parallel in the other arm. How far will the sliding contact be from the end of the metre wire which is connected to the 15Ω resistor?

8.8 An unknown resistance is fitted in one arm of a Wheatstone Bridge and a 3Ω resistor in the other. Balance is obtained when the slider is 25 cm from one end of the metre wire. Give two possible values for the unknown resistance, and explain how you would decide which value is correct.

8.9 A Wheatstone Bridge is made from resistances of 10Ω and 100Ω in its ratio arms and a calibrated resistance box of 0–100 Ω as the third arm (Fig. 8–33). When a 3-metre length of resistance wire is attached to the fourth arm of the bridge, balance is obtained when the resistance box reads 60 Ω. Find the resistance of 1 metre of the resistance wire.

8.10 A galvanometer of unknown resistance is connected in one arm of a Wheatstone Bridge circuit (Fig. 8–34). In the normal galvanometer position a key is fitted. How could you use this circuit to determine when $R_G/R = R_1/R_2$?

8.11 A 2V accumulator is used to supply a 1-metre potentiometer. The reading in the galvanometer is zero with switch S open and the slider at the 75 cm mark (Fig. 8–35). If S is closed, the balance position is at the 60 cm mark. Find the internal resistance of the cell C.

If the accumulator voltage had fallen from 2V to 1.9V just before the second reading was taken, would your *calculated* value of internal resistance be high, low or correct?

8.12 A resistance X is measured on a Wheatstone Bridge consisting of a 10Ω fixed resistance and a 1-metre wire with a sliding contact. What is the value of X if balance is obtained 25 cm from one end as shown in Fig. 8–36? What additional component(s) should be included in a practical Wheatstone Bridge circuit? Why?

How would the measured value of resistance alter if the supply voltage fell to half its original value?

8.13 Consider Fig. 8–37. What will be the t.p.d. across the cell? Will the current increase or decrease as R is increased? Will the t.p.d. then increase or decrease?

8.14 Why are dry cells not used for car batteries?

8.15 A battery consists of two cells connected in series. Each has an e.m.f. of 1.5V and internal resistance of 1Ω. What will be the measured terminal voltage of the battery if the voltmeter used has a resistance of (a) 10Ω, (b) 1000Ω? Which is the better meter?

8.16 (a) A cell of e.m.f. 1.5V and internal resistance 2Ω is connected to a 3Ω resistor. What is the current?
(b) Two such cells are wired in series with the same resistor. Find the current.

8.17 A 2V accumulator has an internal resistance of 0.1 Ω. What value of external resistor will limit the current to 2A? What will be the 'lost volts'?

8.18 A 1.5V cell with a 1Ω internal resistance is connected in a circuit which contains other sources of e.m.f. A high resistance voltmeter across the terminals of the cell reads zero. Give a possible explanation.

8.19 A 12V car battery which is discharged has a total internal resistance of 0.6Ω. A 20V charger is to be used to recharge the battery at 2A. What additional resistance will be required?

8.20 A small gas-lighter element has a resistance of 0.2Ω. You are given five dry cells, each with an e.m.f. of 1.5V and internal resistance of 1Ω. Would you connect the cells in series or parallel to obtain the greatest current through the element? Why is one very large cell normally used for this purpose?

8.21 When a 1.5V dry cell is short-circuited, there is a current of 3A. When a 2V accumulator is short-circuited, there is a current of 200A. What is the internal resistance of each? What harm may be done by short-circuiting an accumulator?

8.22 If you are given a 3V torch battery and a 9V transistor radio battery, how would you obtain the maximum e.m.f.? If the two positive terminals were connected together, what would a voltmeter connected to the two negative terminals read?

If the voltmeter was then removed and the negative terminals of the two batteries were then also joined together, what would happen?

8.23 An ammeter of negligible resistance reads 1A in the circuit shown in Fig. 8–38.

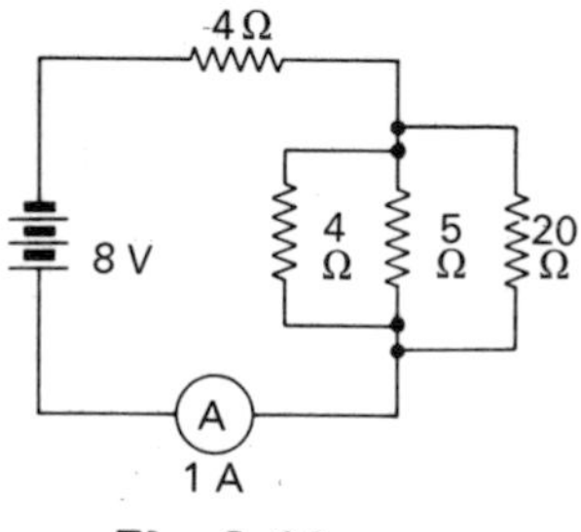

Fig. 8–38

If the e.m.f. of the battery is 8V, what is its internal resistance?

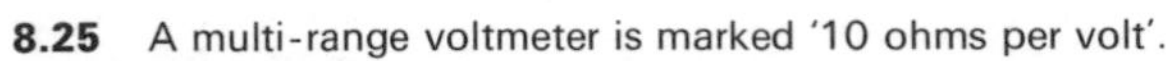

8.24 A moving iron meter reads the potential difference across the terminals of a torch battery as 1.8 V and a moving coil meter reads 2.6 V. Why is this?

8.25 A multi-range voltmeter is marked '10 ohms per volt'.
(a) What will be its resistance on the 200 V range?
(b) What would it read across an H.T. battery which has an e.m.f. of 120 V and an internal resistance of 400 ohms?
(c) What would the same meter read on its 1000 V range?
(d) If a meter marked '100 ohms per volt' were connected across the battery, what would be the reading on the 200 V range?

8.26 Draw a sketch showing how you would wire a bench with 3 pairs of terminals, each giving 12 V from a 12 V battery. If the internal resistance of the battery was 0.2 Ω and a current of 3 A was taken from one of the terminals and 7 A from the other, what would be the p.d. across the third pair of terminals?

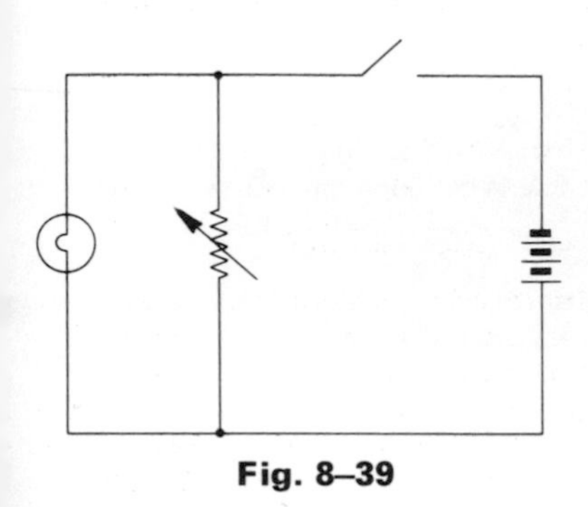
Fig. 8–39

8.27 A dimmer has to be fitted to a quartz-iodine headlamp bulb. The circuit shown in Fig. 8–39 was suggested. Would the lamp be dimmed by such a resistor? Do you consider this to be a satisfactory arrangement? Explain your answer.

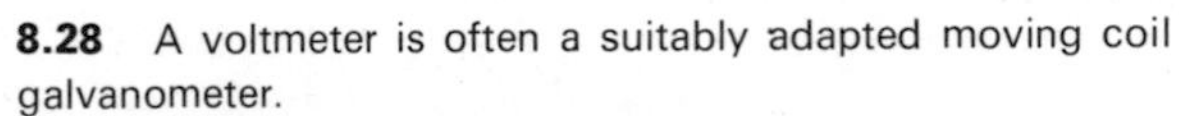

8.28 A voltmeter is often a suitably adapted moving coil galvanometer.
(a) How would you adapt a galvanometer of resistance 5 Ω and full scale deflection current 20 mA to read as a voltmeter for the range 0–3 V?
(b) Why is the p.d. across the terminals of a cell sometimes different from its e.m.f.?
(c) A cell, X, has an e.m.f. of 1.5 V and internal resistance of 5 Ω. An attempt is made to measure its e.m.f. with the voltmeter above. What will the meter read?
What will be the main difference between the voltmeter above and one able to measure the e.m.f. accurately?
(d) The diagram (Fig. 8–40) shows a potentiometer consisting of a 1-metre length of resistance wire suitably mounted over a scale marked off in millimetres. It is arranged to measure the e.m.f. of the cell, X, mentioned above.
(i) Why is this method preferable to the voltmeter method?
(ii) What is the function of the standard cell S?
(iii) What is the function of the resistor R?
(iv) How would you allow for the possible variation of the p.d. of the battery B during the experiment?
(v) Explain whether it is necessary for the meter to be sensitive or accurately calibrated.
(e) If the potentiometer wire has a resistance of 10 Ω, explain how it could be used in conjunction with the cell, X, to measure a small e.m.f. of up to 10 mV.
Sketch the arrangement you would use in order to make an accurate measurement, specifying completely any additional items required.

[S.C.E. (H) 1969]

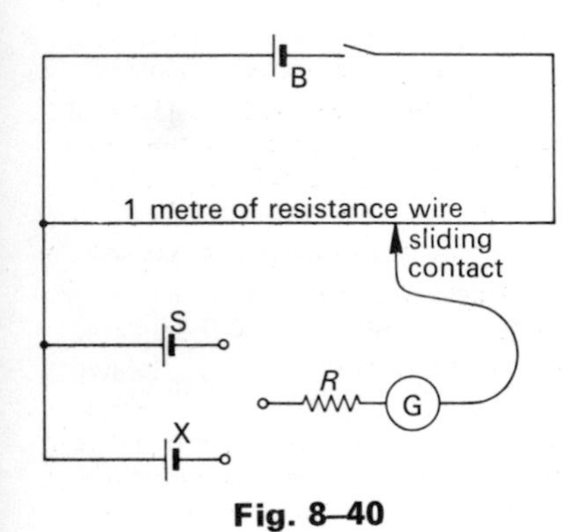

Fig. 8–40

8.29 (a) Fig. 8–41 shows a circuit in which the cell has negligible internal resistance. The circuit is used for measuring the resistance R, of a conductor XY.

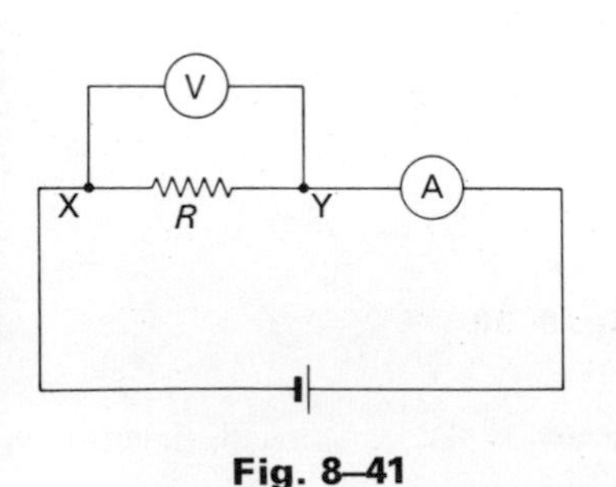

Fig. 8–41

(i) If the voltmeter reading is 3.95 V and the milliamme[ter] reading is 40.3 mA, what value does this give for t[he] resistance R?
(ii) If the voltmeter has a resistance of 5000 Ω, what is t[he] current through the voltmeter?
(iii) What is the value of the current through the resistor [R]?
(iv) What is the true value of the resistance R?
(v) What is the error in the value of the resistance found [in] (i), expressed as a percentage of the true value?
(b) Fig. 8–42 shows a circuit which is to be used to che[ck] readings on the ammeter A, between 0.10 A and 1.00 A.

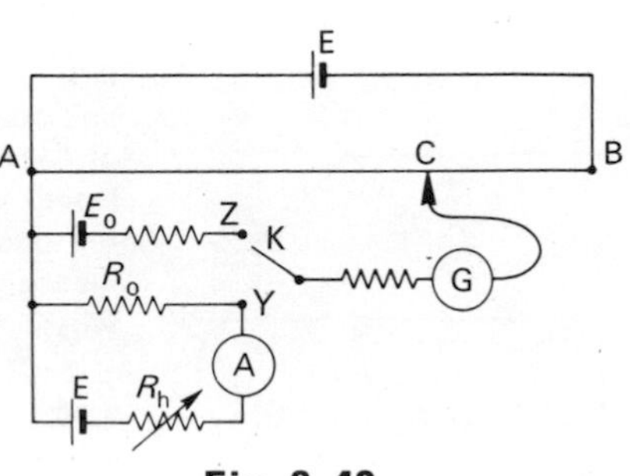

Fig. 8–42

AB is a resistance wire of uniform cross-section. Both c[ells] E are lead-acid accumulators of e.m.f. 2.0 V and the cell [E_0] is a standard cell of e.m.f. 1.02 V. The resistor R_0 [has] resistance 1.00 Ω.
(i) What is the function of the rheostat R_h?
(ii) Over what range of resistance should the rheostat be capable of varying?
(iii) What power should the resistor R_0 be capable [of] dissipating?
(iv) With the switch K connected to Z, the galvanomete[r] shows no deflection when AC = 95.4 cm, and with switch K connected to Y, a similar null-point is fou[nd] when AC = 67.2 cm. What is the current through resistor R_0, correct to the nearest milliampere?

[S.C.E. (H) 19..]

8.30 A bulb, marked '6 V: 0.06 A', is brightly lit wh[en] connected to a battery of four dry cells. Explain w[hy] another bulb, marked '6 V: 36 W' (in good working ord[er]) does not light up at all when connected to the same batt[ery].

[S.C.E. (H) 19..]

8.31 The resistance of one wire, A, is three times as gr[eat] as that of another wire, B.
(a) Compare the rates of dissipation of energy in the t[wo] wires when they are connected *in parallel* across the sa[me] source of supply.
(b) Compare the rates of dissipation of energy in the t[wo] wires when they are connected *in series* across a source [of] supply.

[S.C.E. (H) 19..]

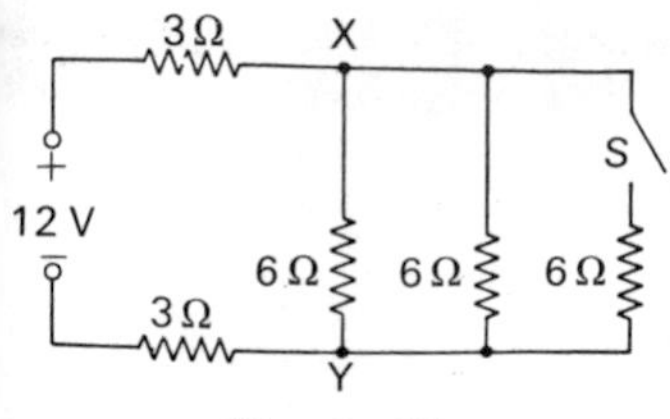

Fig. 8–43

8.32 The circuit shown in Fig. 8–43 consists of a resistance network connected to a 12 V source. The internal resistance of the source is negligible.
(a) Determine the p.d. across XY with (i) S open and (ii) S closed.
(b) When S is closed what is (i) the current from the source and (ii) the power supplied by the source?
[S.C.E. (H) 1971]

8.33 (a) Explain briefly how you would check the approximate current rating of a given specimen of fuse wire. Your answers should include a circuit diagram.
(b) Two lengths of fuse wire are connected in parallel with a space between them. It is found that they melt when the current through each of them is 10 A. When they are twisted together, melting takes place with a current less than 10 A in each. Suggest an explanation for this difference.
[S.C.E. (H) 1971]

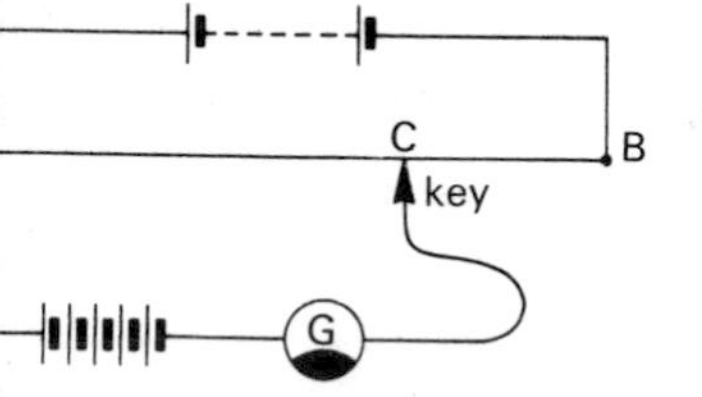

Fig. 8–44

8.34 When five 1.5 V cells are connected, as shown in Fig. 8–44, between the points A and C of a potentiometer wire AB, balance is obtained. AC is found to equal 75 cm.
(a) What is the potential gradient along AB?
(b) How would you protect the galvanometer G in out-of-balance conditions?
(c) If one of the 5 cells were now reversed, explain how far the new balance point would be from A.
[S.C.E. (H) 1971]

8.35 A d.c. motor, with permanent magnetic poles and armature windings of resistance 1 Ω, is connected to a 12 V supply. When the motor is running under normal load conditions, it is found that a current of 4 A flows through the windings.
(a) Explain why the current is not 12 A.
(b) State and explain what would happen to the armature current if the load on the motor were increased.
[S.C.E. (H) 1971]

8.36 (a) Fig. 8–45 shows two circuits commonly used to determine the value of an unknown resistance. Both ammeter and voltmeter are moving coil instruments.

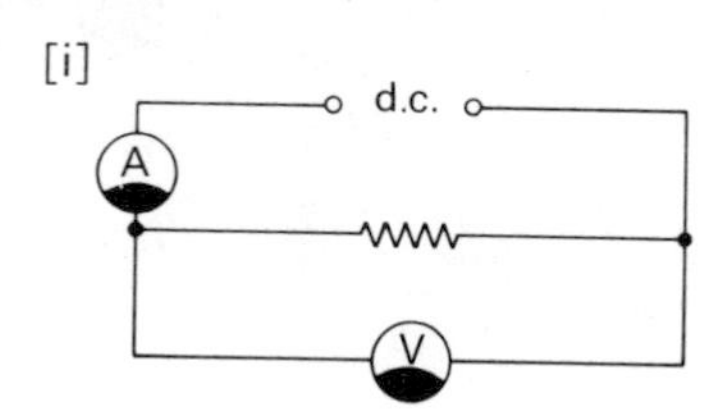

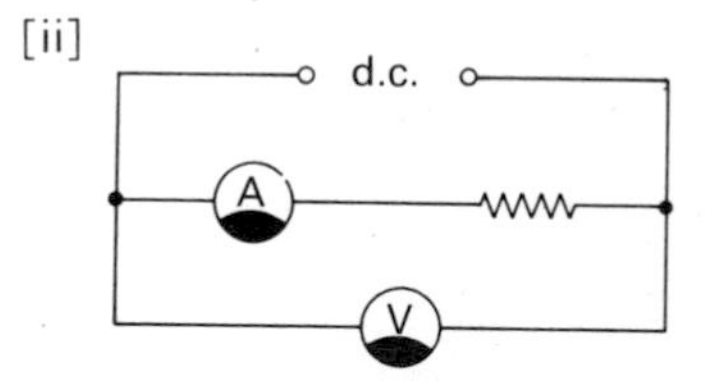

Fig. 8–45

State which circuit you would use if the unknown resistance had a value comparable with the resistance of the voltmeter. Justify your choice, commenting on the position of the ammeter in each case, and the errors likely to result if the other circuit were used.
(b) A galvanometer of resistance 10 Ω gives its maximum deflection for a current of 50 mA. What is the greatest p.d. it can measure if used as a voltmeter (i) alone, (ii) with a series resistance of 190 Ω? In the latter case, what would be the current through the meter if the scale reading were 2 volts?
(c) Three uniform wires XY, YZ, ZX, are joined to form a triangle XYZ (Fig. 8–46).

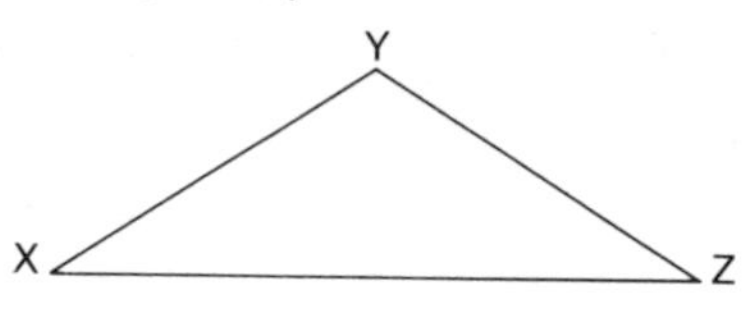

Fig. 8–46

The resistances of XY, YZ, and ZX are 6, 6, and 10 Ω respectively. A p.d. is applied across the resistance XZ. Which point on XZ has the same potential as (i) the point Y and (ii) the mid-point of XY?

The wire YZ is replaced by another of unknown resistance *R*. Explain how, with the aid of a centre-zero galvanometer, connecting wire and a metre rule, you would proceed to find the value of *R*. [S.C.E. (H) 1971]

8.37 (a) Comment on the advantages of a null deflection method as used in a potentiometer circuit.
(b) Discuss the disadvantages of using the brightness of a bulb as a means of measuring electric current.
[S.C.E.(H) 1972]

8.38 An electric toaster has four identical heating elements each of effective resistance 50 Ω connected as shown in the diagram.

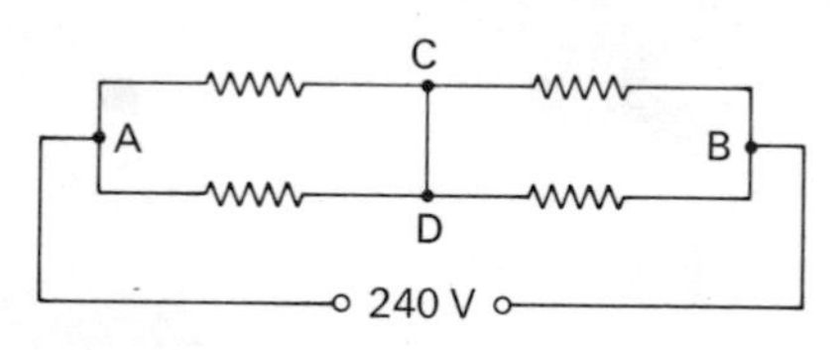

Fig. 8–47

(a) If the element AC breaks, what power is dissipated in each of the remaining elements, assuming their resistances do not vary?
(b) Explain why element AD is likely to fuse if left in use with AC broken.
(c) In practice the resistance of the elements increases with temperature. Explain whether this makes AD more likely or less likely to fuse.
(d) Suggest any practical advantage in having C and D joined. [S.C.E (H) 1972]

9 Measuring Alternating Curren

Fig. 9–1

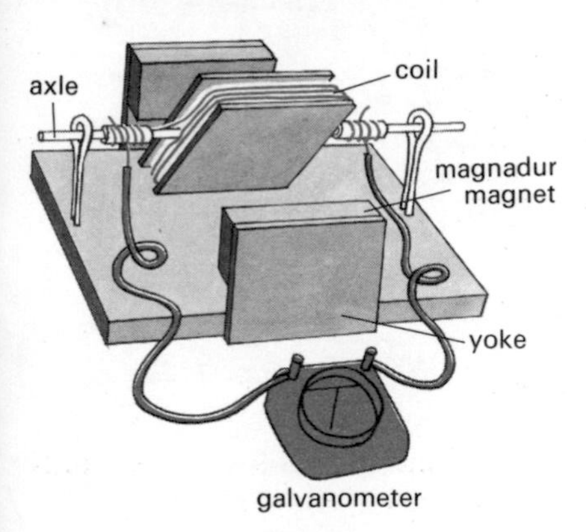

Fig. 9–2

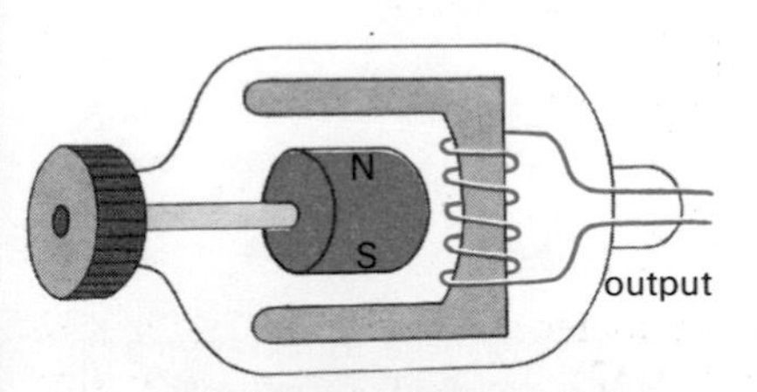

Fig. 9–3

PRODUCING A.C.

In general, when a conductor moves in a magnetic field or a magnet moves near a conductor, an e.m.f. is induced in the conductor. In practice the simplest way of generating a continuous supply of electricity is to rotate a coil in a magnetic field, as in the model alternator illustrated in Fig. 9–2. Alternatively, a magnet can be rotated close to a coil as in the bicycle dynamo (Fig. 9–3). Increasing the speed of rotation increases the amplitude of the e.m.f. and also, of course, the frequency. The output wave form from one of these simple devices is not, however, the pure sine wave we expect from commerically produced alternating current (Fig. 9–4).

MEASURING A.C.

Frequency

The number of complete oscillations per second is called the *frequency*. The unit of frequency is the hertz (Hz) named after the German physicist Heinrich Rudolph Hertz. 1 Hz is 1 cycle per second.

Frequency meters are fairly complicated instruments, and we cannot discuss their operation in this course. Basically they count the number of pulses fed into them in a given time and record the number on a digital display (Fig. 9–5). However, a simple C.R.O. can be used to measure frequency. Suppose, for example, that the mains supply (50 Hz) produces the pattern shown in Fig. 9–4.

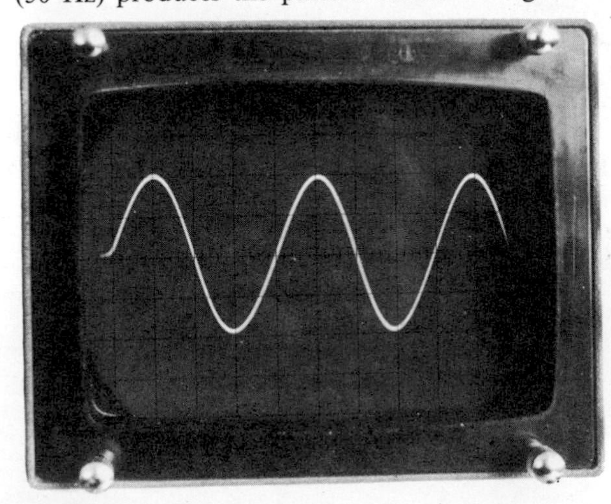

Fig. 9–4

Fig. 9–5

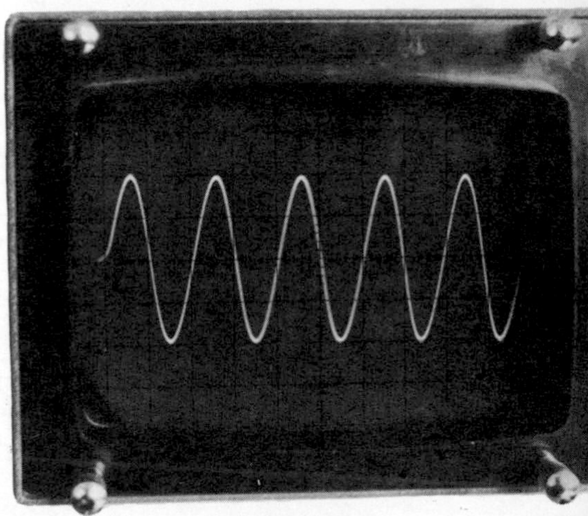

Fig. 9–6

Without altering the time base setting on C.R.O., we can connect another source of unkn frequency to the input terminals. If the new pat produced looks like Fig. 9–6, the unknown quency must be twice the original; that is, it m be 100 Hz.

If a calibrated signal generator and an osc scope are available, the task of measuring freque is even easier. You can simply connect the sou whose frequency is to be measured, to the C.I and adjust the time base to give a reasonable n ber of waves. Then replace the source by the si generator and adjust its frequency until the same number of 'waves' appears on the screen. The unknown frequency is then equal to the frequency reading on the dial of the signal generator.

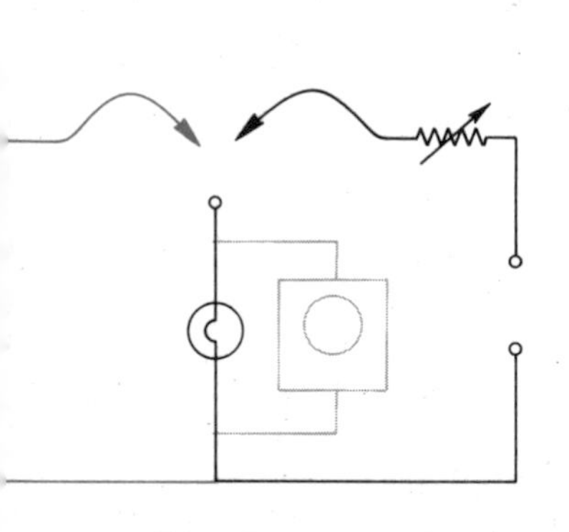

Fig. 9–7

Voltage

Before considering how to measure alternating voltage, we ought to ask exactly what it is we want to measure. If we buy a bulb marked '6 V', we would expect it to light up equally brightly on a 6 V dry battery producing d.c., or a 6 V bicycle dynamo producing a.c. As the 'volt' is defined as one joule per coulomb, we would expect each 6 V source to produce the same number of joules of heat or light for every coulomb of charge flowing through the bulb. The alternating voltage which is capable of producing the same heating effect as the 6 V battery (d.c.) is called the effective or *r.m.s. value* of the alternating voltage.

$$6\text{ V (d.c.)} = 6\text{ V r.m.s. (a.c.)}$$

R.M.S. stands for 'root mean square'. For the moment you can think of it simply as the value of an alternating voltage which will cause heat to be produced in a given resistor at the same average rate as will a direct voltage of that value.

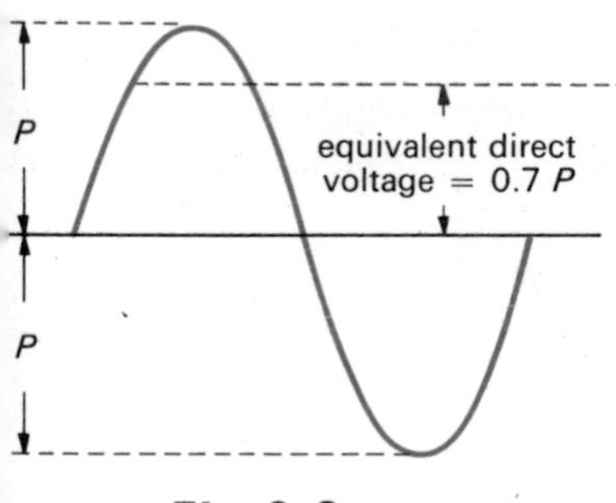

Fig. 9–8

Using the apparatus illustrated in Fig. 9–7, we find that, when the same amount of light is produced by a direct current and an alternating current, the direct voltage is 0.7 times the peak alternating voltage. The effective or r.m.s. value of an alternating voltage is therefore 0.7 peak voltage (Fig. 9–8). This is, however, only valid for sinusoidal (i.e. sine-wave) voltages. *For what kind of alternating voltages would you expect the effective and peak values to be the same? (9.1)*

As practically all the a.c. we normally use is sinusoidal, a.c. voltmeters are calibrated to read the r.m.s. value at 50 Hz. Unless it is otherwise stated, you can assume that when an alternating voltage is mentioned it is the r.m.s. value that is intended.

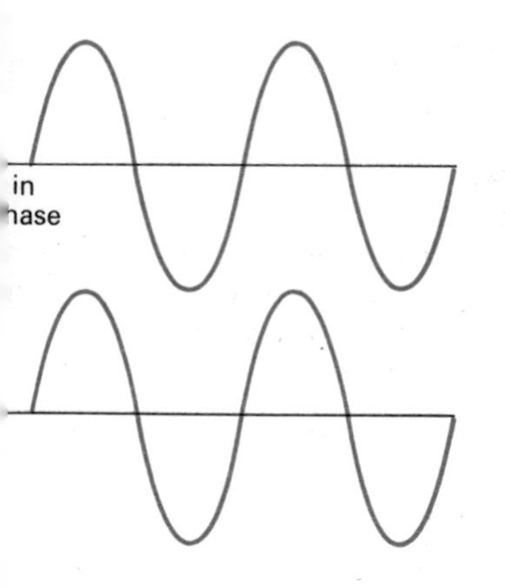

Fig. 9–9

Adding alternating voltages

When two alternating supplies of the same frequency are connected in series, the result can be the sum *or* the difference of these voltages, depending on whether they are in phase (Fig. 9–9) or out of phase (Fig. 9–10). If the supplies shown in Fig. 9–11 are wired in phase, the voltmeter will read 4 V; if they are out of phase, it will read zero. If the supplies are neither exactly in phase or out of phase, the result of the addition or subtraction is not so simply worked out and, if the supplies are of different frequencies, the situation is much more complex.

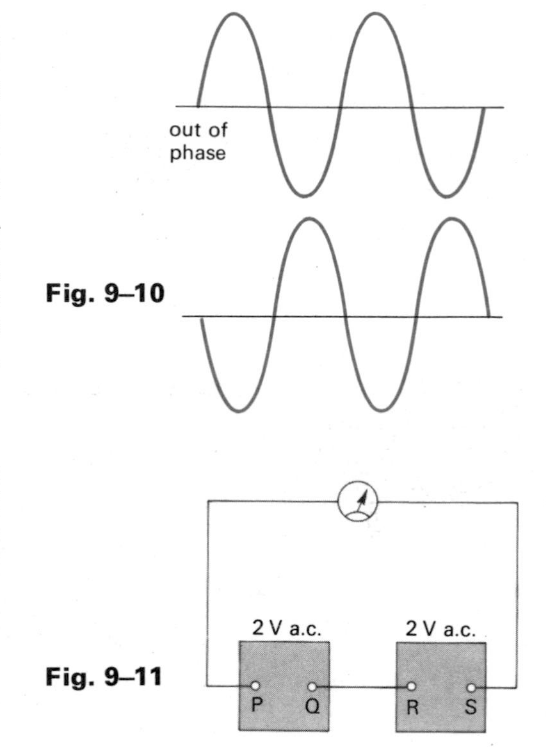

Fig. 9–10

Fig. 9–11

Current

Moving coil meter and rectifier

A normal moving coil meter is not, by itself, suitable for measuring a.c. *Why not? (9.2)*

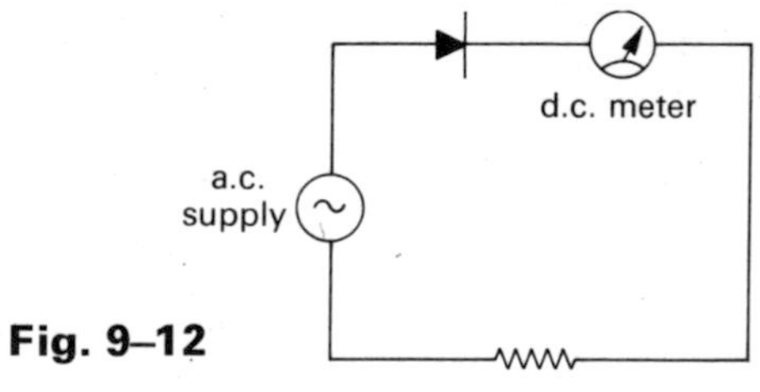

Fig. 9–12

Of course such a meter can be used if the a.c. is rectified first (Fig. 9–12). Such meters are normally calibrated in r.m.s. units.

To allow current to flow through the meter during both half-cycles, a full wave rectifier or bridge rectifier is normally used. This is illustrated in Figs. 9–13 and 9–14. *Can you work out the operation of these circuits?* (9.3)

Most multirange meters (e.g. AVO) have this type of circuit for their a.c. ranges.

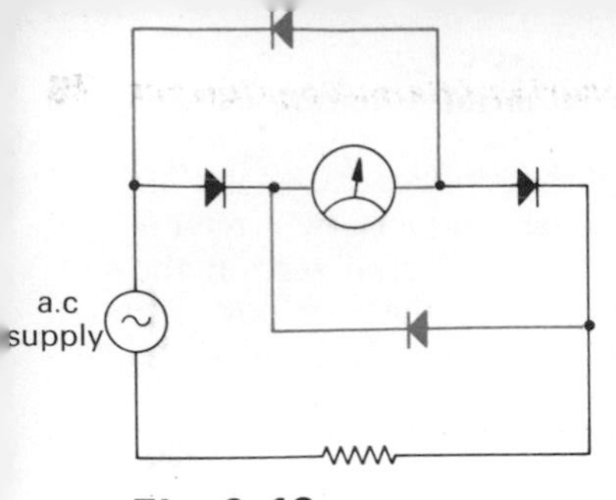

Fig. 9–13

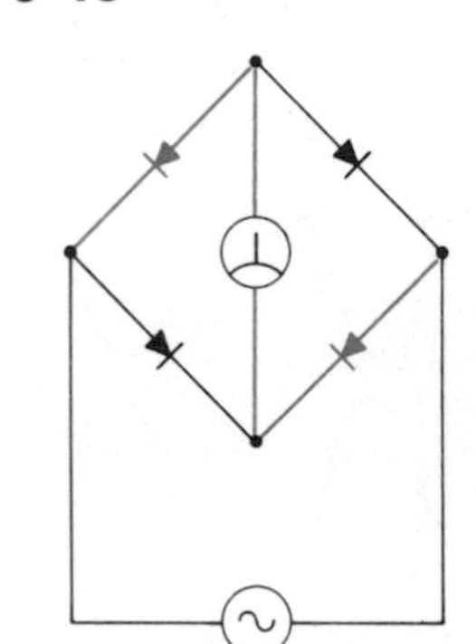

Fig. 9–14

Moving iron meter

Perhaps the cheapest form of a.c. meter is the moving iron meter. The a.c. flows through a coil which then attracts a lump of iron attached to a pointer. As the attraction is independent of the direction of the current, this meter indicates a.c. Fig. 9–15 illustrates a simplified form of moving iron meter. The force exerted on the bar depends on both the strength of the magnetic field of the coil *and* the induced magnetism in the iron. This results in a non-linear scale.

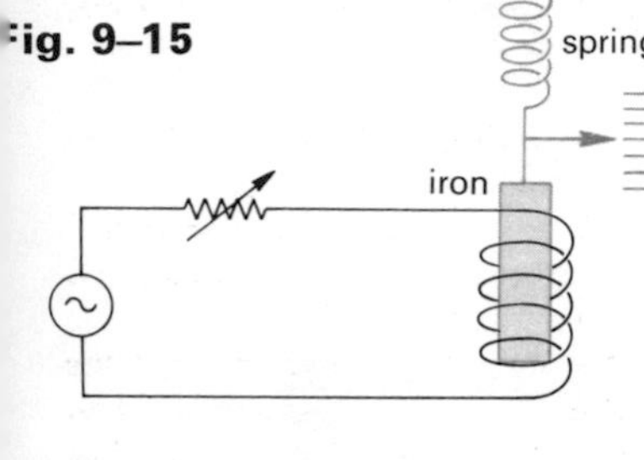

Fig. 9–15

Meters depending on heating effect

The heating effect of a current does not depend on the direction, so that this effect can be used to indicate the strength of an alternating current.

(i) In the hot-wire meter a wire expands when heated by the current (Fig. 9–16). As the heating effect (I^2R) depends on the square of the current, the hot wire ammeter is also a 'square law' instrument.

(ii) In the thermo-junction ammeter the current heats a thermocouple and the output is read in a moving coil meter (Fig. 9–17). The scale is again non-linear.

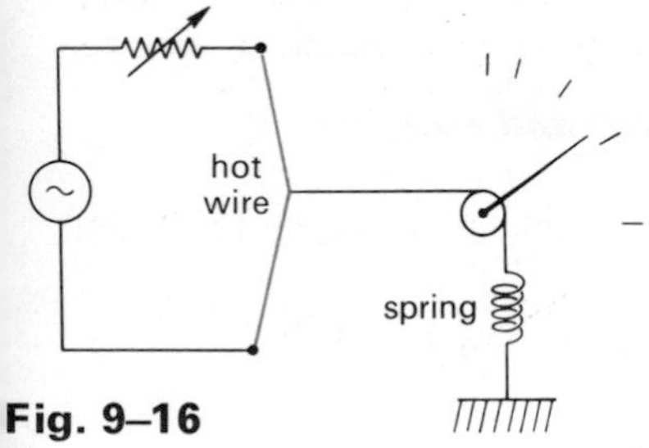

hot wire

spring

Fig. 9–16

In general, a.c. ammeters are calibrated to read the *r.m.s. value* of the current; that is, the value of a.c. which produces heat in a resistor at the same rate as that value of d.c. A 12 V 4 A car headlamp bulb would, for example, produce heat and light at the same rate if it were connected to a 12 V d.c. or a 12 V (r.m.s.) a.c. supply. In the first case the current would be 4 A d.c. and in the second 4 A (r.m.s.) a.c.

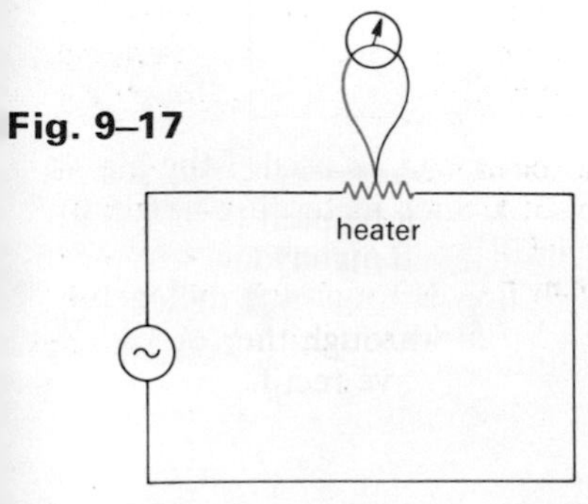

Fig. 9–17

R.M.S. value of current

If a *steady* current I_e flows in a resistance R, the rate of energy transfer is $I_e^2 R$. To find what value of steady current converts energy at the same rate as a sinusoidal alternating current i of maximum (i.e. peak) value I_m, consider the graph on Fig. 9–18.

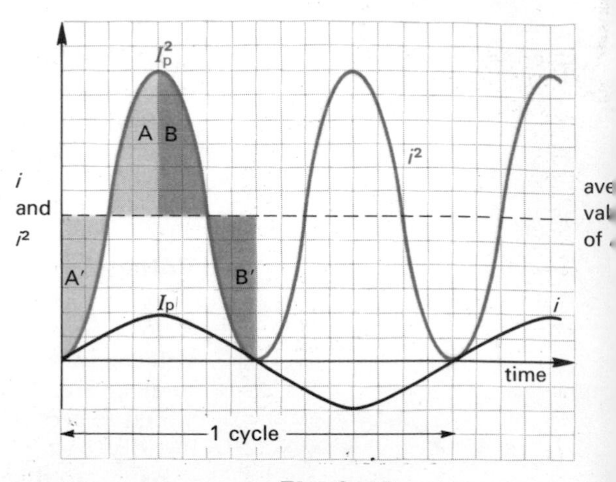

Fig. 9–18

When the instantaneous values of the current are squared, the graph of i^2 is also a sine curve but it has twice the original frequency.

Taken over one complete cycle, the average or mean value of i^2 is $\frac{1}{2}I_m^2$. To see this, we note that area $A = A'$, $B = B'$, etc.

For this alternating current in a resistance R, the average rate of energy transfer over one complete cycle is therefore given by

$$\text{power} = \tfrac{1}{2}I_m^2 R$$

If we call the steady current which would produce heat at the same rate the *equivalent* steady current I_e, then

$$\text{power} = I_e^2 R$$

Thus $$I_e^2 = \tfrac{1}{2}I_m^2$$

$$\Rightarrow I_e = \sqrt{\frac{I_m^2}{2}} = \frac{I_m}{\sqrt{2}} = 0.707 I_m$$

This is called the *root mean square* current

$$I_{\text{r.m.s.}} = 0.707 I_m$$

Power and Energy

The rate at which electrical energy is transformed is measured in joules per second or watts. The volt is defined as a joule per coulomb or a watt per ampere:

$$\text{volts} = \frac{\text{joules}}{\text{coulombs}} = \frac{\text{joules}}{\text{seconds}} \times \frac{\text{seconds}}{\text{coulombs}}$$

$$= \text{watts} \times \frac{1}{\text{amperes}}$$

i.e. watts = volts × amperes

Fig. 9–19

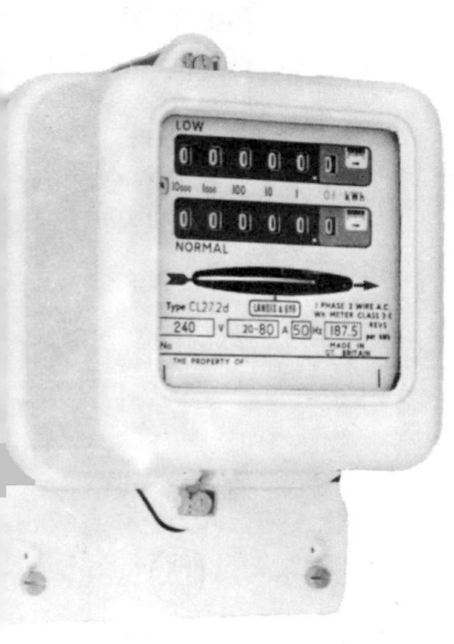

Fig. 9–20

To measure power it is therefore necessary to measure the voltage and the current and to multiply them together. This method is perfectly satisfactory when dealing with direct current as the voltage and current are always in step or in phase. It is not, however, possible to use this method when dealing with alternating currents if the current and voltage are not in phase (see page 88).

Fortunately there is an instrument which enables us to find, in joules, the total amount of energy transformed, so that we can find the power by noting the energy transformed in a certain time. The instrument is called a joulemeter (Fig. 9–19), and we will use it later to find the amount of energy being transformed by various components. The speed at which the aluminium disc in the meter rotates will give us a qualitative idea of power.

The theory behind the joulemeter and its domestic counterpart, the kilowatt-hour meter (Fig. 9–20), is too difficult to discuss here, but this need not prevent our using it. After all, few car drivers know how an internal combustion engine works!

With most electrical household equipment the current and voltage are very nearly in phase, so that we can calculate power, for most practical purposes, as 'volts times amps', using r.m.s. values of voltage and current. The following quantities apply.

Flow rate: current I in amperes
1 A = 1 coulomb per second

Potential difference: p.d. V in volts
1 V = 1 joule per coulomb

Opposition: resistance R in ohms
1 Ω = 1 volt per ampere

Work: energy E in joules
1 J = 1 volt coulomb
= 1 watt second

commercial unit
1 kW h = 3 600 000 joules

Rate of working: power P in watts
1 W = 1 joule per second
= 1 volt ampere

Impedance

As we shall see later the opposition to the flow of alternating current is not as simple as the opposition to direct current. In addition to resistance we have to take into account the effects of capacitors and coils. The opposition they offer to a.c. is called *reactance*. The total opposition resulting from resistance and reactance is called impedance. It is measured in *ohms*.

SUMMARY

A.C. measurement

Frequency	:	C.R.O. frequency meter
Voltage	:	C.R.O. moving coil meter + rectifier
Current	:	moving coil meter + rectifier moving iron meter hot wire meter thermo-junction
Energy	:	joulemeter kW h meter

PROBLEMS

9.4 The oscilloscope trace shown in Fig. 9–4 was obtained on the 5V/cm sensitivity setting. What was the r.m.s. voltage of the input?

9.5 What was the setting of the time base control in seconds per cm, if the input frequency used to produce Fig. 9–4 was 50 Hz?

9.6 (a) A projector lamp which passes 2 A at 50 V is to be operated from 240 V a.c. mains. If the wires to the lamp have a total resistance of 5 Ω and a demountable transformer kit is available having a primary coil of 1200 turns, how many turns should the secondary coil have?

Assuming the transformer is perfectly efficient, find the current taken from the mains. If, in practice, the transformer is not perfectly efficient, state two possible reasons for this.
(b) Write a short note explaining the principles of operation of any type of meter suitable for measuring the current taken by the transformer.

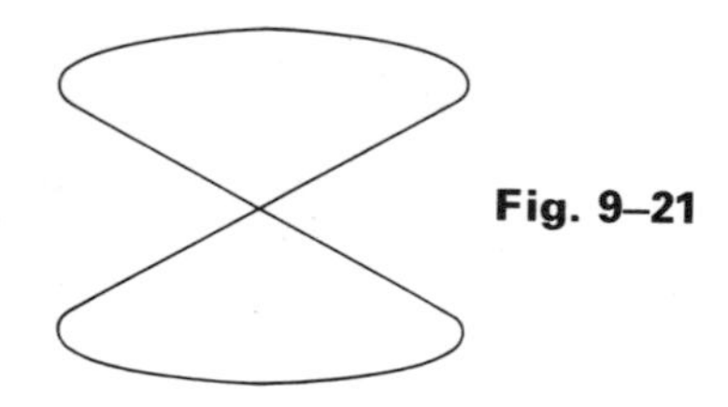

Fig. 9–21

(c) The diagram (Fig. 9–21) shows the pattern on the screen of an oscilloscope with a signal of frequency 50 Hz applied to the X-plates. What is the frequency of the signal on the Y-plates?
(d) Devise an experimental method by which the frequency of the mains supply might be measured, using a watch or clock as the standard of time. Explain how you would deduce the result from the measurements which you propose. [S.C.E. (H) 1969]

9.7 (a) Convert 1 kW h into joules.
(b) An electric kettle is designed to operate with a 240 V a.c. supply. Due to overloading of the supply, the voltage falls temporarily to 220 V.
(i) Calculate the percentage drop in power output assuming that the resistance of the heating element is unchanged.
(ii) If the variation in resistance with temperature is taken into account, explain whether the actual drop in power output will be greater or less than that calculated in (i).
(iii) Explain whether it would cost more to boil water at the lower voltage.
(c) In a diode valve the potential difference between anode and cathode is 150 V. The current through the valve is 12 mA.
(i) What is the energy, in joules, of each electron as it arrives at the anode? (You will find the charge on the electron given in your data book.)
(ii) In what form is this energy?
(iii) What becomes of this energy when the electron arrives at the anode?
(iv) How many electrons arrive at the anode in 1 second?
(v) From your answers to (i) and to (iv) calculate the rate, in watts, at which energy is dissipated at the anode.
(vi) Show that you get the same answer for the power dissipation at the anode by consideration of the anode-cathode voltage and the current through the valve. [S.C.E. (H) 1970]

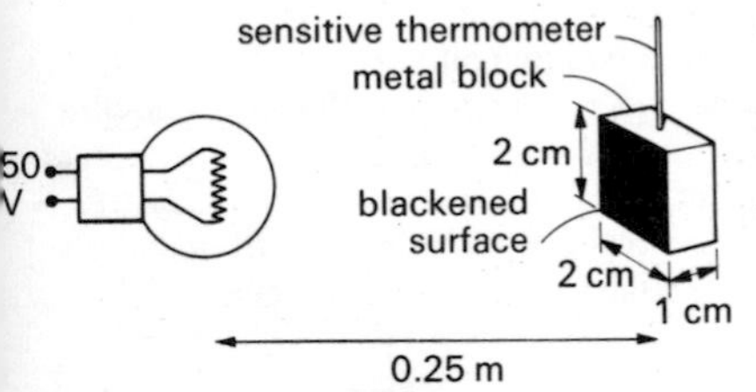

Fig. 9–22

9.8 The apparatus shown in the sketch (Fig. 9–22) shows a pupil's attempt to measure the efficiency of a light bulb marked 250 W: 250 V.
The metal block measured 2.0 cm × 2.0 cm × 1.0 cm and weighed 0.036 kg. It was blackened on the side facing the bulb and set 0.25 m from the centre of the bulb. The bulb was switched on for 4 minutes, and it was found that the temperature of the block rose from 18.20 °C to 19.70 °C.
(a) Calculate the quantity of heat absorbed by the block in 4 minutes, taking the specific heat capacity (specific heat) of the metal as 400 J kg^{-1} K^{-1}.
(b) Estimate, therefore, the total heat energy radiated from the bulb in 1 second. (Surface area of a sphere is $4\pi r^2$ and $\pi = 3.14$.)
(c) If it is assumed that the energy not radiated as heat is converted into light, how much energy is converted into light in 1 second?
(d) Hence calculate the percentage efficiency of the bulb as a light source.
(e) State *three* factors which have not been taken into consideration, and say how each one would affect the result.
(f) Explain why the side of the metal block facing the bulb is blackened, and why it would be inadvisable to blacken the other sides as well.
[S.C.E. (H) 1970]

9.9 Fig. 9–23 shows a power supply, an oscilloscope and a transformer.

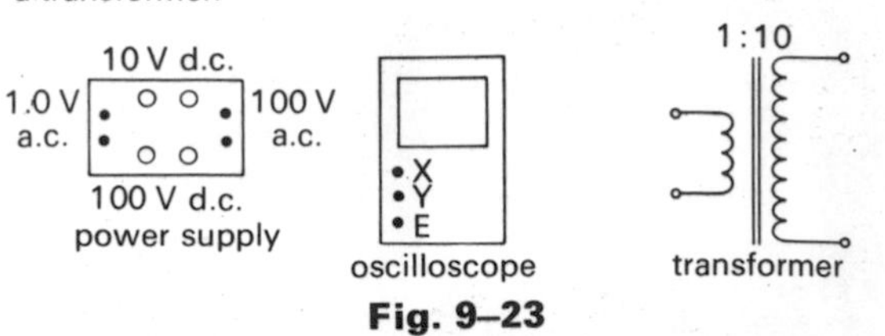

Fig. 9–23

(a) By means of a circuit diagram, show how you would connect the power supply, the oscilloscope and the transformer to display 10 V a.c. on the oscilloscope.
(b) The a.c. voltages shown on the power supply are root-mean-square values. Show, by a sketch, the wave form to be seen on the oscilloscope when the 100 V a.c. terminals are properly connected to it. Indicate the approximate value of the peak voltage shown.
(c) Draw a circuit diagram to show how you would connect the power supply and the oscilloscope to the rectifier shown in Fig. 9–24, to display a rectified voltage.

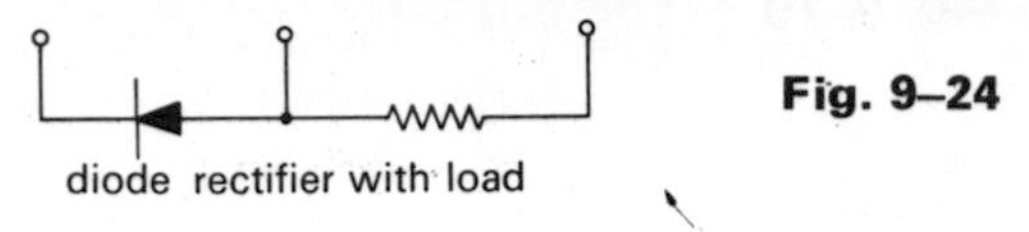

Fig. 9–24

Show what the wave-pattern would look like.
(d) The 1.0 V a.c. output from the power supply is used as input signal for the amplifier shown in Fig. 9–25.

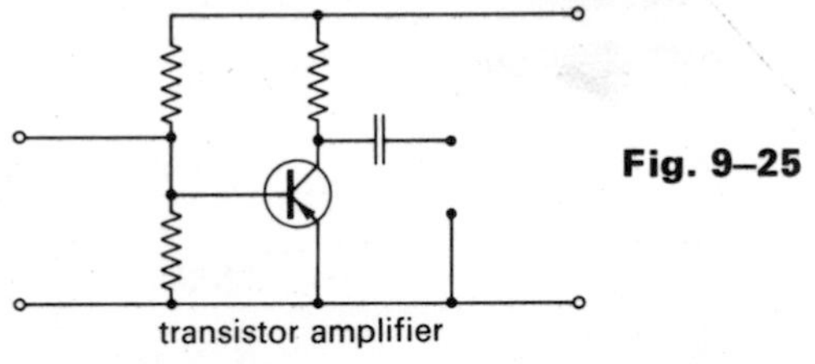

Fig. 9–25

(i) Draw a circuit diagram, using the power supply, the amplifier and the oscilloscope to display the amplified voltage on the oscilloscope.
(ii) Explain how you would use the oscilloscope to measure the gain of the amplifier. [S.C.E. (H) 1970]

9.10 The circuit shown in Fig. 9–26 can be used to compare the r.m.s. and peak values of an alternating voltage.

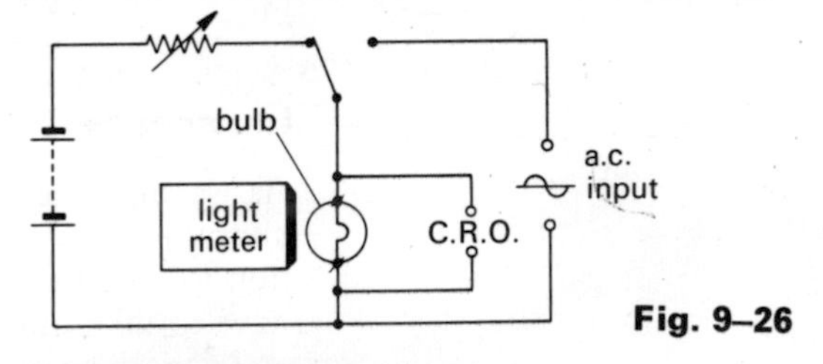

Fig. 9–26

(i) What is the purpose of
A. the two-way switch,
B. the rheostat, and
C. the light-meter?

10 Storing Electrical Energy

THE CAPACITOR

Fig. 10–1

During the early part of the eighteenth century, there were many attempts to construct machines which would produce large electric charges. Francis Hauksbee used a rotating glass sphere which he rubbed with a woollen cloth and later Otto von Guericke constructed a similar machine with a sulphur ball. One obstacle to producing a large charge was the absence of any device for storing charge. Various attempts had been made without success and then, in 1745, a suitable device was discovered by scientists in Germany and in Holland.

In one experiment a bottle of water was used to see if it would hold electric charges. An electric machine consisting of a rotating sphere produced the electric charges, which were conducted away by a metal bar suspended on silk threads. The water-filled bottle was fitted with a cork and a nail through the cork led the charge into the water. The Abbé Nollet was one of the first men to hold such a bottle to the end of the conducting bar (Fig. 10–1). He received an unexpected and substantial shock! Later experiments showed that the charge was not stored in the water. The accumulation of charge seemed to depend on the glass between the water and the Abbé's hand. Soon, glass bottles were being coated inside and outside with metal foil and even greater shocks produced. These were called Leyden jars (Fig. 10–2), after the Dutch town in which they were developed.

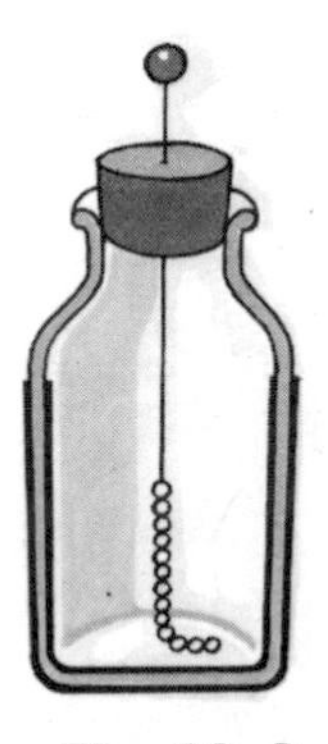

Fig. 10–2

The Leyden jar was regarded by many as an interesting gimmick rather than as an important scientific discovery. Such tricks as firing a cannon (Fig. 10–3) or shocking a few hundred monks at one time (Fig. 10–4) were more fun than puzzling over the storing of electrical energy!

Fig. 10–3

It was later discovered that even the glass could be dispensed with provided some other means could be found of keeping the two metal foils apart. The device formed is called a *capacitor*. When a capacitor is charged, the two plates (foils) carry equal and opposite charges. When we speak of the 'charge on a capacitor', we are referring to the size of the charge on *either* plate.

When a deep can is placed on an electroscope and a charged electrophorus plate is lowered into

Fig. 10–4

the can, the leaves diverge. The induced charges are indicated in Fig. 10–5(b). If the plate is removed, the leaves fall and the plate remains charged (c). If, however, the plate is allowed to touch the bottom of the inside of the can before it is removed,

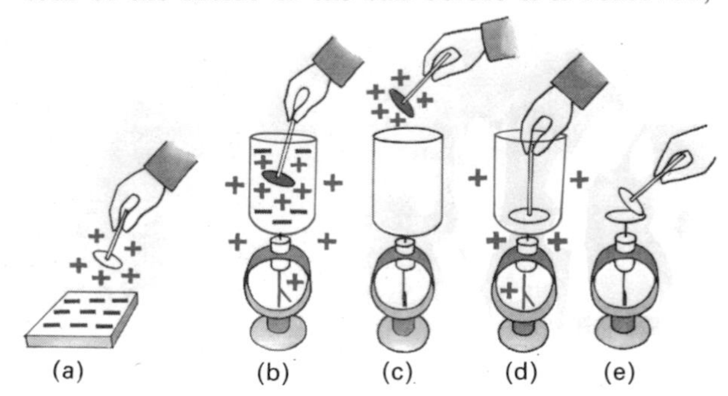

Fig. 10–5

nearly all the charge on the plate is transferred to the can (d). A test using another electroscope will confirm that there is no longer any measurable charge on the plate (e).

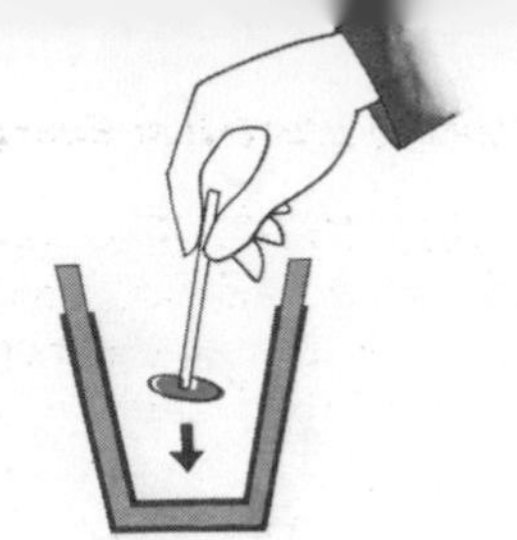
Fig. 10–6

If we use a Leyden jar capacitor made from two cans separated by an insulator, we can transfer all the charge on an electrophorus plate to the inside of the inner can in a similar way (Fig. 10–6).

Although the electroscope is used in Fig. 10–5 as a device to indicate charge, it is really measuring the *potential difference* between the leaves and the case. It is true, of course, that the greater the charge on the leaves, the greater will be the p.d. between them and the case. But the deflection of the leaves tells us nothing about the amount of charge *on a body* connected to the electroscope. A large charged sphere and a small charged sphere could produce identical deflections (Fig. 10–7), yet there could be a thousand times more charge on the large sphere. The p.d. between each sphere and earth would, however, be the same.

Fig. 10–7

The situation is analogous to placing one mercury thermometer in a bucketful of warm water and another thermometer in a thimbleful of warm water. There is much more internal energy (heat) in the bucket than in the thimble but both thermometers read the same. They indicate temperature and not internal energy. Yet it is true that the greater the internal energy *of the mercury* in a thermometer, the greater will be the reading.

CHARGE AND P.D.

To measure the p.d. between the plates of a capacitor, an electroscope can be used once it has been calibrated against a suitable moving coil voltmeter (Fig. 10–8). If an image of the leaf is projected on to a blackboard, a more accurate calibration can be obtained.

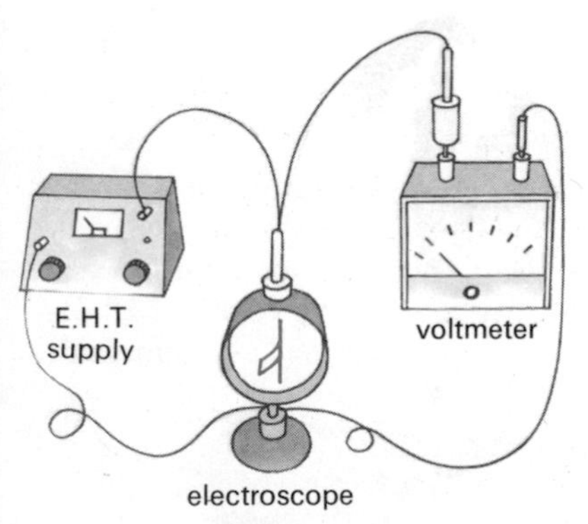

Fig. 10–8

To find out how the p.d. across the capacitor changes as more and more charge is added, we can use a Leyden jar capacitor connected to the calibrated electroscope (Fig. 10–9).

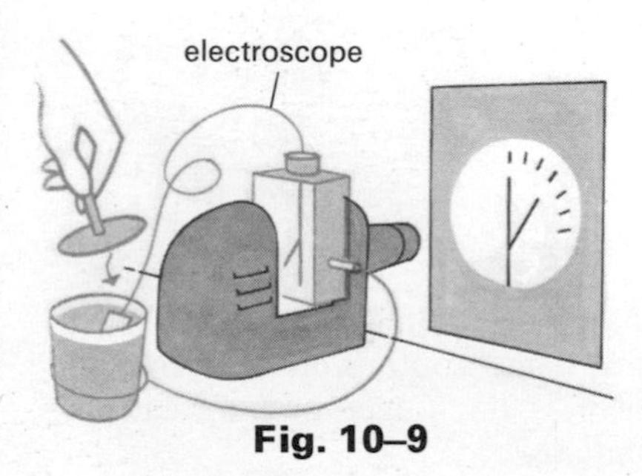

Fig. 10–9

Assume that every time a fully charged electrophorus disc is lifted off the charged plate, it carries the same quantity of charge. Assume also that all this charge can be transferred to the inside of the Leyden jar capacitor. If then we count the number of 'strokes', that is, the number of times we transfer charge from the electrophorus to the capacitor, we can take the number of strokes as an indication of the charge on the capacitor. The results obtained show that the p.d. across the capacitor V is directly proportional to the number of charging strokes, that is, to the charge Q. In other words, the ratio Q/V is constant. We use this ratio as a measure of *capacitance*. If 1 coulomb of charge produces a p of 1 volt between the plates, the capacitance 1 coulomb/volt. This is called a *farad* (F).

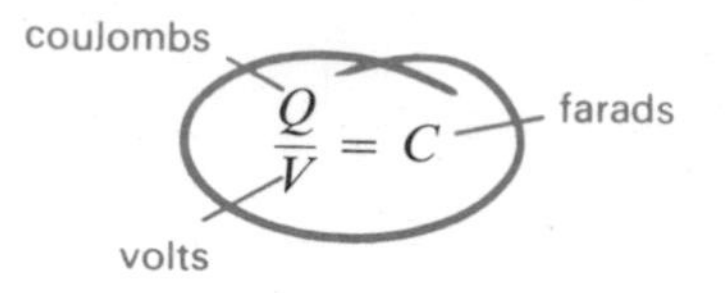

$$\frac{Q}{V} = C$$

The electroscope can now be used to compare capacitance of one capacitor with that of t similar capacitors in parallel (Fig. 10–10).

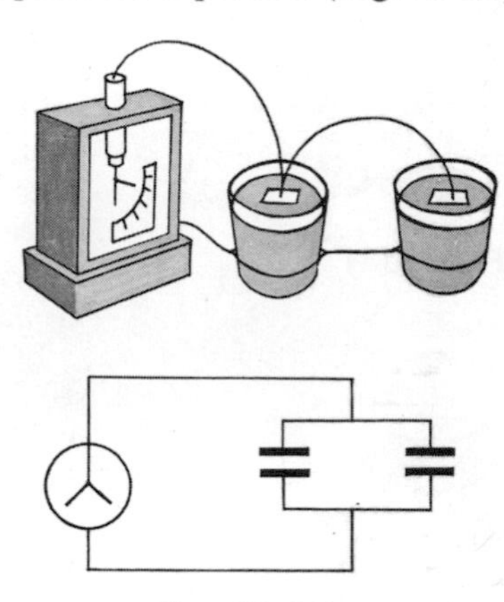
Fig. 10–10

Let us assume that a charge Q will produce a p of V across one capacitor. *If twice as many charg strokes* ($2Q$) *are needed to produce the sa deflection* V *across the two capacitors in paral what is the total capacitance?* (10.1)

When the same two capacitors are wired series (Fig. 10–11) the same deflection is obtair with half the original number of charging stro ($\frac{1}{2}Q$). *What is the total capacitance of the t capacitors in series?* (10.2)

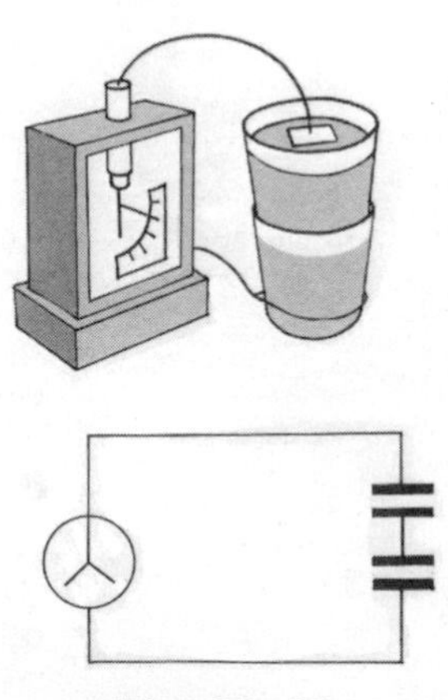
Fig. 10–11

THE PARALLEL PLATE CAPACITOR

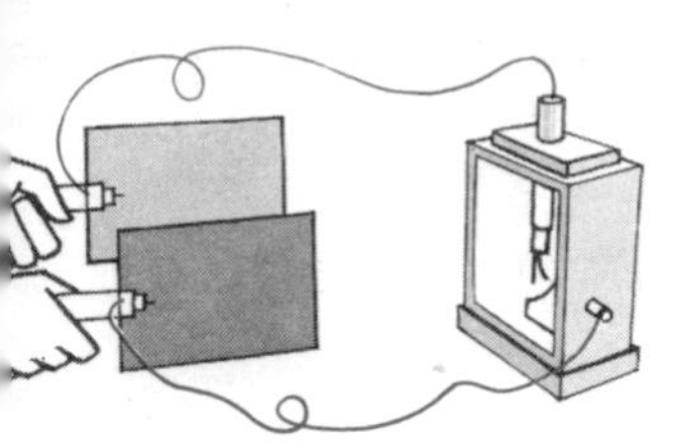

Fig. 10–12

For most practical purposes capacitors consist of two metal plates separated by an insulator of some kind. They are used in all radio and TV sets and in computers and electronic flash guns. Capacitors are also used as suppressors on electric motors and in car engines. Using the relationship $C = Q/V$, the factors affecting the capacitance of a parallel plate capacitor can be investigated. A parallel plate capacitor can be made from two metal plates on insulating handles. Once charged, the p.d. across the plates can be measured with an electroscope (Fig. 10–12). As the charge on the plates will not be affected by moving them relative to one another, an increase in p.d. V will indicate a reduction in the capacitance C.

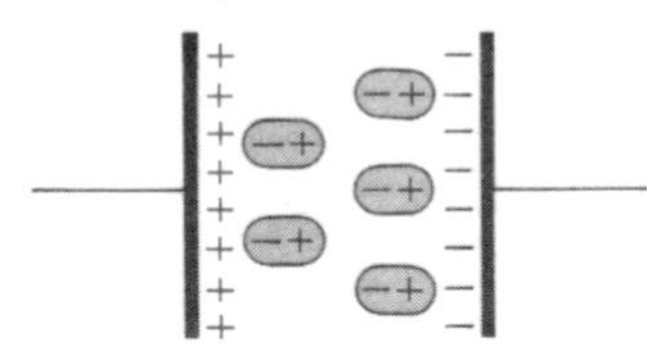

Fig. 10–13

This experimental arrangement may be used to show that the capacitance *increases* as

(a) the area of overlap increases,
(b) the plates are moved closer together or,
(c) slabs of paraffin wax, polythene or perspex (i.e. dielectrics) are introduced between the plates.

The Dielectric

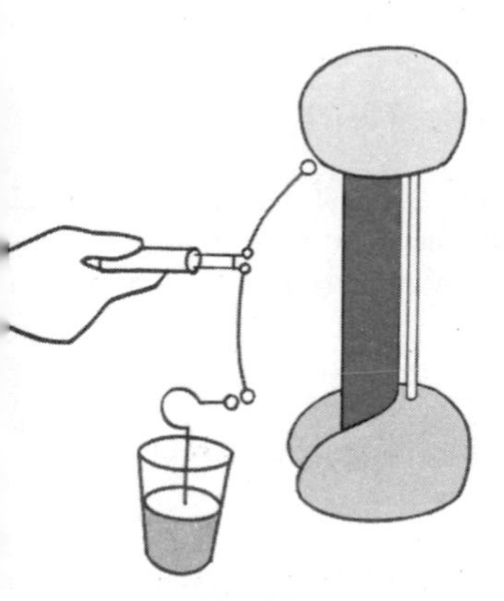

Fig. 10–14

We have seen that the capacitance of two parallel plates depends on the material between the plates. When a dielectric is introduced, the plates can hold more charge for every volt of potential difference across them; that is, the capacitance (coulombs per volt) is increased. There is no flow of charge in the dielectric, which is an insulator, but the molecules are either re-aligned (polar molecules) or their charge distribution is re-arranged (non-polar molecules) as shown in Fig. 10–13. The positively charged nuclei of the atoms can be displaced slightly towards the negative plate and the negatively charged electrons can be shifted slightly towards the positive plate. Energy is required to do this and the capacitor stores this energy in the form of dielectric strain. This can be demonstrated by charging a demountable Leyden jar with a Van de Graaff generator (Fig. 10–14). The jar can then be dismantled and the two metal cups touched together so that they are completely discharged. When the jar is reassembled, an impressive spark can be produced!

If the insertion of a slab of dielectric material doubles the capacitance of two plates previously separated by air, the dielectric material is said to have a *dielectric constant* of 2. If it trebles the capacitance its dielectric constant is 3, and so on. If k is the dielectric constant, A the overlapping area of the plates, and d the distance between them, the capacitance is found to be proportional to kA/d.

Here are some approximate values for common dielectrics.

Material	*Dielectric constant*
Air	1
Waxed paper	2
Mica	7
Ceramics	6–4000

Many of the capacitors in radio, television, and other electronic equipment use these dielectrics. Fig. 10–15 shows some of them. As the farad is too large a unit for most purposes, we use microfarads ($1\ \mu F = 10^{-6}$ F), nanofarads ($1\ nF = 10^{-9}$ F), and picofarads ($1\ pF = 10^{-12}$ F).

Fig. 10–15

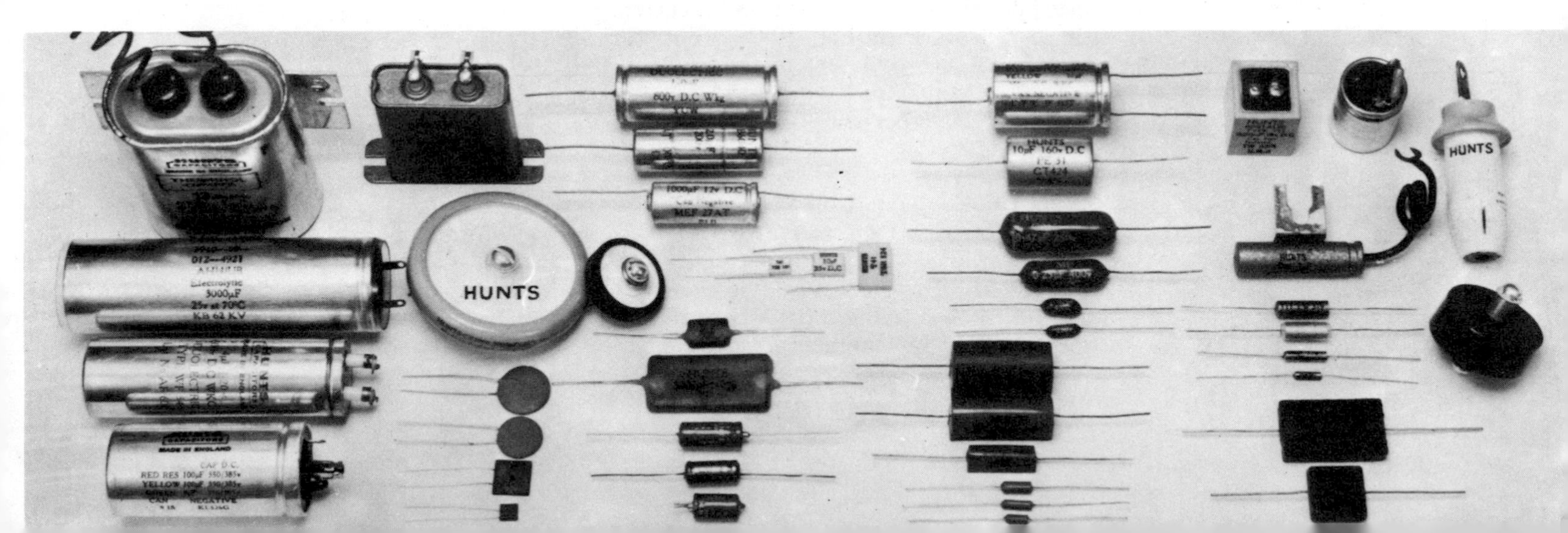

CHARGING FROM A D.C. SOURCE

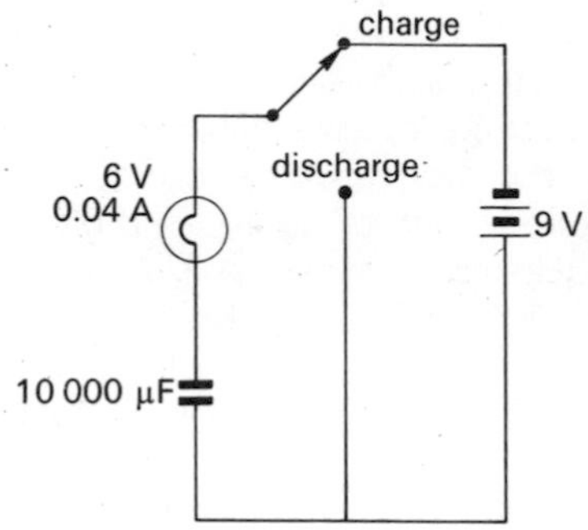

The charge on a capacitor is the product of p.d. and capacitance ($Q = VC$). Until now we have considered large voltages of several thousand volts which can be relatively easily measured using an electroscope. The capacitances have been small. However, we could store a reasonable quantity of charge with *low* voltages, provided we had a *large* capacitance. To obtain this we can use *electrolytic capacitors* (Fig. 10–16). The dielectric is a very thin film of aluminium oxide formed on the *positive* aluminium plate. The negative plate consists of an electrolyte and another aluminium plate. Such capacitors must not be charged in the reverse direction, as the oxide layer will decompose. An electrolytic capacitor can have a capacitance of several thousand microfarads. If such a capacitor is charged from a 9 V transistor battery, it will store enough energy to light a bulb for a second or so (Fig. 10–17).

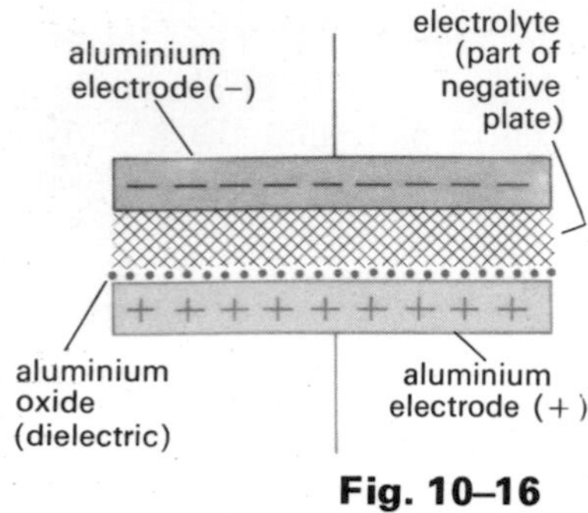

Fig. 10–16

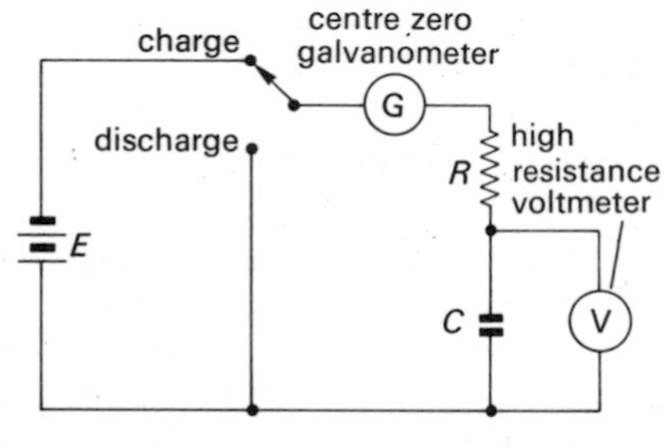

Fig. 10–17

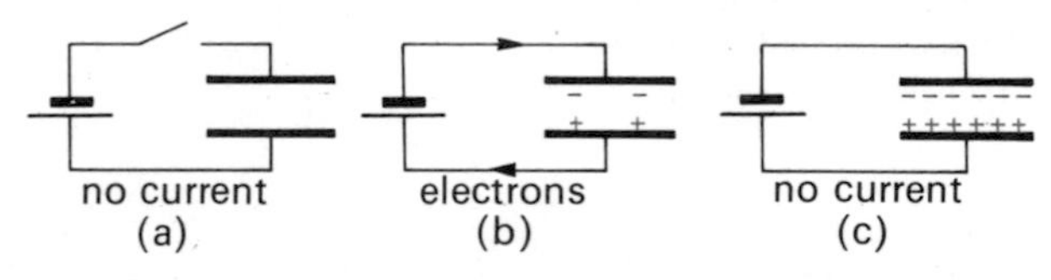

Fig. 10–18

When a capacitor is connected as shown in Fig. 10–18(a), no electrons flow and there is no p.d. across the plates of the capacitor. When the switch is closed as in Fig. 10–18(b), electrons flow in the direction shown and the capacitor is charged. When the p.d. across the plates is equal to the e.m.f. of the cell, the current stops (Fig. 10–18(c)). Note however that as the current falls exponentially, it takes, theoretically, an infinite time to stop.

A circuit such as that illustrated in Fig. 10–19 can be used to investigate changes in the charging current and the p.d. across a capacitor. As the capacitor is being charged up, the current falls rapidly and the p.d. across the capacitor increases. During discharge through the resistor, current and voltage fall together (Fig. 10–20). The rates of charge and discharge depend on the product RC. If either of these quantities is increased, the charging and discharging process will take longer.

Fig. 10–19

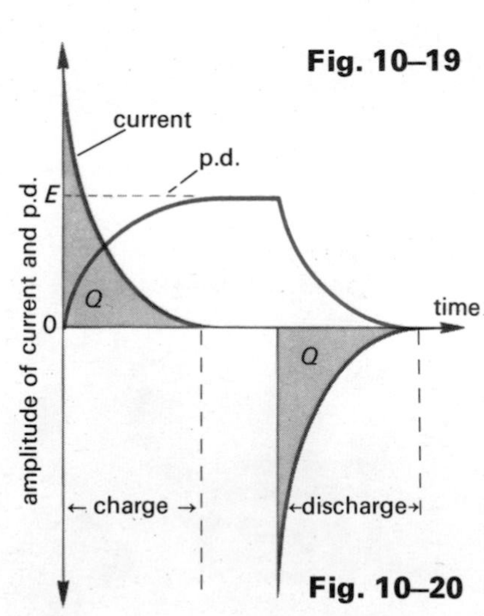

Fig. 10–20

Charging with a Constant Current

Normally the charging current for a capacitor varies as shown in Fig. 10–20. However, it is possible to keep this current nearly constant by continuously adjusting a variable resistance in se with the capacitor (Fig. 10–21). If the curren

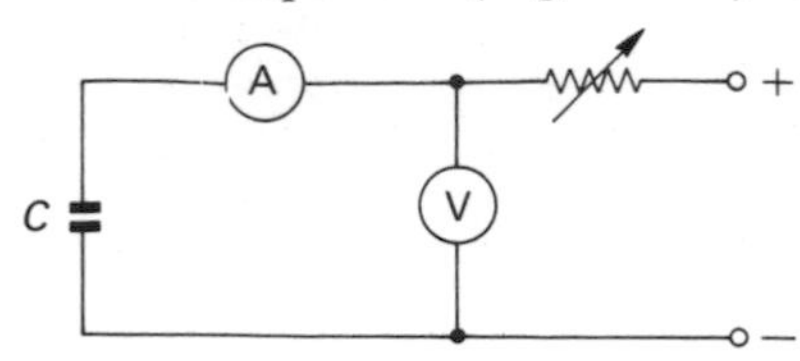

Fig. 10-

held constant in this way over a period of time, charge It transferred to the capacitor will be p portional to the time. We can therefore use *time* measure of charge transferred.

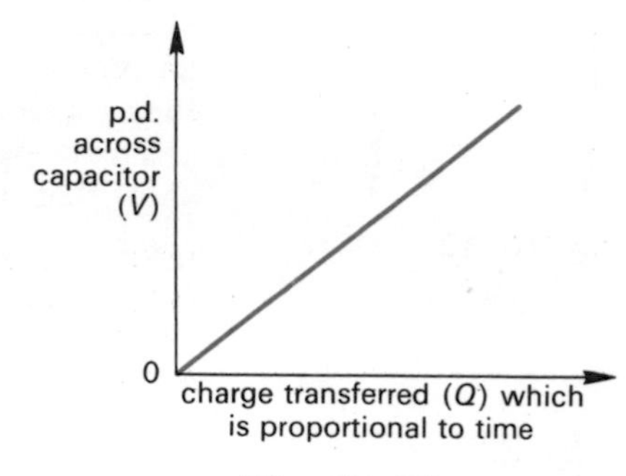

Fig. 10–22

A graph of the results is shown in Fig. 10–22 it is a straight line, the ratio Q/V is constant ar a measure of the *capacitance* of the capacitor.

CAPACITORS IN SERIES AND PARALLEL

The total capacitance of two capacitors wire parallel (page 78) could have been predicted

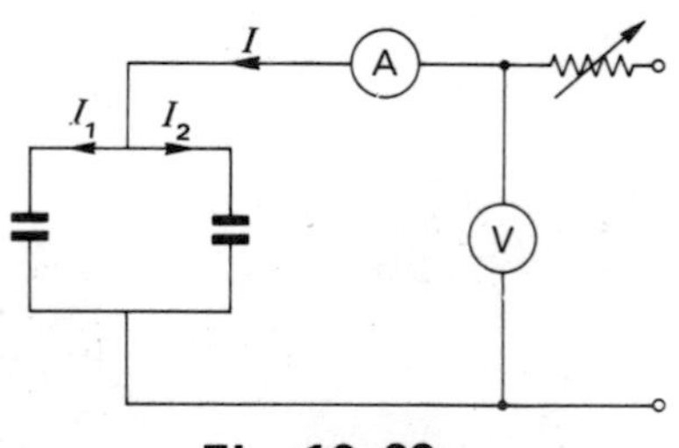

Fig. 10–23

considering the charging current (Fig. 10–23). I gine that the charging current I is kept constant, example by continuously varying a rheostat, u after a time interval t, the p.d. across both cap tors is V.

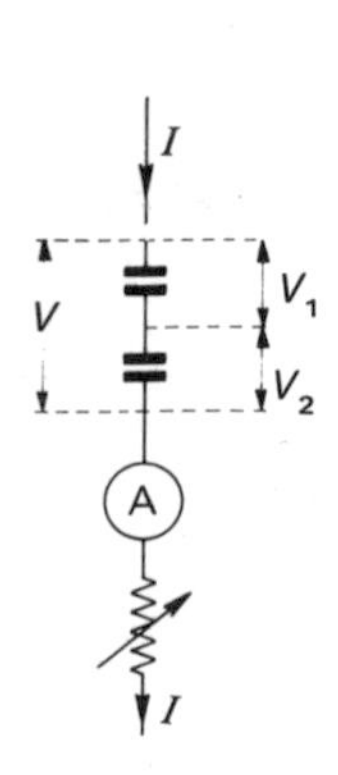

Fig. 10–24

$$\text{Total capacitance } C_T = \frac{Q}{V} = \frac{It}{V}$$

also $$C_1 = \frac{I_1 t}{V}$$

and $$C_2 = \frac{I_2 t}{V}$$

but $$I = I_1 + I_2$$

therefore, multiplying each term by $\frac{t}{V}$ we have

$$\frac{It}{V} = \frac{I_1 t}{V} + \frac{I_2 t}{V}$$

and so for capacitors in parallel

$$C_T = C_1 + C_2$$

With capacitors in series the charging current I is the same at every part of the circuit (Fig. 10–24). After a time interval t, the charge passed is It, and the total potential difference V is equal to the sum of the p.ds across C_1 and C_2

$$\text{Total capacitance } C_T = \frac{Q}{V} = \frac{It}{V}$$

also $$C_1 = \frac{It}{V_1}$$

and $$C_2 = \frac{It}{V_2}$$

but $$V = V_1 + V_2$$

therefore, dividing each item by It we have

$$\frac{V}{It} = \frac{V_1}{It} + \frac{V_2}{It}$$

and so for capacitors in series

$$\frac{1}{C_T} = \frac{1}{C_1} + \frac{1}{C_2}$$

BALLISTIC GALVANOMETER

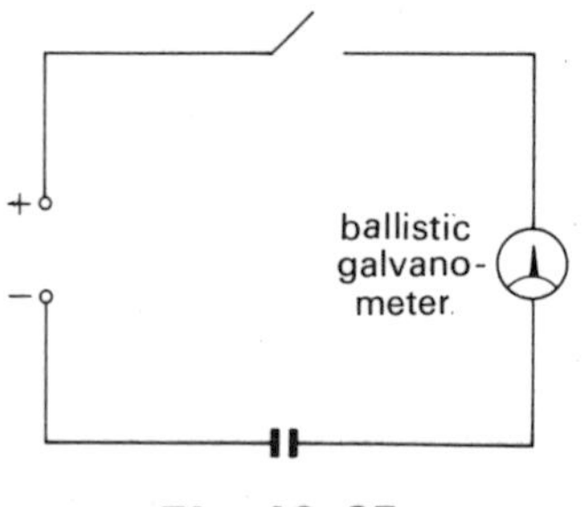

Fig. 10–25

A capacitor can be charged very quickly by a short 'burst' of charge, provided there is very little resistance in the circuit. If this pulse flows through a sensitive galvanometer (Fig. 10–25), and if the needle is still rising after the pulse has died down, then the *maximum deflection* is proportional to the *total charge Q* which has been transferred to the capacitor. This type of meter is called a ballistic galvanometer. It can be used to compare capacitances.

ENERGY STORED IN A CAPACITOR

When a spring is being stretched, the force exerted is directly proportional to the extension. The energy stored in the stretched spring is $\frac{1}{2}Fd$ where F is the final value of the force and d the extension (see page 26).

When a capacitor is being charged, the p.d. across the plates is directly proportional to the charge (Fig. 10–26). As p.d. is measured in *joules per coulomb* and charge in *coulombs*, the product p.d. × charge will be expressed in *joules*. The total energy stored in the capacitor can be represented by the area under the graph, that is $\frac{1}{2}QV$, where Q is the charge on the capacitor and V the *final* voltage.

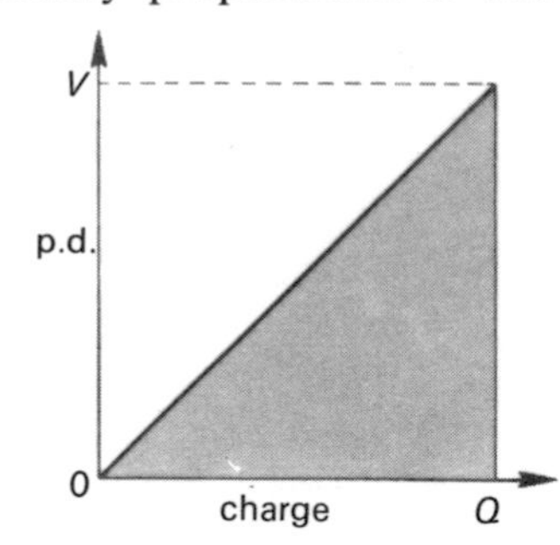

Fig. 10–26

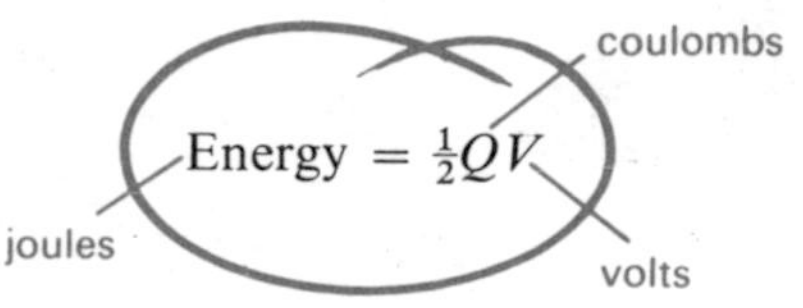

From $Q = VC$ we have

$$\text{energy stored} = \tfrac{1}{2}QV = \tfrac{1}{2}CV^2 = \tfrac{1}{2}Q^2/C$$

Electronic flash guns operate from the energy stored in a capacitor. In atomic research, capacitors are used to give large pulses of current for short periods. The capacitors shown in Fig. 10–27 store several megajoules of energy.

Fig. 10–27

SUMMARY

Capacitors

When a capacitor is connected to a d.c. supply, the charging current reaches its maximum value immediately. As the charge builds up on the plates, the charging current decreases.

Capacitance $C = \frac{Q}{V}$

Energy stored $= \frac{1}{2}QV = \frac{1}{2}CV^2 = \frac{1}{2}\frac{Q}{C}$

Capacitors in series $\frac{1}{C_T} = \frac{1}{C_1} + \frac{1}{C_2} + \frac{1}{C_3} \cdots$

Capacitors in parallel $C_T = C_1 + C_2 + C_3 \ldots$

PROBLEMS

10.3 A 100 μF (10^{-4} F) capacitor is charged to 50 V. How much charge has been transferred from one plate to the other? If it were discharged completely in 2 milliseconds, what would be the average current?

10.4 An electroscope has a capacitance of 10^{-10} F. What charge will it hold if it is charged to 1200 V?

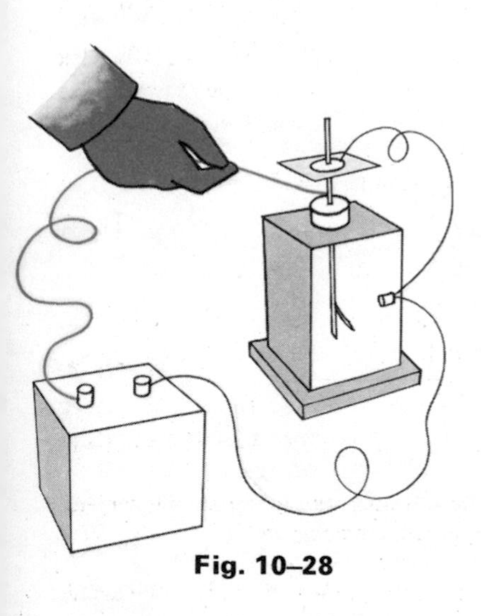

Fig. 10–28

10.5 A thin uncharged polythene sheet lies on the plate of an electroscope. An electrophorus plate is then placed on top of it and connected by a wire to the electroscope case. One terminal of a 9 V battery is also taken to the case of the electroscope, and a wire from the other terminal is touched momentarily on to the electroscope plate (Fig. 10–28). What happens when the electrophorus plate is raised? Why? Has the system gained energy, other than gravitational potential energy, when the plate is raised? If so, where has it come from?

10.6 Why do the leaves of an electroscope fall slightly if you bring your hand near the plate?

(a)

(b)

(c)

Fig. 10–29

10.7 Fig. 10–29 (a) shows three capacitors in series. Fig. 10–29 (b) shows them pushed close together so that their inside plates coincide. Fig. 10–29 (c) shows the two central plates removed. If the central plates are thin, the set up in Fig. 10–29 (c) is equivalent to that in Fig. 10–29 (a). Do you think the total capacitance of three capacitors in series is greater or less than that of each individual capacitor?

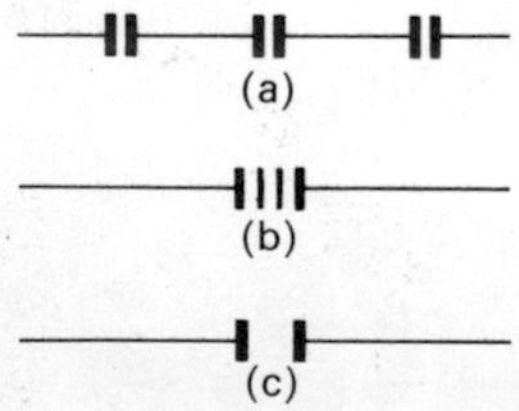

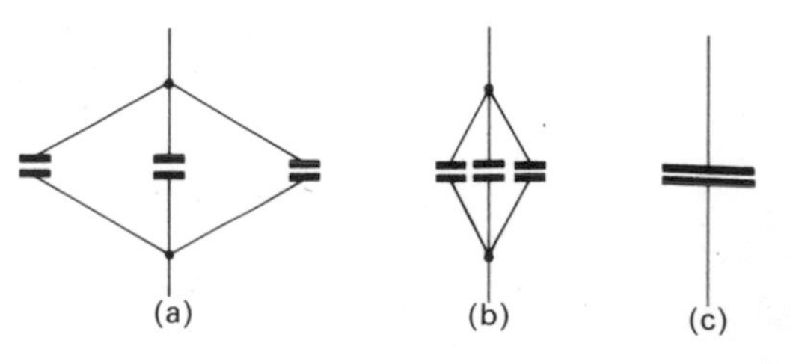

Fig. 10–30

10.8 Fig. 10–30 (a) shows three capacitors in parallel. Fig. 10–30 (b) shows them pushed together and Fig. 10–30 (c) shows them so close that the plates touch. The set up in Fig. 10–30 (c) is equivalent to that in Fig. 10–30 (a). Do you think the total capacitance of three capacitors in parallel is greater or less than that of each individual capacitor.

10.9 A parallel plate capacitor has air as its dielectric. How would its capacitance be altered if

(a) a thin metal sheet were held by an insulator midway between the capacitor plates?

(b) a thick metal block were held by an insulator between, but not touching, the capacitor plates?

(c) a block of paraffin wax were inserted between the plates?

10.10 In a radio tuning capacitor air is often used as the dielectric (Fig. 10–31). A tuning capacitor with its vanes

Fig. 10–31

closed is connected to an electroscope and charged. What will happen to the deflection of the electroscope leaf when the capacitor's vanes are opened? Explain the answer you give.

10.11 When the switch in Fig. 10–32 is closed, electrons

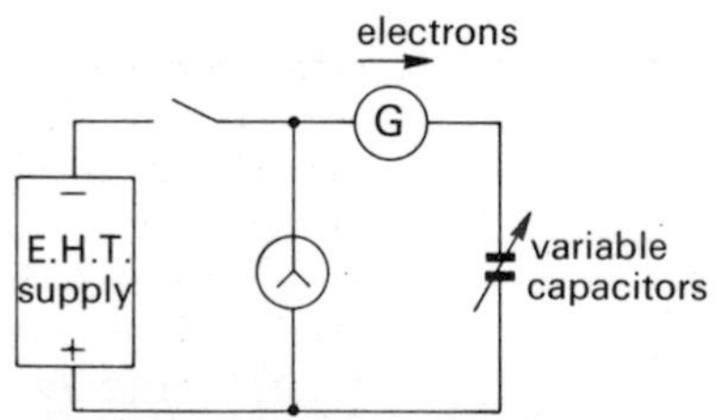

Fig. 10–32

flow in the direction indicated, thus charging up the variable capacitor set with its vanes half open. When the switch is opened, the electroscope indicates a steady p.d.

across the capacitor. What will happen to the p.d. if the capacitor vanes are now closed? Will there be any indication in the sensitive galvanometer? With the vanes still fully closed, the switch is again closed. Describe what you would expect to happen. Finally with the switch still closed the vanes of the capacitor are opened. Will this cause any change in the p.d. across the capacitor or any flow of charge through the galvanometer?

10.12 The circuit shown in Fig. 10–33 is wired up. The switch is then closed and the rheostat R continuously adjusted to maintain a steady current of 1 mA. It is found

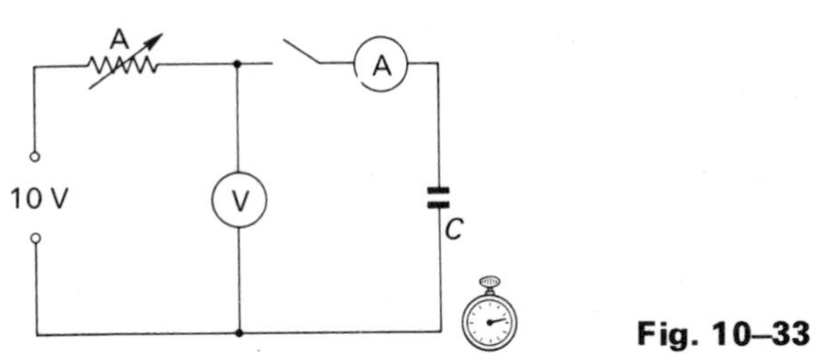

Fig. 10–33

that it takes 20 seconds for the p.d. across the capacitor to reach 5 V. Find the capacitance of C. Draw rough graphs showing how you think the p.d. across the capacitance would vary with (a) time and (b) charge.

10.13 A 4 μF and a 6 μF capacitor are connected in series to a 10 V battery. Find the total capacitance, the charge on each, and the p.d. across each capacitor.

10.14 A capacitor of 10 μF has a p.d. across it of 40 V. If it is discharged in 0.2 seconds, find (a) the charge stored in the capacitor and (b) the average value of the current during discharge.

10.15 A switch which vibrates to and fro 50 times a second is used to connect a capacitor alternately to a 10 V

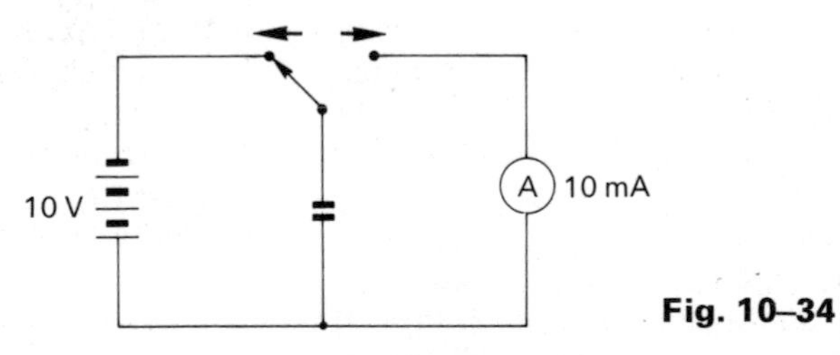

Fig. 10–34

battery and a milliameter (Fig. 10–34). The meter indicates that an average current of 10 mA is passing. Assuming that the capacitor is fully charged and then fully discharged each time the contacts are made, find:
(a) the charge passing through the meter every second,
(b) the charge passing through the meter every $\frac{1}{50}$th second,
(c) the number of coulombs given to the capacitor during each charge,
(d) the p.d. across the capacitor when fully charged, and
(e) the value of the capacitance.
(Hint: current is the rate of flow of charge.)

10.16 If a capacitor is charged to 1000 V and then discharged through a 100 Ω resistor, a certain amount of heat is produced. If a 50 Ω resistor had been used, would the heat produced have been (a) the same, (b) half, (c) a quarter, (d) double or (e) four times as great?

10.17 What quantity of electricity is required to charge a 2 μF capacitor to a p.d. of 100 V? The 2 μF capacitor is in the form of a cylinder about 5 cm long and 1 cm in diameter. Describe its possible internal construction. The outside is marked with the value of its capacitance and also the information '250 V working'. What might happen if a much higher voltage were applied? [S.C.E. (H) 1969]

10.18 (a) 'A capacitor has a capacitance of 4 μF.' Explain what is meant by this statement.
(b) Upon what factors does the capacitance of a parallel-plate capacitor depend?
(c) A parallel-plate capacitor is charged using a battery and is then isolated from the battery. A slab of insulating material, such as mica, is slipped between the plates. How does this affect (i) the charge held, (ii) the potential difference between the plates, and (iii) the capacitance?
(d) If the capacitor in (c) had remained connected to the battery while the slab was being inserted, what effect would the insertion of the slab have had on (i) the potential difference between the plates, (ii) the charge held, and (iii) the capacitance?
(e) The switch K is made to vibrate between contacts X and Y at a rate of 50 complete vibrations per second. The voltmeter V reads 10.0 V and the milliammeter A reads 8.00 mA (Fig. 10–35).

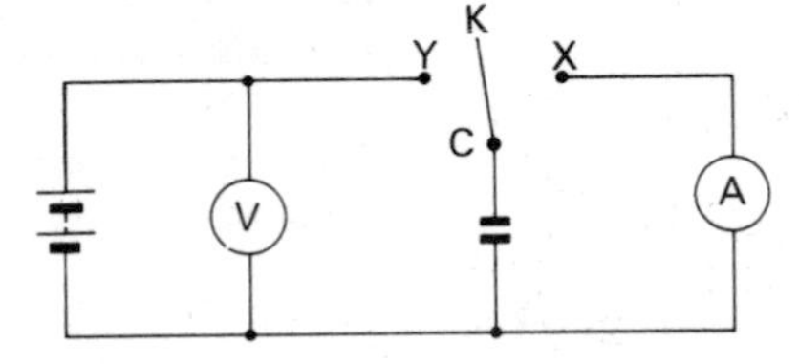

Fig. 10–35

(i) How much charge passes through the milliammeter in 1 second?
(ii) How much charge passes through the milliammeter each time the switch makes contact with Y?
(iii) What is the capacitance of the capacitor C?
(f) Explain how an alternating current can flow in the circuit shown in Fig. 10–36 in spite of the space between the plates of the capacitor. [S.C.E. (H) 1970]

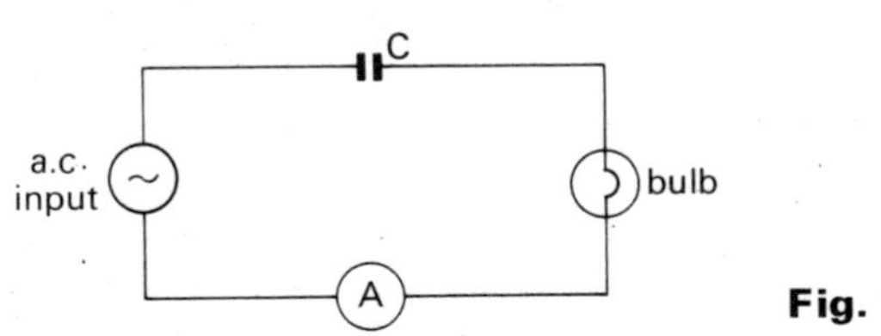

Fig. 10–36

11 Opposition to Alternating Current

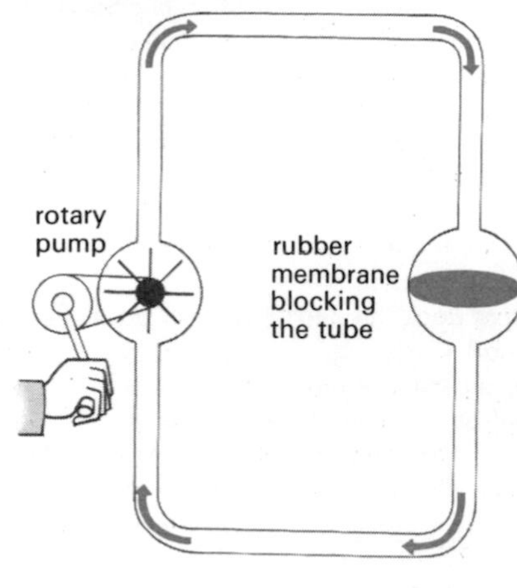

Fig. 11–1

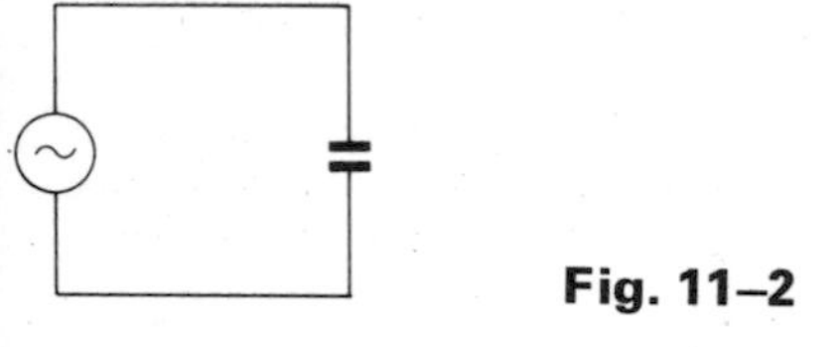

Fig. 11–2

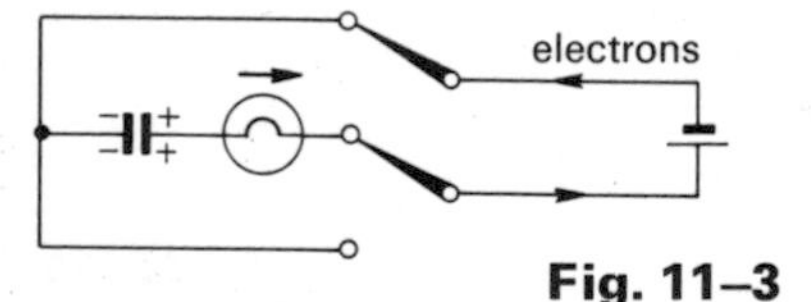

Fig. 11–3

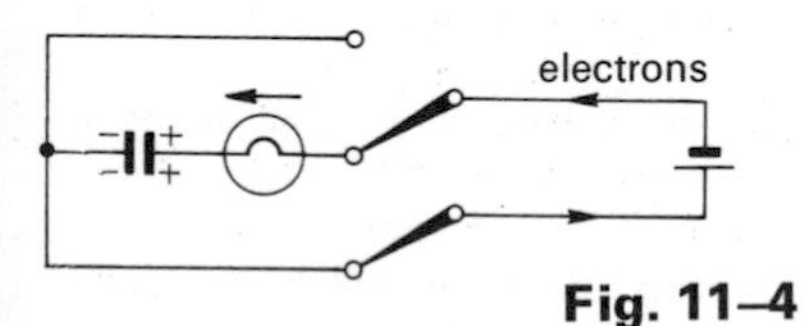

Fig. 11–4

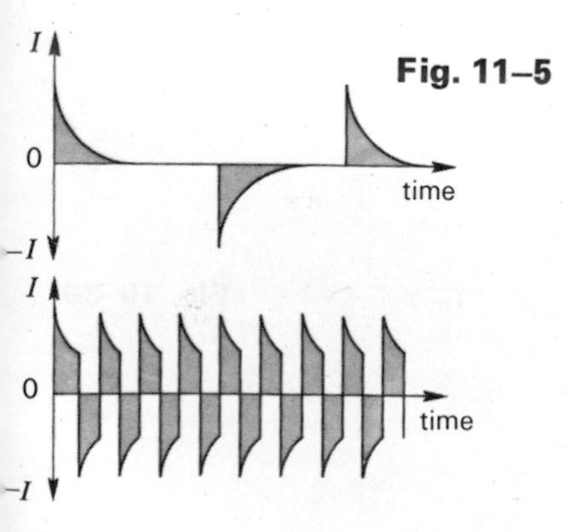

Fig. 11–5

RESISTANCE *R*

In direct current circuits the only opposition to the steady flow of charge is *resistance*. The component responsible is called a *resistor* and its property *resistance*. The opposition to the flow of charge is measured by the ratio V/I and this also is labelled *resistance*.

Whenever current flows through a resistor, heat is produced from electrical energy. It is this ability to obtain heat from electrical energy, rather than the opposition to the flow of current, which enables us to identify resistance in an a.c. circuit.

CAPACITANCE *C*

Water in the model illustrated in Fig. 11–1 can be made to surge to and fro by turning the handle first one way and then the other. Although the water is always moving it never flows *through* the rubber membrane.

In the electrical circuit, Fig. 11–2, an alternating voltage makes charges surge to and fro. The capacitor is charged in one direction, discharged and then charged in the opposite direction. Yet no charges flow *through* the capacitor. If a bulb were inserted in the circuit it would glow, as the heating effect of a current does not depend on the direction of flow.

You can produce a primitive kind of a.c. using a battery and reversing switch. When the contacts are switched to the position shown in Fig. 11–3, the bulb flashes as the capacitor is being charged. Once it is fully charged, no current flows. By switching over to the position shown in Fig. 11–4, the capacitor is discharged and then charged in the opposite direction, and again the bulb flashes. If the switch is operated rapidly, the bulb will appear to glow continuously. The higher the frequency of switching the higher will be the current. You can see why this should be so from Fig. 11–5. At a low switching frequency, there are long periods during which there is no current. At a high frequency, the current I is never allowed to fall to a low value and so the effective current is much greater. The higher the frequency, the greater will be the shaded area during a given time interval, and thus the greater will be the effective current.

If the capacitor in Fig. 11–3 were replaced by resistor, the switching frequency would not affe

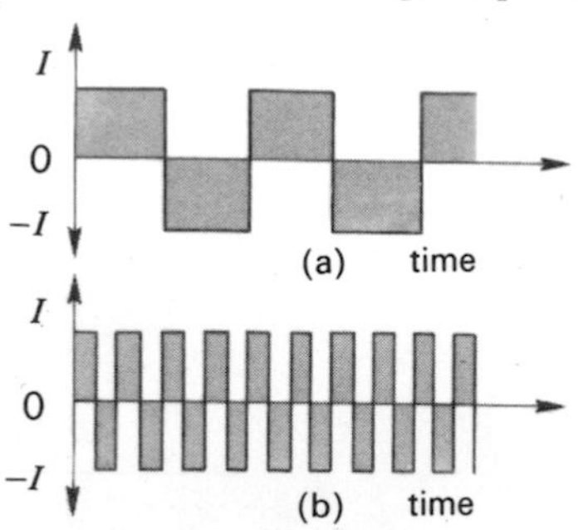

Fig. 11-

the current. The total shaded areas during a giv time interval and thus the currents are the same Figs. 11–6(a) and (b).

Capacitive Reactance

Similar results can be obtained using the slow a. generator, which produces *constant amplitude* a. at different frequencies (Fig. 11–7). This allows to study the effect of changes in frequency witho changes of amplitude complicating the issue. with the reversing switch, we find that the curre through a resistor does not depend on frequenc With a capacitor the current increases as t frequency increases.

In a capacitor the opposition to current is call the reactance or, more correctly, *capacitive r actance* (X_C). It is measured by the ratio of t applied voltage to the current.

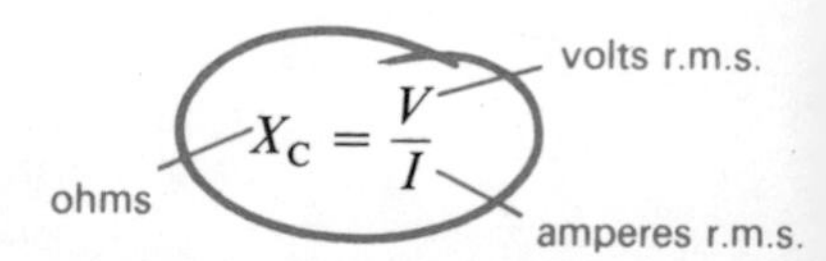

Fig. 11–7

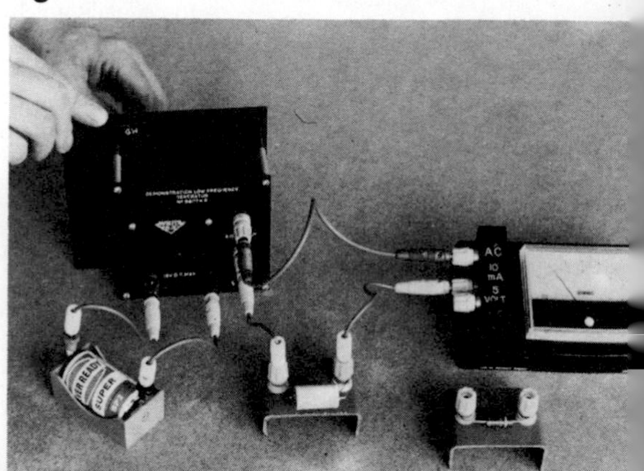

Fig. 11–8

By inserting various capacitors and combinations of capacitors in the circuit shown in Fig. 11–8, you can study capacitive reactance qualitatively. *To measure reactance, what other instrument would be required? (11.1)*

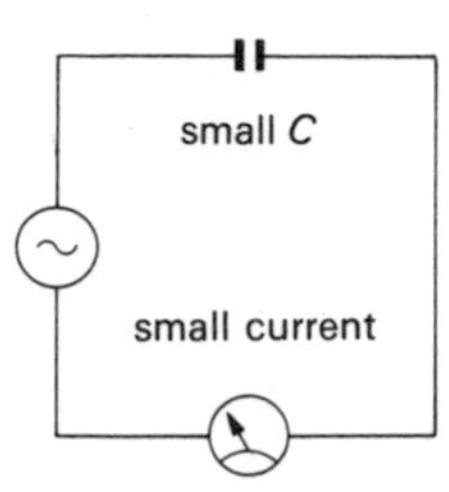

Fig. 11–9

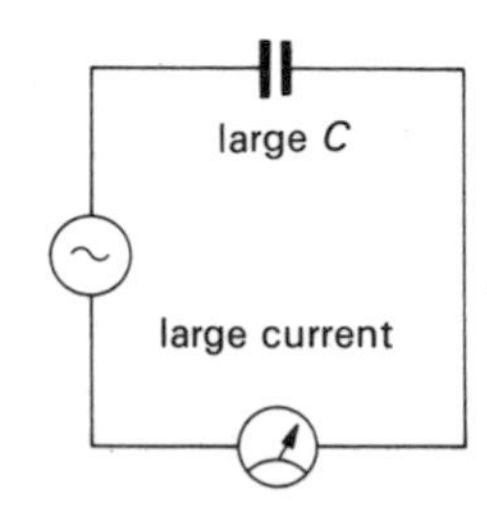

Fig. 11–10

Figs. 11–9 and 11–10 indicate the currents obtained when different capacitors are used. You can see that the bigger the capacitor, the bigger the current, and so the smaller the reactance.

The experimental results on reactance can be summed up by saying that, when the frequency or capacitance is *increased*, the reactance is *decreased*.

$$f\uparrow \quad X_C\downarrow$$
$$C\uparrow \quad X_C\downarrow$$

Accurate experiments show that the reactance is inversely proportional to each of these quantities. The capacitive reactance is, in fact, given in SI units by

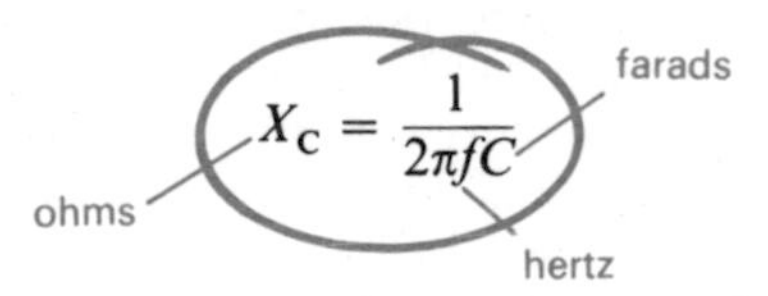

Capacitors in Series and Parallel

In the experiments illustrated in Figs. 11–11, 11–12, and 11–13, medium sized capacitors are used. Two such capacitors in parallel behave like a *big* capacitor and two capacitors in series behave like a *small* capacitor. Notice that reactances behave like resistors; that is, when connected in series the total reactance is greater and when connected in parallel the total reactance is smaller.

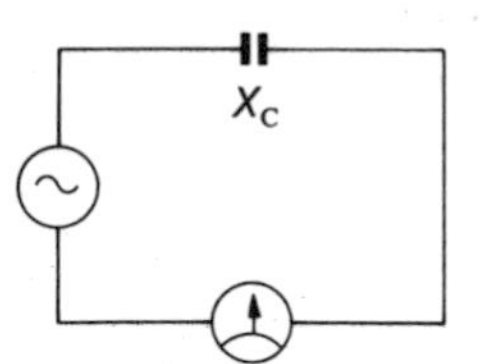

Fig. 11–11

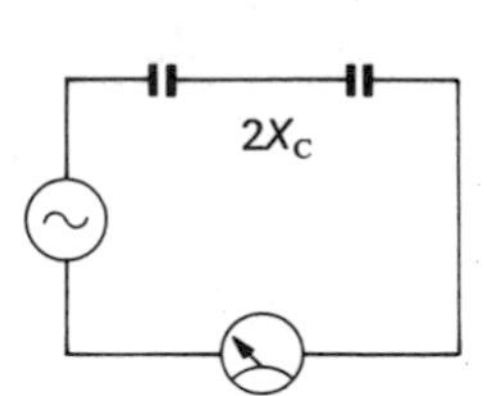

Fig. 11–12

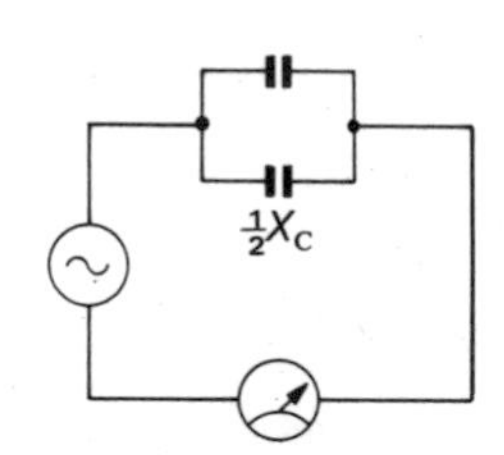

Fig. 11–13

Capacitors in parallel

$$C_T = C_1 + C_2 + C_3$$

but

$$\frac{1}{X_T} = \frac{1}{X_1} + \frac{1}{X_2} + \frac{1}{X_3}$$

Capacitors in series

$$\frac{1}{C_T} = \frac{1}{C_1} + \frac{1}{C_2} + \frac{1}{C_3}$$

but

$$X_T = X_1 + X_2 + X_3$$

INDUCTANCE

You have seen earlier in the course that, whenever there is a change in the magnetic flux linked with a conductor, an e.m.f. is induced in the conductor. Moreover the direction of this induced e.m.f. is such that it always tends to oppose the change which causes it (Lenz's law). For example, when a magnet is plunged into a closed loop of wire, the induced current in the wire sets up a magnetic field which tends to slow down the moving magnet (Fig. 11–14).

Fig. 11–14

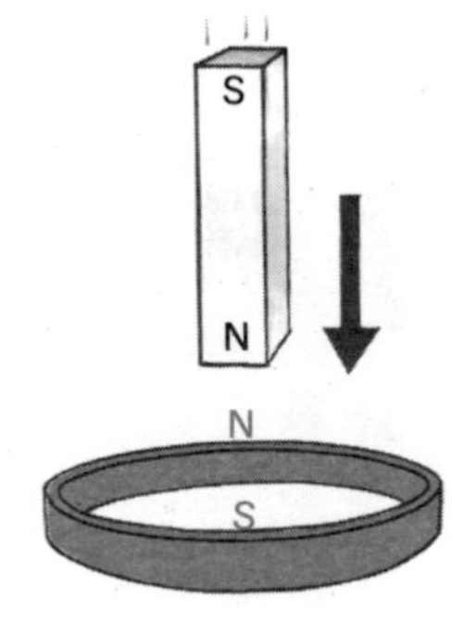

A current in a conductor produces a magnetic field in and around the conductor. The magnetic flux which is linked with the circuit will change if the current changes. An opposing e.m.f. is therefore induced in the conductor if the current in it changes. Furthermore, the size of this induced e.m.f. is proportional to the rate of change of flux.

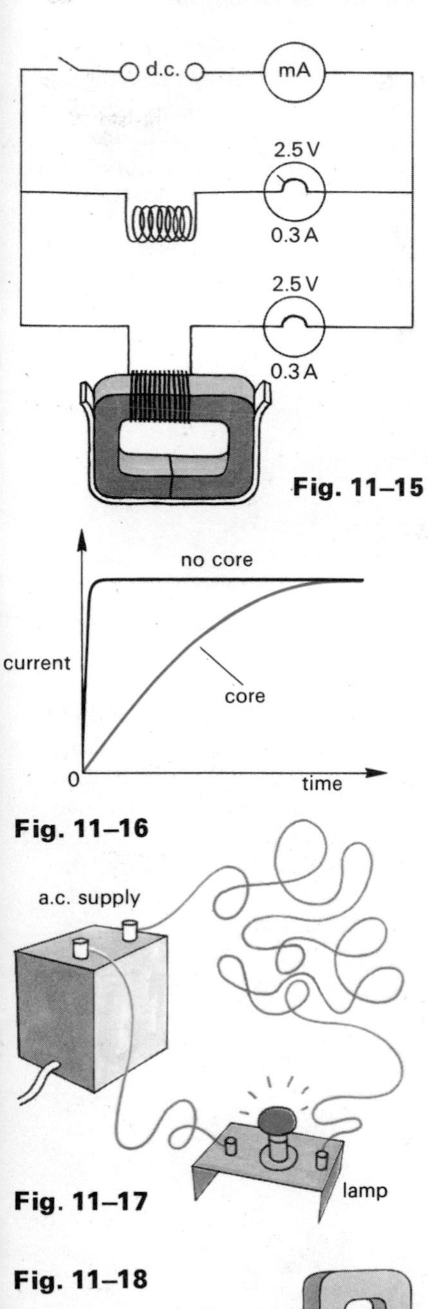

Fig. 11–15

Fig. 11–16

Fig. 11–17

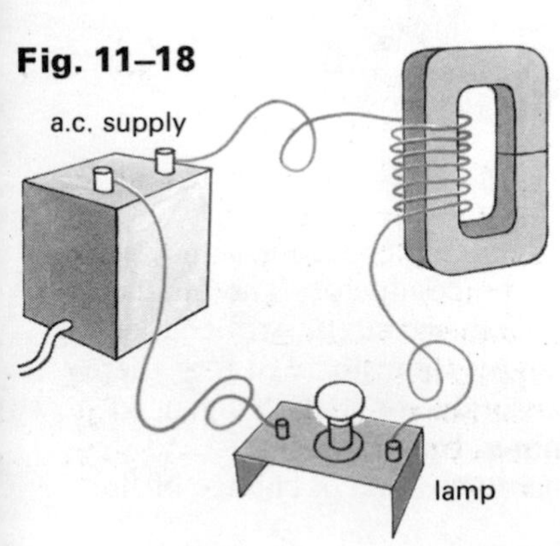

Fig. 11–18

The resistance in each branch of the circuit shown in Fig. 11–15 is the same, yet, when the switch is closed, one bulb lights up before the other. The only difference is that one coil is wound round an iron core and the other is not. In the coil on the core there is a greater rate of change of flux and therefore a greater induced e.m.f. As this e.m.f. *opposes* the rising current, it takes longer for the current in this coil to build up. This is shown in Fig. 11–16.

With a.c., the current is always changing and a back e.m.f. will always oppose the direction of the change. If the alternating voltage is increasing, the back e.m.f. will tend to slow down the rate of increase in current. *What would you expect to happen if the battery in Fig. 11–15 was replaced by an a.c. supply? (11.2)*

If a length of insulated wire is connected to a lamp and an a.c. supply, the lamp lights up (Fig. 11–17). If, however, we wind the same wire on to an iron core, we discover that, as the number of turns increases, the brightness of the lamp decreases (Fig. 11–18). The current has therefore decreased and so the opposition to the flow, i.e. the reactance, must have increased.

This property of a coil carrying a varying current to produce a varying magnetic field which in turn produces an opposing e.m.f. is called the *inductance* L of the coil. The unit of inductance is the henry H, named in honour of the American physicist Joseph Henry (1797–1878), who discovered this property.

Henry's career in physics started when, as a boy of sixteen, he chased a rabbit under the floorboards of a church! He lost the rabbit but discovered in the church library a book called *Lectures on Experimental Philosophy*. This so aroused his curiosity that he went back to school to study mathematics and science. Later he discovered electromagnetic induction just before Faraday, but Faraday published his results first. Henry also laid the foundations for the telegraph but Wheatstone, after a long conference with Henry, cashed in on it!

If a current in a coil is changing at a rate of one ampere per second and the induced e.m.f. is 1 volt, the inductance is 1 henry. The value of L is given by

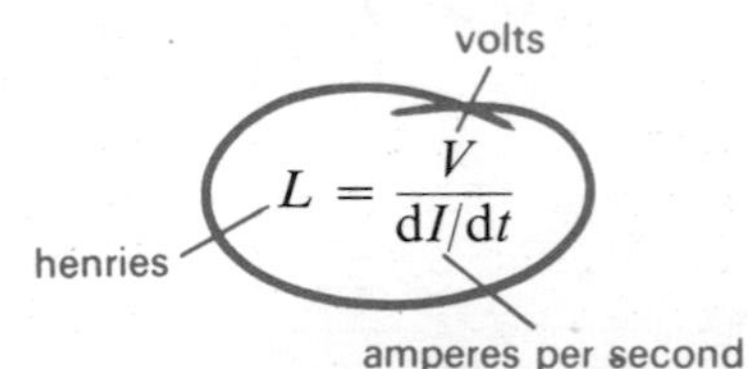

$$L = \frac{V}{\mathrm{d}I/\mathrm{d}t}$$

where $\mathrm{d}I/\mathrm{d}t$ means 'rate of change of current'.

The inductance of a coil depends on many fa tors. Your experiments have shown that it increas when the number of turns is increased and when iron core is used. If two C-cores are clamp together to form a complete magnetic circuit (Fi 11–18), the inductance is greatly increased. As t current in an a.c. circuit decreases as the inductan increases, the reactance of a coil must *increase* the inductance *increases*.

To study the effect of frequency on inducti reactance, we can use the slow a.c. generator or t

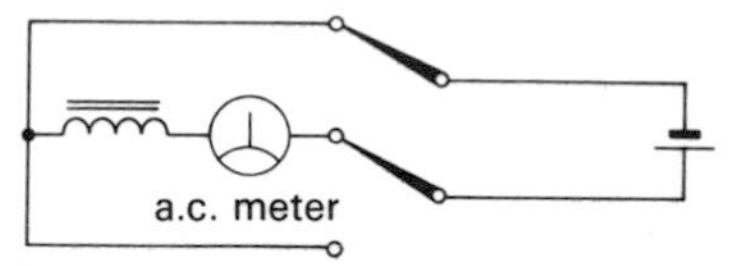

Fig. 11–19

switching circuit of Fig. 11–19. If the current estimated by using an a.c. meter, it is found decrease as the frequency increases; that is, t reactance *increases* as the frequency *increases*. Th is not difficult to understand if we consider t

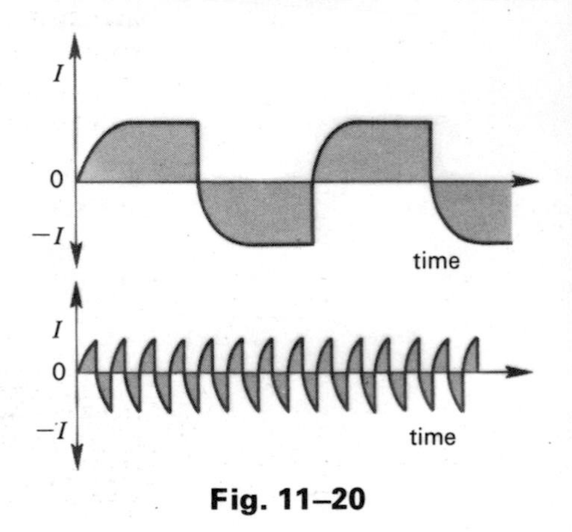

Fig. 11–20

graphs of Fig. 11–20. The current at low switchi frequencies is shown first. Although it takes tir to build up, a current is at its maximum value fo large part of each cycle. At high frequencies t current does not have time to build up to maximum value before the supply is reversed a so the effective current is much smaller.

Summarizing these results we have

$$L\uparrow \quad X_L\uparrow$$
$$f\uparrow \quad X_L\uparrow$$

Accurate experiments show that the reactance directly proportional to each of these quantiti

The inductive reactance is, in fact, given in SI units by

$$X_L = 2\pi f L$$

ohms — hertz — henries

If two inductors were connected in series what would be the total inductance in the circuit? (11.3) Would the current be greater or less than it would be with only the inductor in the circuit? (11.4)

OUT OF STEP

When dealing with a d.c. circuit we did not have to consider whether or not the current and the voltage were 'in step', as they both had steady values. Current and voltage are changing continuously in alternating current and so we ought at least to ask whether or not these two quantities are always 'in step'. When a p.d. is applied to a coil, does the charge flow immediately? Does the current reach its maximum at exactly the same instant as the voltage applied? The current through a component can be measured with a milliammeter and the p.d. across it can be indicated by the deflection of the spot on a C.R.O. (Figs. 11–21 and 11–22). Using this apparatus, we can see whether or not current and voltage are in phase.

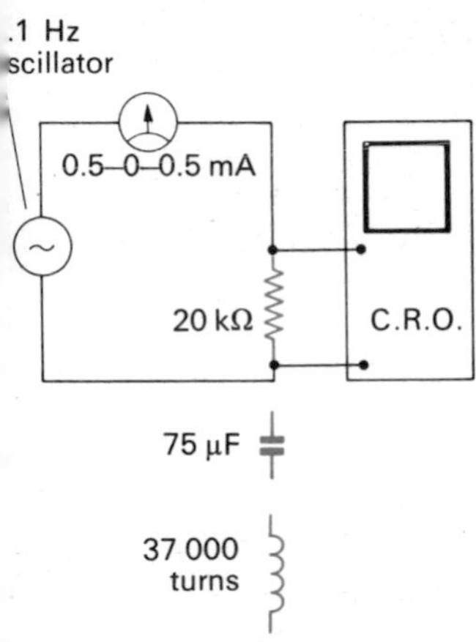

Fig. 11–21

Fig. 11–22

Phasors

Vectors, or phasors, as they are more properly called, can be used to represent the phase angle between the current through and the p.d. across various components. Arrowheads can be used to indicate the ends of the lines, but they are not necessary as they do not denote the *directions* of the current or voltage. With the resistor R the current and voltage are in phase (Fig. 11–23). With the capacitor C the current reaches its peak *before* the voltage; that is, the current *leads* the voltage by 90° (Fig. 11–24). In the inductor L the current takes time to build up and the voltage reaches its peak first. In a pure inductor, that is, one with no resistance, the current would lag behind the voltage by 90° (Fig. 11–25). If a circuit contains both resistance and capacitance (Fig. 11–26), the current leads the *input* voltage by less than 90° (Fig. 11–27). If the circuit has resistance and inductance (Fig. 11–28), the current lags behind the *input* voltage by less than 90° (Fig. 11–29).

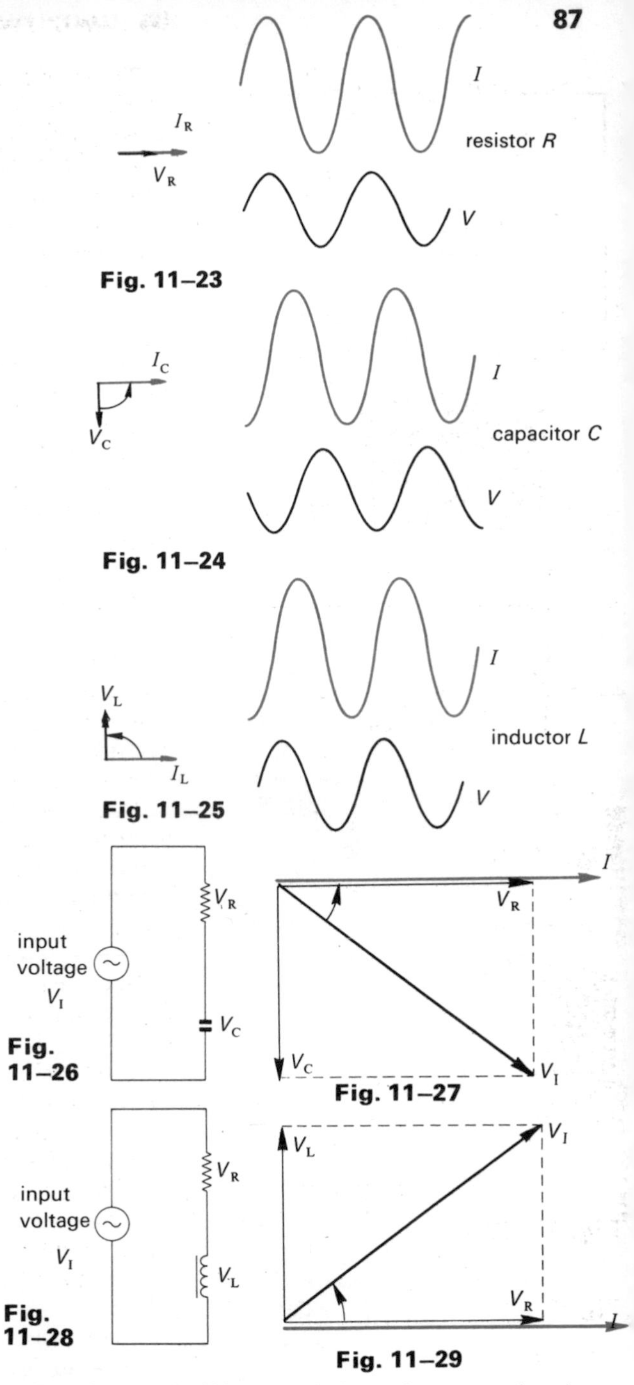

Fig. 11–23

Fig. 11–24

Fig. 11–25

Fig. 11–26

Fig. 11–27

Fig. 11–28

Fig. 11–29

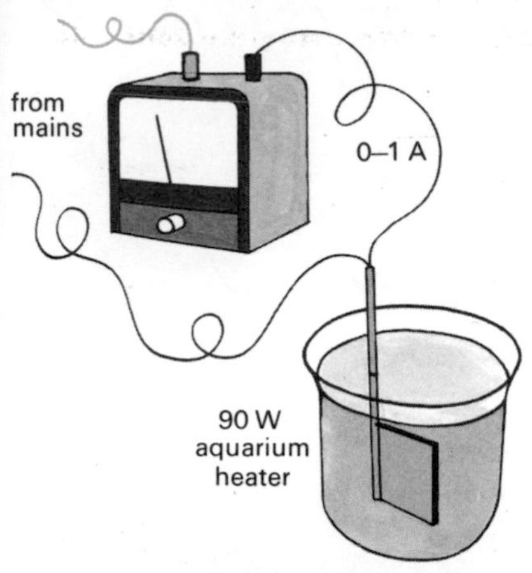

Fig. 11–30

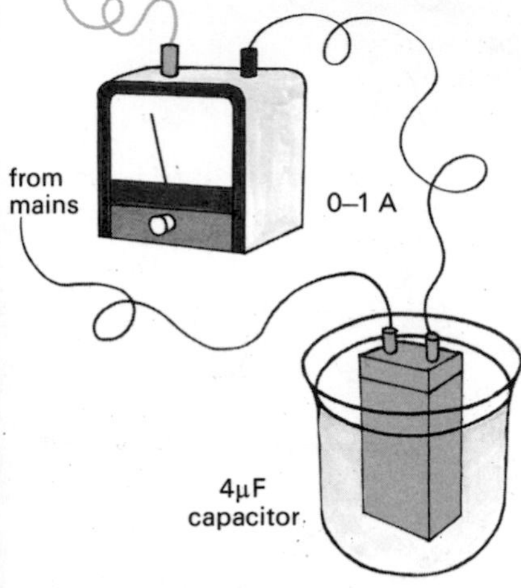

Fig. 11–31

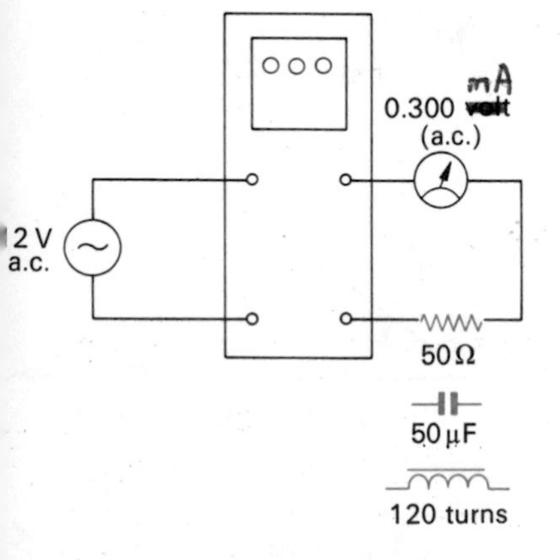

Fig. 11–32

POWER IN A.C. CIRCUITS

It is possible to have the same alternating current in two components, e.g. a resistor and a capacitor, and to have the same p.d. across them, yet one quickly warms up and the other does not (Figs. 11–30 and 11–31). The energy transferred per second (power) is not the same, even though the product 'amperes × volts' is the same.

In d.c. circuits the current and voltage are steady and thus in phase. We can find the power by multiplying current and voltage together. As we have seen that current and voltage are not necessarily in phase in an a.c. circuit, we shall now consider the power in various components.

In the apparatus illustrated in Fig. 11–32 we can use a joulemeter to measure the total energy transfer. This can then be compared with the product VIt where V is 12 V, I the current reading in the meter, and t the time in seconds during which the current flows. The results show that VIt indicates the energy transfer (heat) only in the case of the resistor. No heat is produced in a purely reactive circuit and therefore we cannot use VI to measure the power or VIt to calculate the energy transfer in such circuits.

The power *at any instant* can be found by multiplying the current and the voltage. In the accompanying graphs this product is plotted in red. The shaded area represents the energy being taken from the supply (above the line) and the energy being fed back into the source of supply (below the line).

Resistance

Power is always *positive* (Fig. 11–33); that is, energy is always being taken from the source to produce heat.

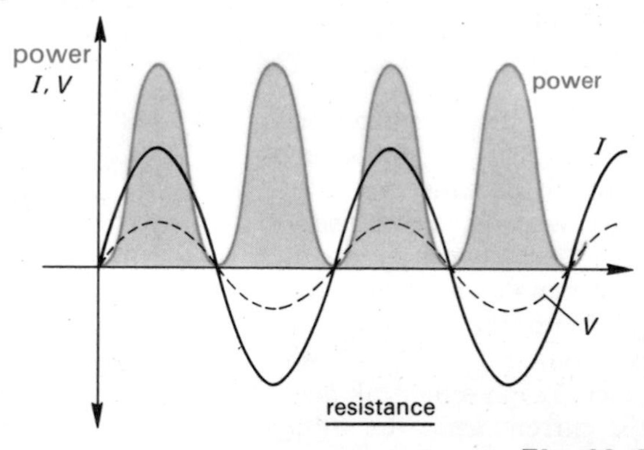

Fig. 11–33

Capacitance

Positive and negative half-cycles of the power curv are equal. On average *no* energy is being taken fro the source. The energy taken from the source o each positive half-cycle is stored in the capacito This energy is returned to the source during th negative half-cycle (Fig. 11–34).

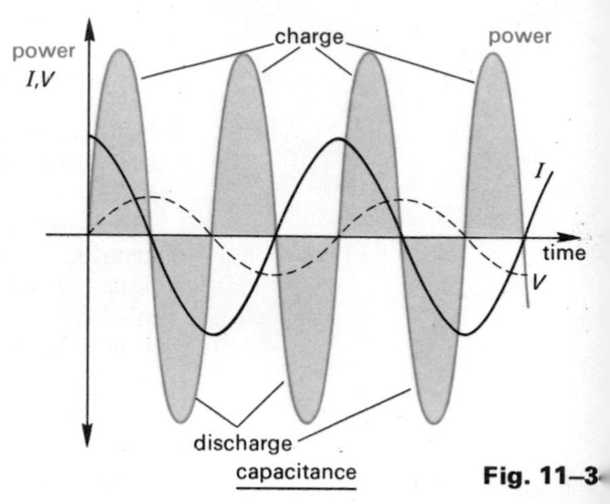

Fig. 11–3

Pure Inductance

Energy is stored in the magnetic field during th positive half-cycle of the power curve and returne to the source of supply when the field collapse On average *no* energy is taken from the sourc (Fig. 11–35). In practice an inductor always ha some resistance and so there will always be som energy transfer (heat). In other words, the phas lag is never quite 90° in a real inductor.

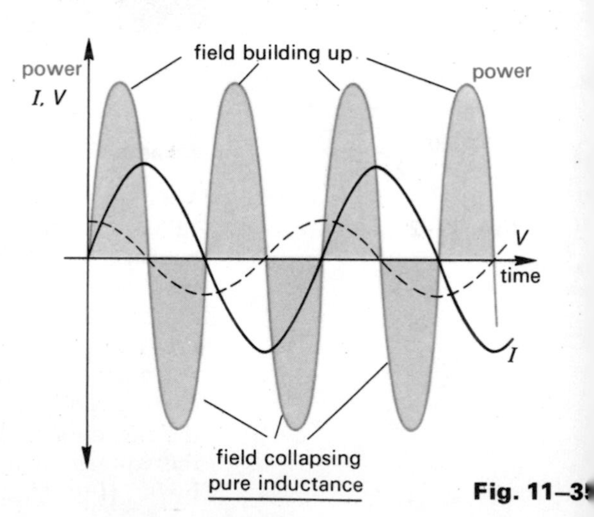

Fig. 11–3

SUMMARY

Production

A.C. may be produced by a vibrating magnet near a conductor or a vibrating conductor in a magnetic field. A rotating magnet near a conductor or a conductor in a magnetic field can also produce a.c.

Transformers

$$\frac{V_p}{V_s} = \frac{t_p}{t_s} \text{ approximately}$$

$$\frac{I_p}{I_s} = \frac{t_s}{t_p} \text{ approximately}$$

V_p = p.d. across primary
I_p = current through primary
t_p = number of turns in primary
V_s = p.d. across secondary
I_s = current through secondary
t_s = number of turns in secondary

Capacitors

As the capacitance is *increased*, the reactance *decreases*.
As the frequency is *increased*, the reactance *decreases*.
The current *leads* the voltage by 90°.
Capacitive reactance (measured in ohms)

$$X_C = \frac{V}{I} = \frac{1}{2\pi fC}$$

Inductors (pure)

D.C. When the switch is closed the current takes time to build up, owing to the inertial effect of the back e.m.f. induced by the changing magnetic field.

A.C. As the inductance is *increased*, the reactance *increases*. As the frequency is *increased*, the reactance *increases*. The current *lags* behind the voltage by 90°.
Inductive reactance (measured in ohms)

$$X_L = \frac{V}{I} = 2\pi fL$$

Resistors

A.C. Resistance is not affected by frequency. The current and the voltage are in phase.

Power

Electrical energy is transformed (to heat) only by the *resistive* component of an a.c. circuit.

PROBLEMS

11.5 Discuss the current in the circuit of Fig. 11–36 if

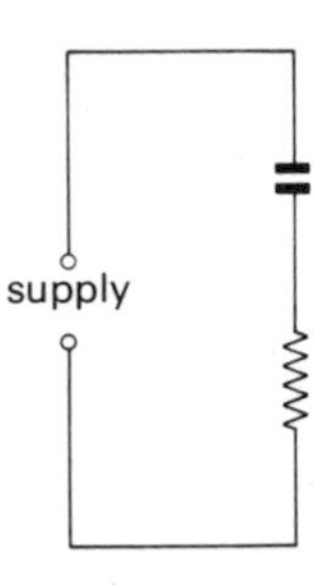

Fig. 11–36

(a) the supply is a 2 V accumulator,
(b) the supply is 2 V 50 Hz a.c.,
(c) the frequency of the 2 V a.c. supply is increased.

11.6 If the capacitor in Fig. 11–36 is replaced by an iron-cored inductor, discuss the current which would flow when
(a) the supply is a 2 V accumulator,
(b) the supply is 2 V 50 Hz a.c.,
(c) the frequency of the 2 V a.c. supply is increased,
(d) the supply is 2 V 50 Hz and the iron core is removed from the inductor.

11.7 Discuss as quantitatively as possible various ways of operating a 12 V 24 W bulb from 240 V a.c. mains. Consider also the practical factors involved.

11.8 In Fig. 11–37, L is a coil of wire of many turns wound on a soft-iron core.

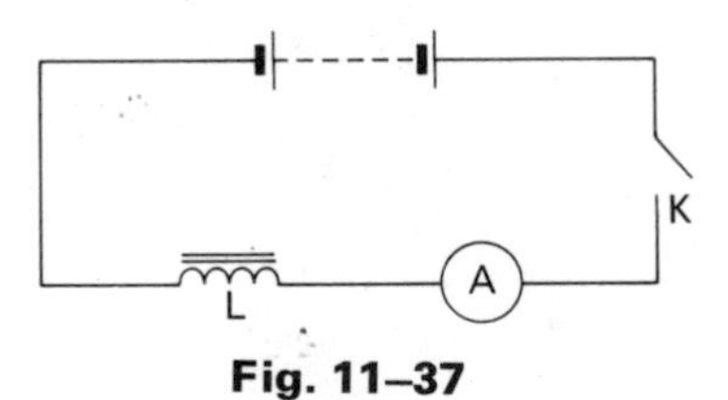

Fig. 11–37

(a) Why, when the switch K is closed, does the current take a little time to grow to its final value?
(b) Explain why, when the switch K is subsequently opened, there is a spark at the switch contacts.
(c) If the experiments were repeated without the iron core, what differences, if any, would be noticed in each case?
[S.C.E. (H) 1970]

12 Oscillations

Fig. 12–1

Bagpipes and bells produce vibrations. In fact we are surrounded by regular to and fro motions. We see by light vibrations, we hear by sound vibrations, we are heated by infra-red vibrations, and we are entertained by electromagnetic vibrations. We live because of our regular heartbeats, and even the atoms we are made of are continually oscillating.

In an oscillatory system energy is continually changing from one form to another; for example, kinetic energy may change to potential energy and back to kinetic energy and so on. The number of complete to and fro movements per second is called the frequency f and the time for one complete swing is the period T.

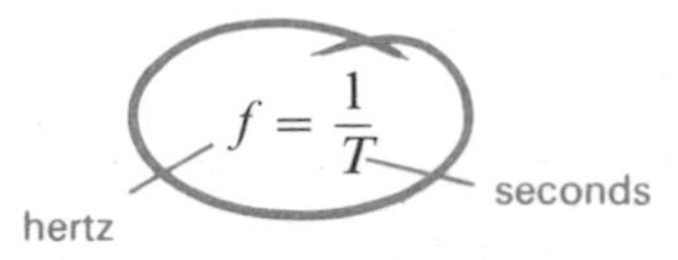

A pendulum is one of the simplest forms of oscillator. Fig. 12–3 shows a strobe photograph of such a pendulum. It is taken with the camera at rest. If, however, the camera is moved vertically *at a constant speed* while the strobe picture of the pendulum is being taken, the result turns out like Fig. 12–4. The pendulum bob traces out a sine curve.

Fig. 12–3

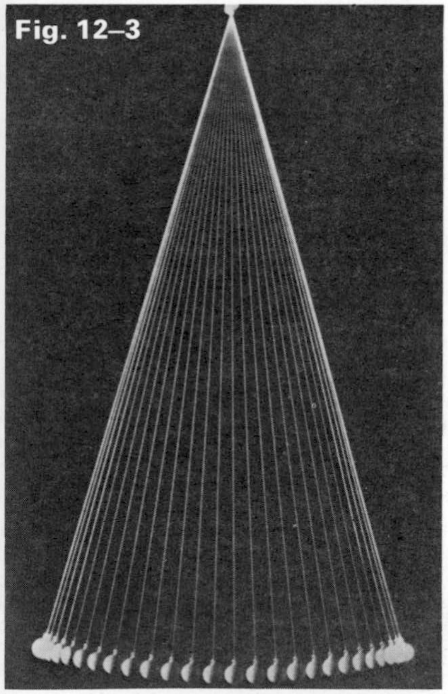

Fig. 12–4

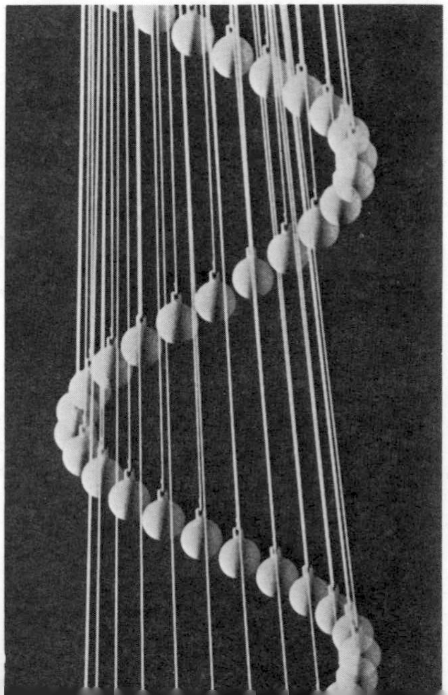

Fig. 12–2

If a heavy mass is attached to the end of a vertic coil spring, the mass will oscillate up and dow Fig. 12–5 shows a strobe picture of this oscillati taken with a camera moving horizontally at constant speed. The mass, which again traces o a sine curve, is said to be moving with *simp harmonic motion.*

When a guitar string is plucked, it vibrates to a fro with a gradually decreasing amplitude. T oscillation is said to be *damped.* The energy ori nally given to the string is 'lost' in overcomi internal friction (the string becomes warme external friction (by its moving through the air t air is warmed), and radiation (sound is produce A vibrating springboard, a swinging lamp and oscillating trampoline are other examples

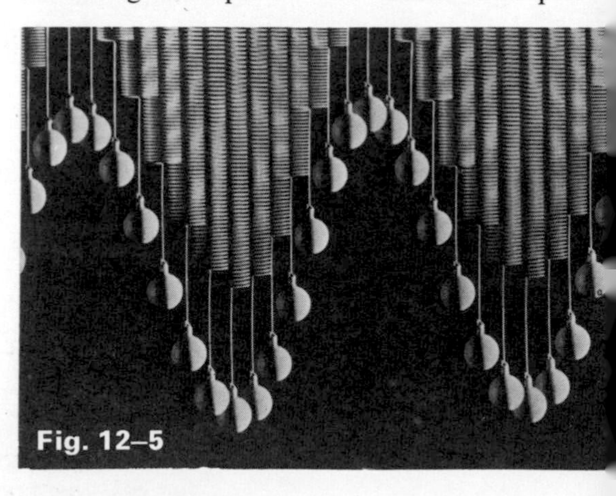

Fig. 12–5

Fig. 12–6

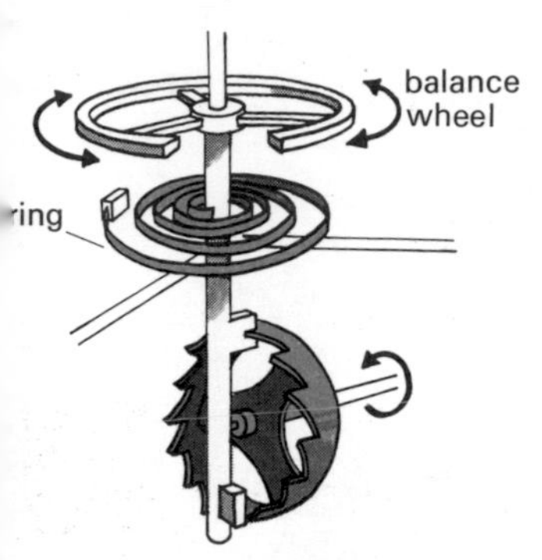

Fig. 12–7

Fig. 12–8

damped oscillations. In the suspension system of a car, unwanted oscillations are quickly damped by the shock absorbers, which are deliberately designed to absorb energy from the oscillations.

If oscillations are to be maintained, then energy must be continually fed into the system at the correct frequency. The vibrations of a pendulum can, for example, be maintained by a mechanical device called the anchor escapement. The escape wheel, driven by a clockwork motor or falling mass, provides the 'make-up' energy at the right moment. Perhaps you can work out its operation from Fig. 12–6, which illustrates an excellent Russian-built working model. In a watch, a hairspring and balance wheel replace the pendulum (Fig. 12–7).

Hacksaw blades of different lengths have different natural frequencies of vibration. This can be simply illustrated with the apparatus shown in Fig. 12–8. If the handle is made to vibrate at various frequencies, one blade will start to swing to and fro vigorously when the frequency of the forced vibrations is the same as the natural frequency of that blade. This phenomenon is called *resonance* and the frequency at which it occurs is called the *resonant frequency* f_0. It is also possible to maintain oscillations if the driving frequency is $\frac{1}{2}f_0$, $\frac{1}{3}f_0$, etc. For example, giving a child's swing a push every second or third swing will keep it going.

If the red ball in Fig. 12–9 is set swinging to produce 'forced vibrations', which pendulum would you expect to resonate with it? (12.1)

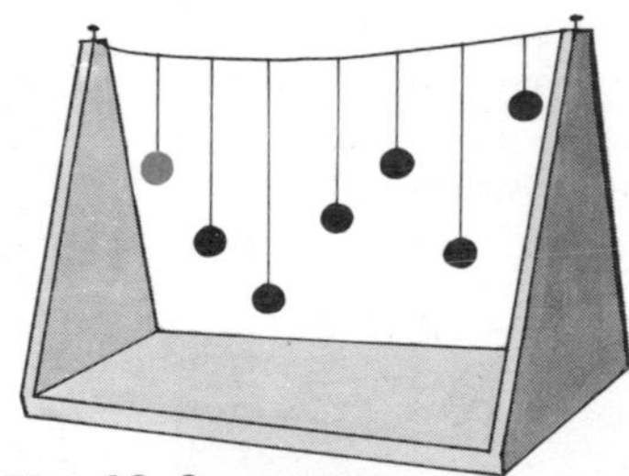

Fig. 12–9

In many modern cars you can estimate the speed quite accurately by the noise of various rattles! (Fig. 12–10). When the frequency of the engine corresponds to the natural frequency of some particular part of the car, resonance is produced. The off-side door might vibrate around 40 km per hour, the heater fan at 50 km per hour, and the exhaust pipe at 60 km per hour.

Fig. 12–10

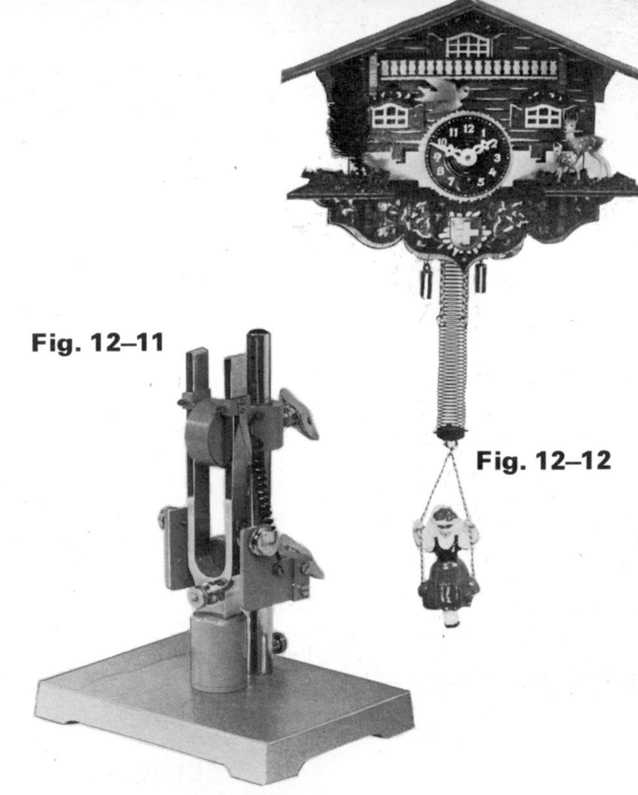

Fig. 12–11

Fig. 12–12

More carefully planned resonances are produced in the electrically-maintained tuning fork shown in Fig. 12–11 or in the Swiss clock controlled by vertical oscillations of a mass on the end of a helical spring (Fig. 12–12). Energy is fed in from a clockwork motor to maintain the oscillations.

Amplitude and Frequency

The apparatus shown in Fig. 12–13 can be used to study the amplitude of the torsional oscillations of a flywheel attached to a clock spring, when energy is fed in at different frequencies. Similarly, the amplitude of the standing wave on a stretched wire can be studied when the driving frequency is

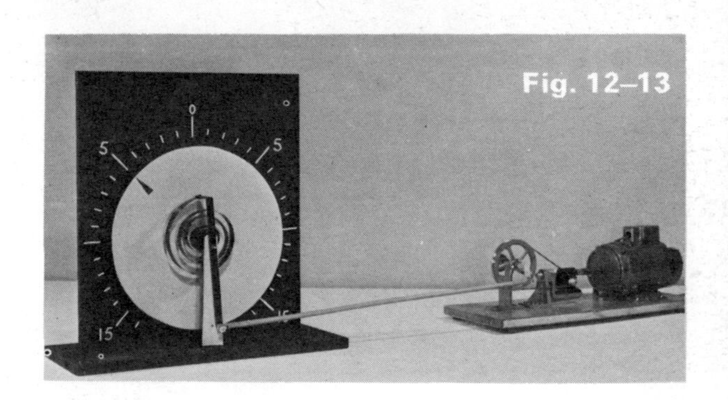

Fig. 12–13

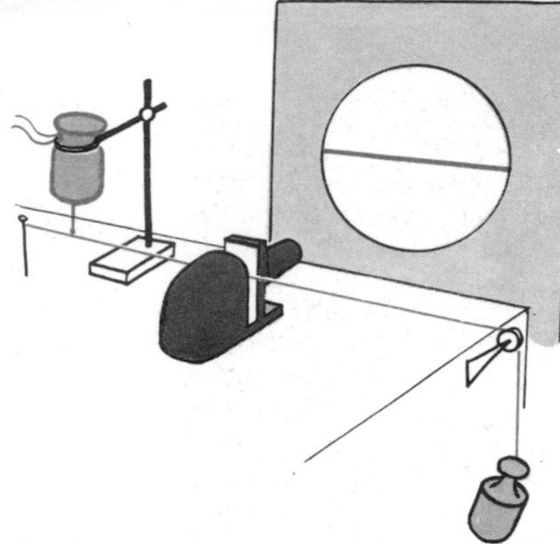

Fig. 12–14

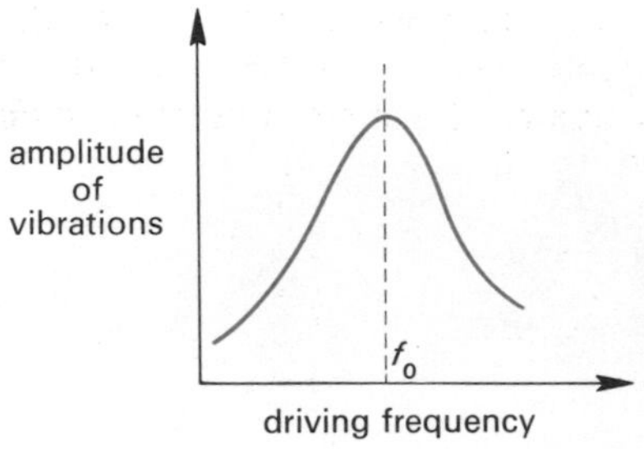

Fig. 12–15

altered (Fig. 12–14). A calibrated signal generator is used to drive the vibrator and the image of the wire is projected on to a screen for ease of measurement. In each case the amplitude increases to a maximum value at resonance, that is, when the driving frequency is equal to the natural frequency of the system (Fig. 12–15).

ELECTRICAL OSCILLATIONS

In today's world, high speed radio and television waves enable us to hear and see events as they take place a quarter of a million miles away on the surface of the moon (Fig. 12–16). In more down to earth scientific jargon, audio and video information is transmitted at 3×10^8 m/s by electromagnetic radiation generated by electrical oscillations.

Mechanical oscillations are associated with continuous changes of energy from one form to another. In the apparatus shown in Fig. 12–13 kinetic energy in the rotating flywheel is used to wind a spring. When the flywheel comes to rest, energy is stored in the spring. The spring th unwinds and loses its potential energy as the f wheel again gains kinetic energy. There is then continuous process:

$$\text{K.E.} \rightarrow \text{P.E.} \rightarrow \text{K.E.} \rightarrow \text{P.E. etc.}$$

We can study electrical oscillations with t apparatus shown in Figs. 12–17 and 12–18. Wh the switch is in the position shown, the capacitor charged. If the switch is moved to the centre po tion, energy is stored in the capacitor (potent energy). Moving the switch to the left allows t charged capacitor to drive charges through the c (kinetic energy), thus building up a magnetic fie

Fig. 12–16

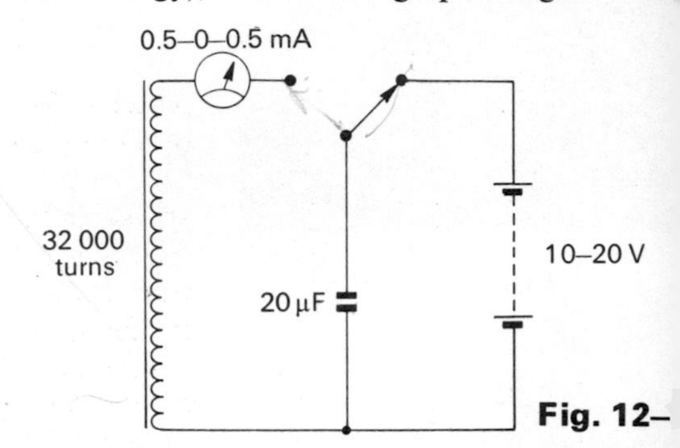

Fig. 12–

Fig. 12–18

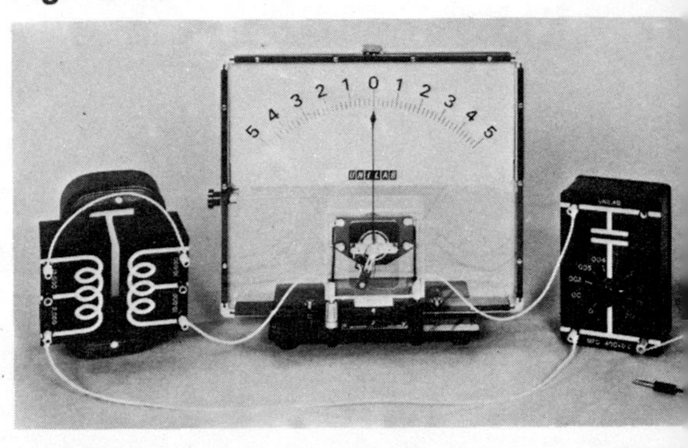

As the discharging current decreases, the magne field collapses and an e.m.f. is induced in the co This e.m.f. tends to keep the current flowing in t same direction, so that the capacitor is aga charged but with the opposite polarity. This and fro movement of charge continues until all t energy has been transformed into heat and rad tion. This process is illustrated in Fig. 12–19. I C.R.O. is connected across the circuit, the volta

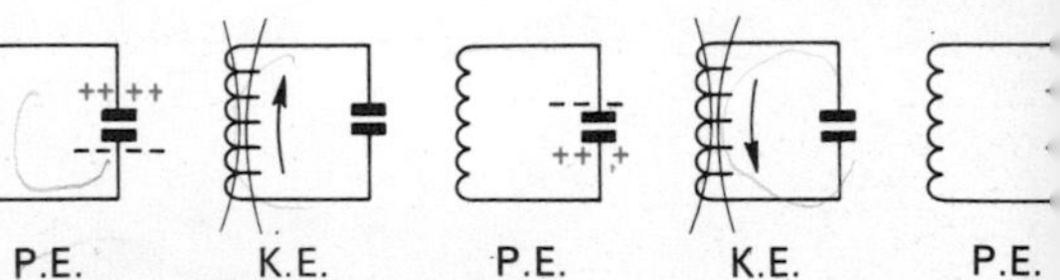

Fig. 12–19

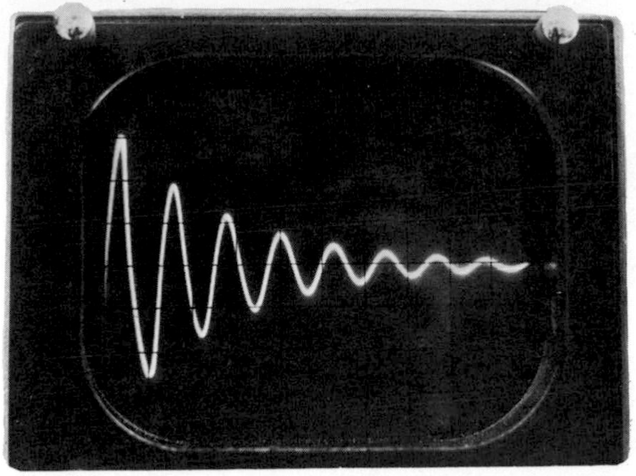

Fig. 12–20

pattern shown in Fig. 12–20 is produced. It is clearly of a damped oscillation.

A mechanical analogue of this tuned circuit is shown in Fig. 12–21. The stretched elastic in which energy is stored causes the trolley to move. When the elastic has lost all its potential energy, the trolley is moving quickly. As it moves, the elastic is stretched in the opposite direction. The kinetic energy disappears and potential energy reappears in the stretched elastic. The to and fro motion continues until all the energy has been transformed to heat and sound.

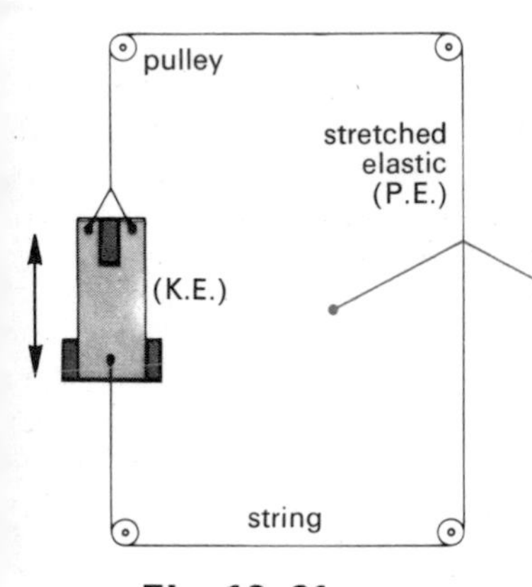

Fig. 12–21

Oscillator Frequency

If a smaller capacitance is used in the tuned circuit of Fig. 12–19, it will store less charge and will therefore discharge more rapidly. Similarly, a smaller inductance will allow the current to build up more rapidly and decrease more rapidly. Reducing either the capacitance or the inductance should therefore produce faster oscillations. Experiments show that this is so.

$$C\downarrow \quad f\uparrow$$
$$L\downarrow \quad f\uparrow$$

Frequencies of the order of one hertz are produced by the components shown in Fig. 12–18. For frequencies used in a radio receiver—e.g. 1 MHz—a coil of a hundred turns and a capacitor of a few hundred picofarads could be used (Fig. 13–29, page 104). At about 150 MHz the coil might consist of a single turn, and the capacitor might be made of two small semi-circular plates (Fig. 12–22).

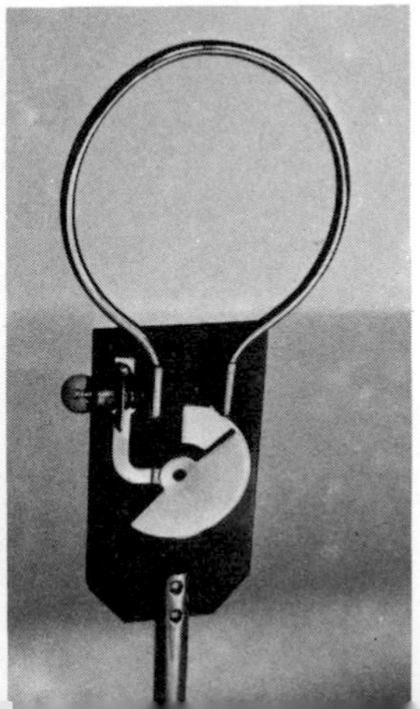

Fig. 12–22

MAINTAINED OSCILLATIONS

In an oscillatory circuit, electrical energy produces heat as the current surges through the wires. Some energy is also radiated as electromagnetic waves. If oscillations are to be maintained at a constant amplitude, energy must be fed into the circuit at the correct frequency. If, for example, energy could be fed from a battery at just the right time during each oscillation, losses in the tuned circuit could be made up (Fig. 12–23). If a signal from the tuned circuit could be used to operate a switch at the appropriate moment, the oscillations could be maintained. Fortunately a simple amplifier, such

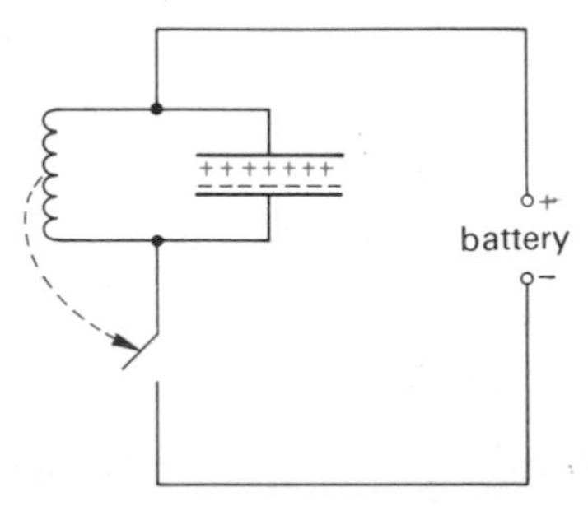

Fig. 12–23

as a triode valve or transistor, can be used to supply the necessary energy automatically. The information telling the valve when to conduct and allow pulses of current to flow comes from a coil coupled to the tuned circuit (Fig. 12–24). Oscillations will be maintained only if the pulses come at the right time. If the connections to the coupling coil are reversed, oscillations will stop. This corresponds to giving a swing a push as it comes towards you, thus slowing it down instead of keeping it going.

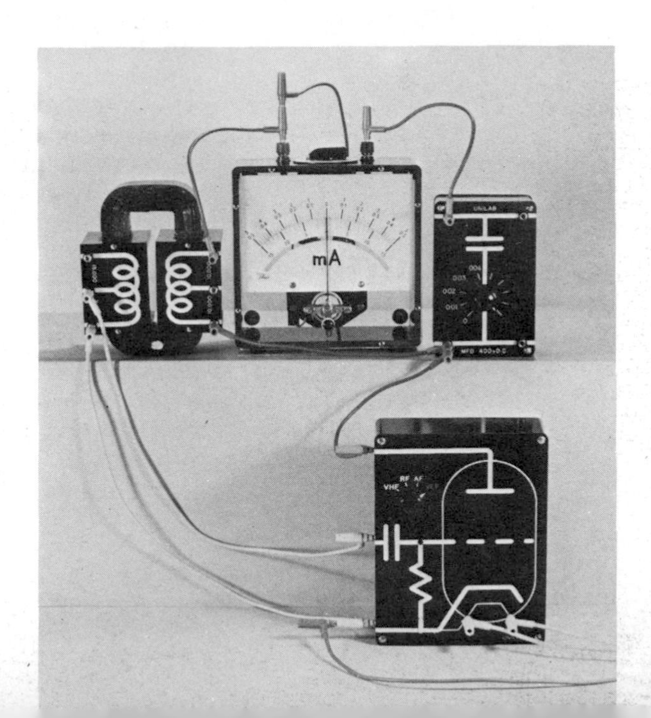

Fig. 12–24

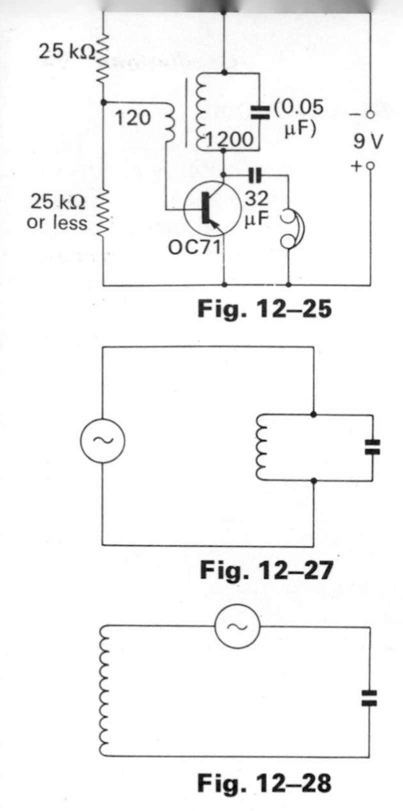

Fig. 12–25

Fig. 12–27

Fig. 12–28

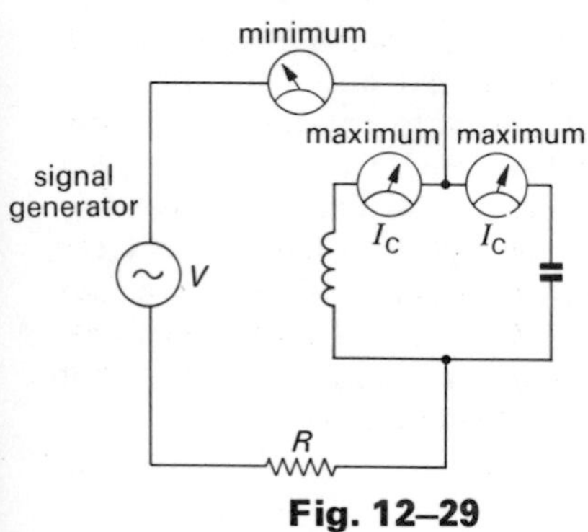

Fig. 12–29

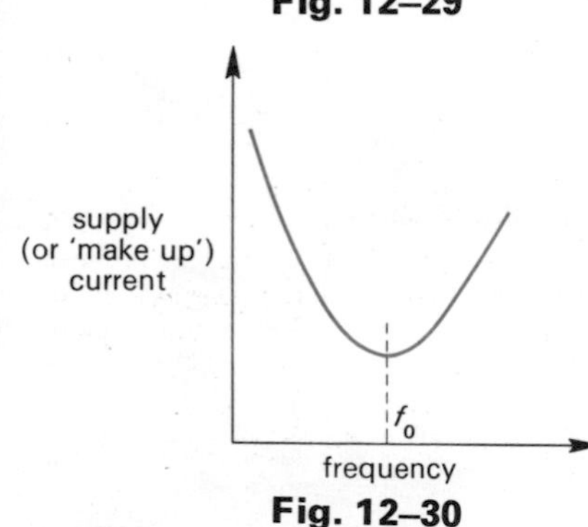

Fig. 12–30

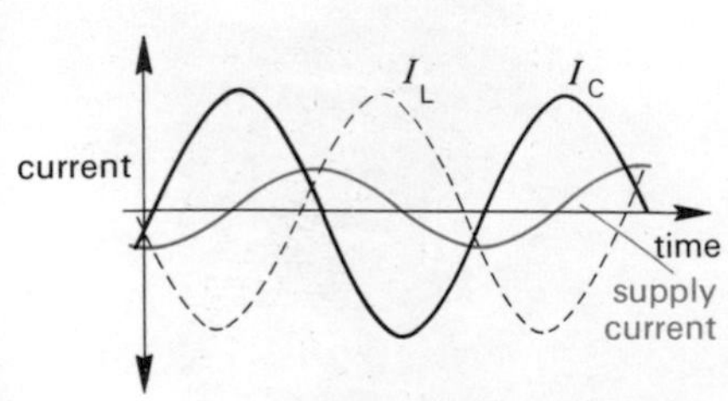

Fig. 12–31

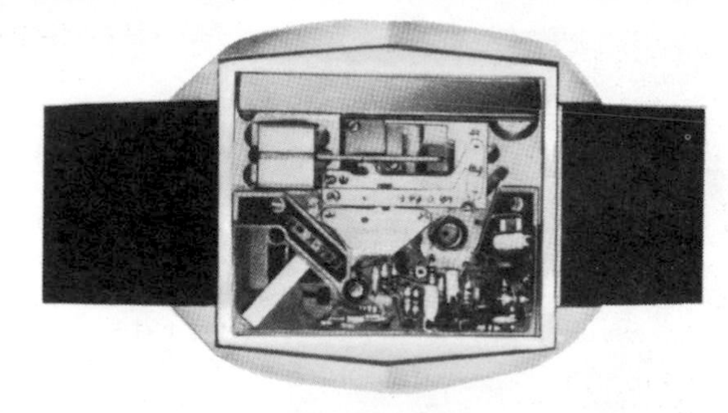
Fig. 12–26

A triode valve is used in Fig. 12–24 to maintain slow oscillations, and a transistor is used in Fig. 12–25 to produce audio oscillations. Electrically maintained oscillations controlled by a quartz crystal are used in the wrist watch illustrated in Fig. 12–26. Such watches are claimed to be accurate to within one minute a year. Although they are extremely expensive at present, the price is likely to topple with the mass production of suitable integrated circuit chips.

Parallel and Series Circuits

When energy is fed into a tuned circuit as in Fig. 12–27, the circuit is called a parallel tuned circuit. If the supply is in series with the two components as in Fig. 12–28, the circuit is called a series tuned circuit. In each case *maximum* current flows in the tuned circuit when the input is supplied at the natural frequency of the circuit. The current taken from the source is, however, very different in the two cases.

If a signal generator is connected via a resistor to a parallel circuit (Fig. 12–29), the supply current can be measured at different frequencies. The variation of the current is shown in Fig. 12–30. At resonance the supply current is minimum and the current in the tuned circuit *maximum*. Moreover, we find that the currents in each branch of the parallel circuit are the same at resonance. Such a circuit is a current amplifying device. This might look at first sight as if we were getting 'something for nothing'. Remembering, however, that the current in the coil (I_L) is nearly 180° out of phase with the current in the capacitive branch (I_C), we see that the *resultant* supply current can be very small (Fig. 12–31). In fact this small 'make up' current is needed only to account for the losses due to heating and radiation. From the signal generator's point of view, the impedance of the tuned circuit (V/I) is *greatest* at resonance—which explains why such a circuit is sometimes called *rejector circuit*. If it were possible to have inductor with no resistance and if there were radiation losses, the impedance would be infini

There is always likely to be some other resistan in the circuit (R in Fig. 12–29) and the p.d. acr this resistance will be least at resonance. *Wh (12.2)* The p.d. across the tuned circuit w therefore be greatest at resonance.

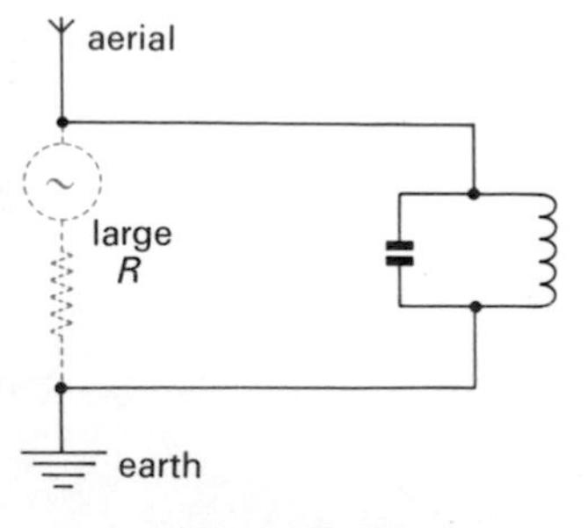

Fig. 12–32

The aerial circuit of a simple receiver may considered as a source of e.m.f. with a high interr resistance (shown red in Fig. 12–32). At resonan there will therefore be a large voltage across t tuned circuit.

In a series circuit we have the opposite eff (Fig. 12–33). The current is the same in every p of the circuit and will reach a maximum at reso ance (Fig. 12–34). The impedance (V/I) is then le

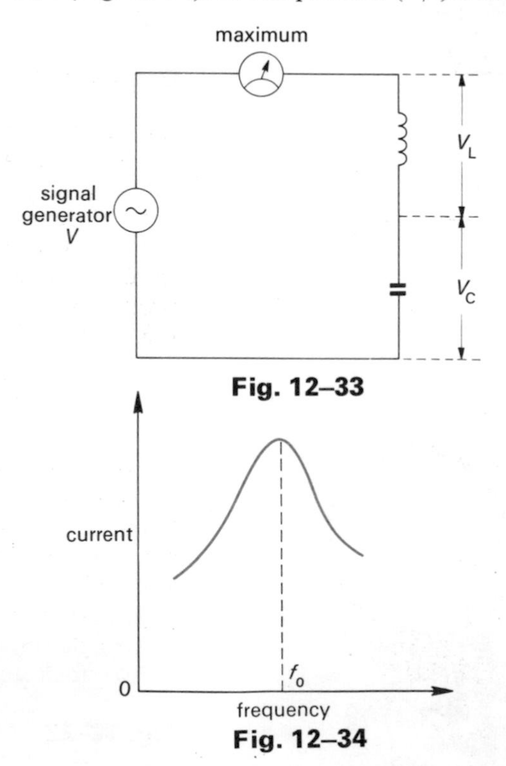

Fig. 12–33

Fig. 12–34

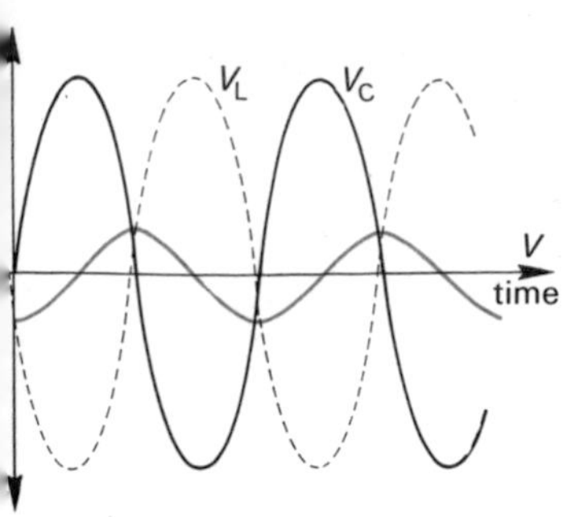

Fig. 12–35

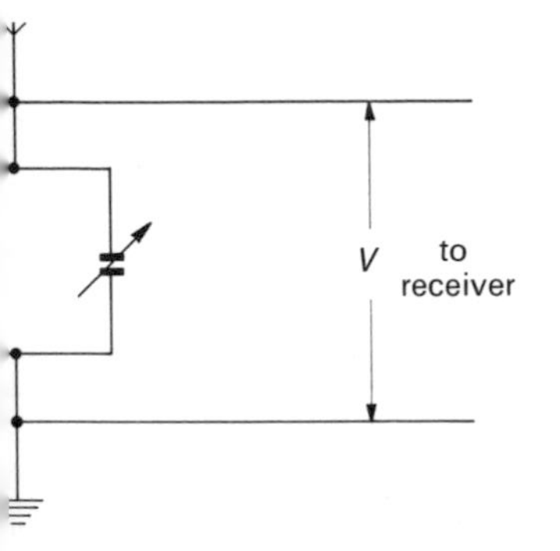

Fig. 12–36

and the circuit is called an *acceptor circuit*. Now, however, the p.d. across the inductor and the capacitor will each be *greater* than the supply voltage; but as these voltages are again nearly 180° out of phase the *resultant* voltage is equal to the supply voltage (Fig. 12–35). A series tuned circuit is a voltage amplifying device.

TUNING IN

The circuits we have considered so far have each had a fixed natural frequency of oscillation and we have altered the supply frequency to study their response. The frequency of a radio signal is, however, determined by the transmitter. Radio 1 (247 m), for example, transmits in Scotland on a frequency of 1.214 MHz. Nothing we can do with the receiver will alter this frequency. When we 'tune in' a receiver we have to alter the tuned circuit wired to the aerial so that *its* natural frequency is adjusted to the frequency of the incoming signal (Fig. 12–36). This can be done by altering the capacitance of a small variable capacitor, or by altering the inductance as a ferrite core is moved in and out of the coil. When this has been done, the peak voltage V at the required frequency should be greater than the peak voltages at all other frequencies, even though stronger signals at other frequencies are being picked up by the aerial. There are, of course, limits to this. If you happen to have a BBC transmitter in your back garden, your listening is going to be rather restricted.

Reactance at Resonance

We have already stated, without proof, that the reactance of a capacitor is given by $1/2\pi fC$ and for an inductor by $2\pi fL$. Our experiments have shown that at resonance the currents are equal in the capacitor and the inductor and that the voltages across each component are also equal. The reactance of each component to the resonant frequency must therefore be the same. This information enables us to calculate the resonant frequency f_0.

$$2\pi f_0 L = \frac{1}{2\pi f_0 C}$$

$$\Rightarrow f_0^2 = \frac{1}{(2\pi)^2 LC}$$

$$\Rightarrow f_0 = \frac{1}{2\pi\sqrt{LC}}$$

hertz
farads
henries

SUMMARY

Mechanical oscillations

Normally all oscillations—longitudinal, transverse, and torsional—are damped because of internal friction, external friction and radiation.

In a vibration or oscillation, energy is continually changing from K.E. to P.E. to K.E. to P.E., etc.

To maintain oscillations, energy is fed into a system at its natural frequency (f_0) or at (f_0/n) where n is an integer.

Standing waves

Standing waves are produced by the interference of incident and reflected travelling waves. Such waves often radiate other travelling waves. For example, a vibrating air column or wire radiates sound waves.

Electrical oscillations

A capacitor stores electrical energy (P.E.). If the capacitor is discharged through a coil, the energy appears in the magnetic field through the coil while charge is flowing (K.E.). When the field collapses, the capacitor is again charged up in the opposite direction (P.E.).

A parallel (rejector) circuit has maximum impedance at f_0. It is a current amplifying device.

A series (acceptor) circuit has minimum impedance at f_0. It is a voltage amplifying device.

The resonant frequency of a tuned circuit *increases* as C or L are *reduced*.

$$f_0 = 1/2\pi\sqrt{LC}$$

PROBLEMS

12.3 Devise a method of maintaining oscillations in *one* of the following (a) a vibrating hacksaw blade, (b) a mass vibrating on a helical spring, (c) a torsional pendulum, (d) a ball running to and fro on a curved rail, or (e) a simple pendulum.

12.4 Discuss energy changes in each of the following (a) a torsional pendulum, (b) a guitar string.

12.5 Why should a mains-operated ticker timer be tuned to a natural frequency of 50 Hz? If the length of the strip could not be altered, how could you reduce its natural frequency? Draw a rough graph showing how the amplitude of the vibration of a ticker timer varies with input frequency.

Fig. 12–37

Fig. 12–38

Fig. 12–39

12.6 Before diving a girl bounces up and down on the end of a springboard (Fig. 12–37). How could a stout man make the same springboard vibrate naturally at the same frequency?

12.4 With what frequency would you push a child's swing to cause it to reach a large amplitude (Fig. 12–38)? Would any other frequency produce similar results?

12.8 You may have bounced up and down on your bed-springs or on a trampoline (Fig. 12–39). Do you think the mass of the person jumping will affect his frequency?

12.9 How is the frequency of a simple pendulum affected by (a) increasing the mass of the bob, (b) increasing the length of the string, (c) increasing the amplitude of swing from 5° to 10°?

When has the bob its maximum (i) speed, (ii) kinetic energy, (iii) potential energy?

12.10 Copy the following statements and complete them. In a parallel tuned circuit

(a) at f_0 the current from the supply is at its . . . .
(b) at f_0 the impedance of the circuit is at its . . . .
(c) below f_0 . . . . current goes through L than C.
(d) above f_0 . . . . current goes through L than C.

12.11 Copy the following statements and complete them. In a series tuned circuit

(a) at f_0 the current is at its . . . .
(b) at f_0 the voltages across L and C are at their . . . .
(c) at f_0 the impedance of the circuit to the supply is . . . .
(d) below f_0 the voltage across L is . . . . than across C
(e) above f_0 the voltage across L is . . . . than across C.

12.12 A coil and capacitor are connected in the circuit shown in Fig. 12–40.

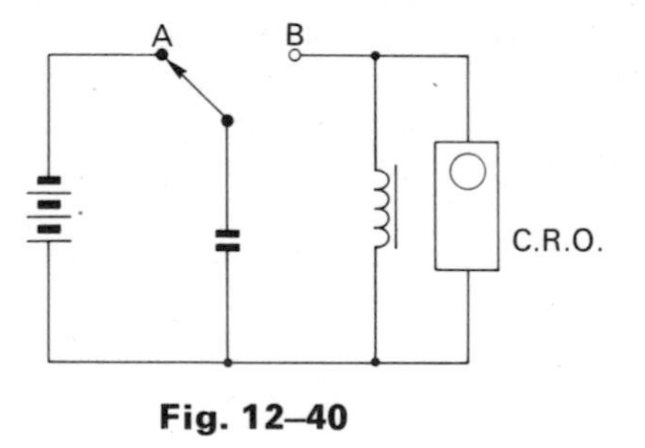

Fig. 12–40

If the C.R.O. has a long persistence tube, sketch the pattern produced when the switch is thrown to position B.

12.13 A constant voltage a.c. supply is connected to 'black box' as shown in Fig. 12–41.

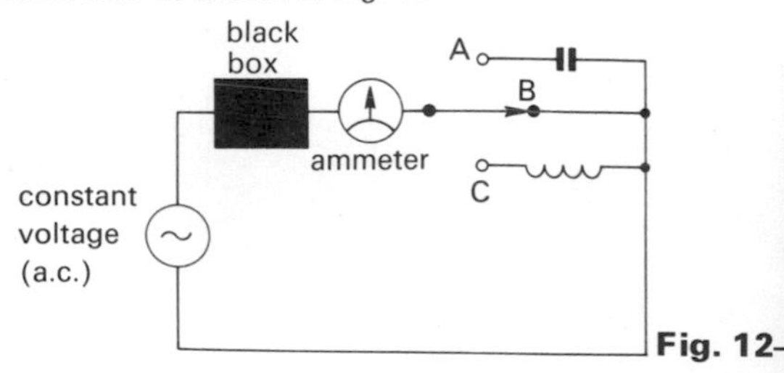

Fig. 12–

When the switch is moved from B to A the current reduced. When moved from B to C it increases. What cou be in the black box?

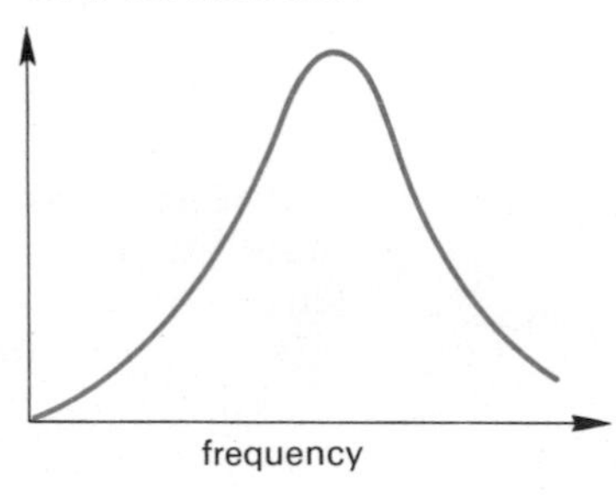

Fig. 12

12.14 Could Fig. 12–42 represent the variation of curre or impedance in (a) a series, (b) a parallel circuit?

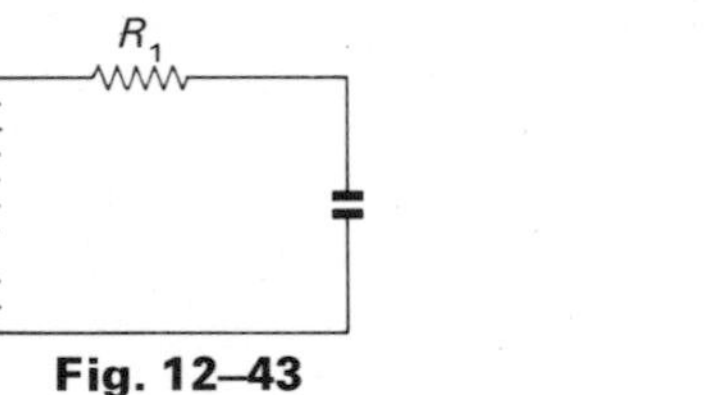

Fig. 12–43

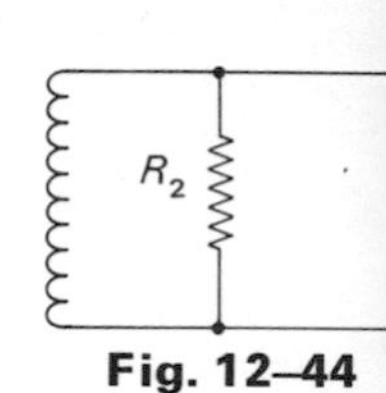

Fig. 12–44

12.15 The damping of a tuned circuit depends on t resistance in the circuit. Should R_1 (Fig. 12–43) and (Fig. 12–44) be large or small to produce the least dampin

12.16

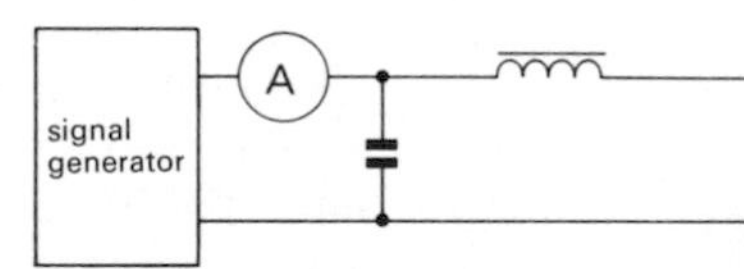

Fig. 12–

Draw a graph showing how the reading on the meter Fig. 12–45 would change as the frequency of the consta voltage source was increased. Do the same for the me readings in Fig. 12–46.

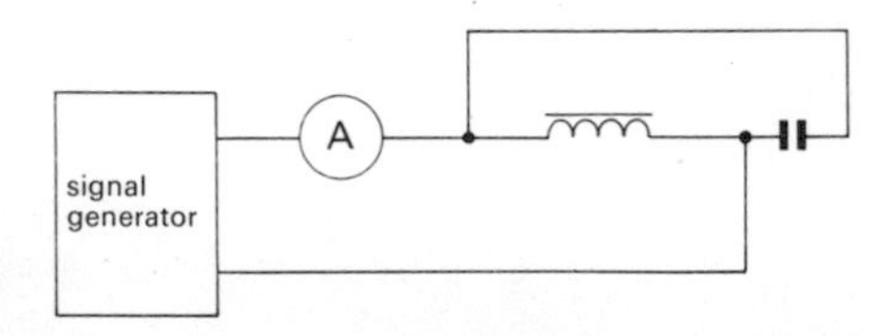

Fig. 12

12.17 Fig. 12–47 shows the variation of inductive reactance with frequency, and the variation of capacitive reactance with frequency.

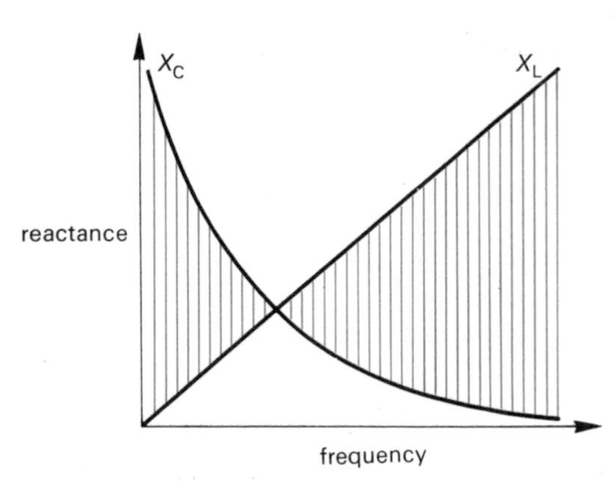

Fig. 12–47

The red lines in Fig. 12–47 represent the difference at various frequencies, that is, the resultant reactance at these frequencies. Plot a graph showing how the resultant reactance varies with frequency. Where is the resonant frequency indicated? Does this graph represent the resultant reactance of a series or a parallel circuit?

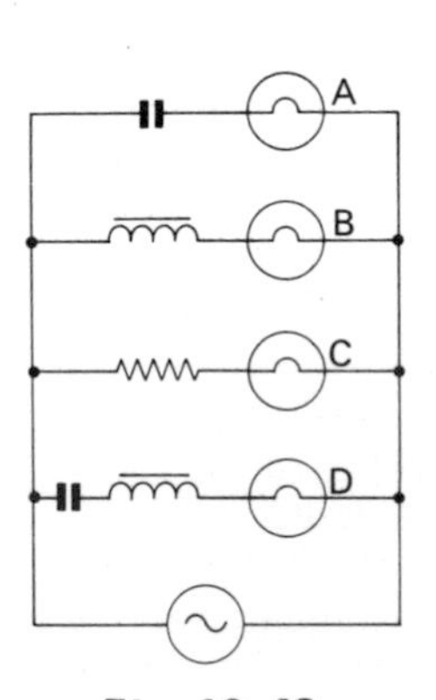

Fig. 12–48

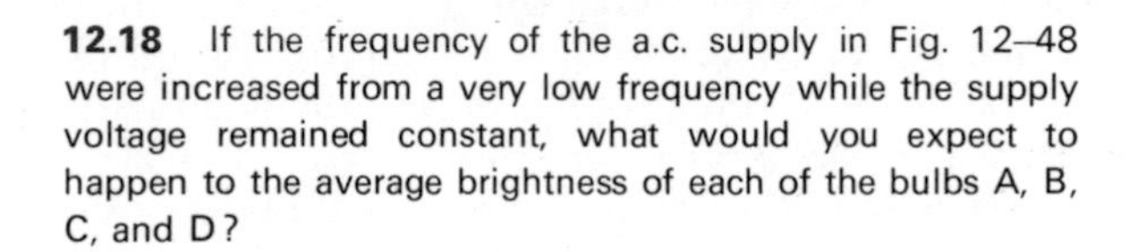

12.18 If the frequency of the a.c. supply in Fig. 12–48 were increased from a very low frequency while the supply voltage remained constant, what would you expect to happen to the average brightness of each of the bulbs A, B, C, and D?

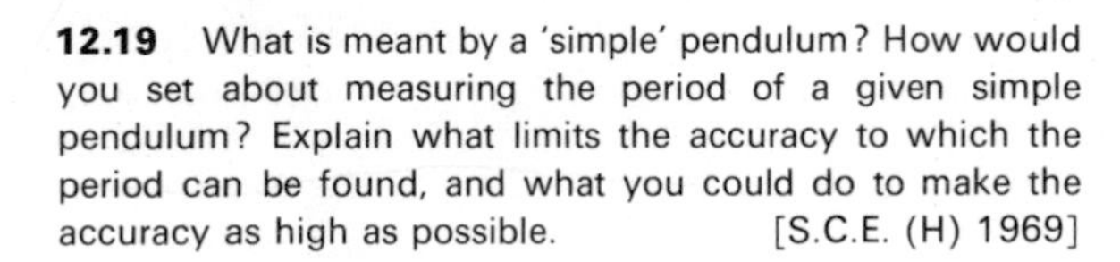

12.19 What is meant by a 'simple' pendulum? How would you set about measuring the period of a given simple pendulum? Explain what limits the accuracy to which the period can be found, and what you could do to make the accuracy as high as possible. [S.C.E. (H) 1969]

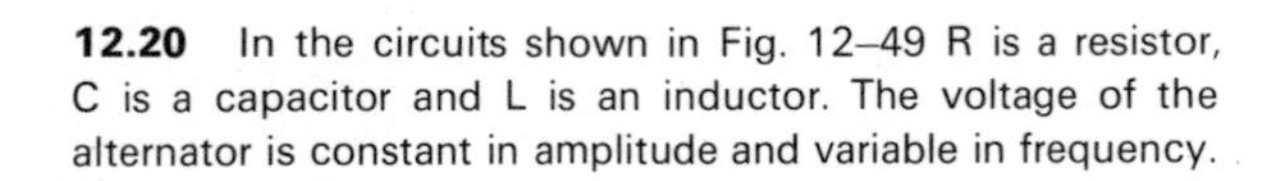

12.20 In the circuits shown in Fig. 12–49 R is a resistor, C is a capacitor and L is an inductor. The voltage of the alternator is constant in amplitude and variable in frequency.

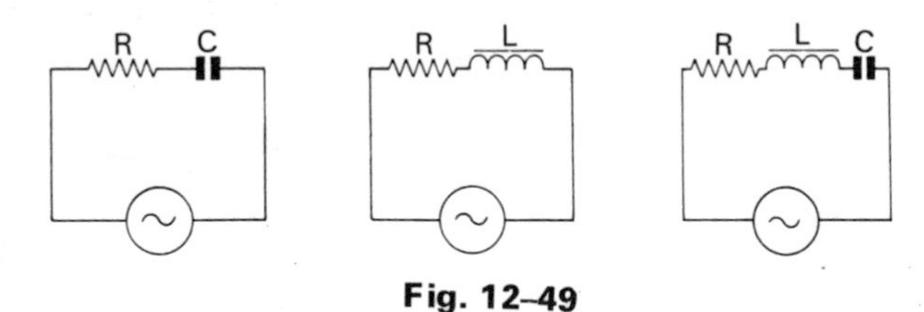

Fig. 12–49

Discuss briefly, in each case, how the temperature of the resistor might change as the frequency of the alternator is increased from zero. [S.C.E. (H) 1970]

12.21 Fig. 12–50 shows a form of vibrator.

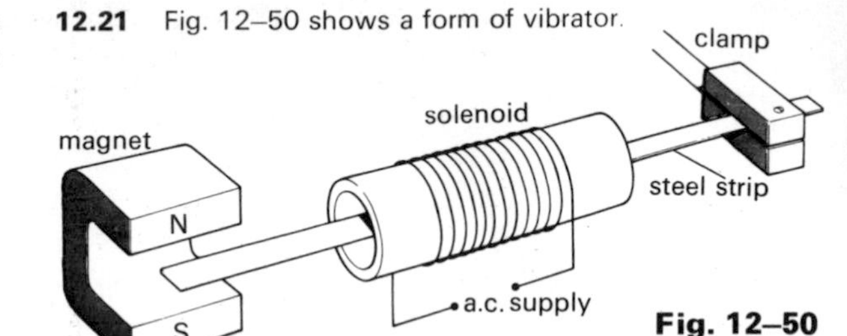

Fig. 12–50

(a) Explain why the steel strip vibrates.
(b) If the supply frequency is 50 Hz, what is the frequency of vibration of the steel strip?
(c) How would you adjust the apparatus to obtain the maximum amplitude of vibration? (S.C.E. (H) 1971]

12.22 In the circuit shown in Fig. 12–51, L has a large inductance and R is a variable resistor. After S is closed R is adjusted until the identical bulbs are equally bright. S is now opened.

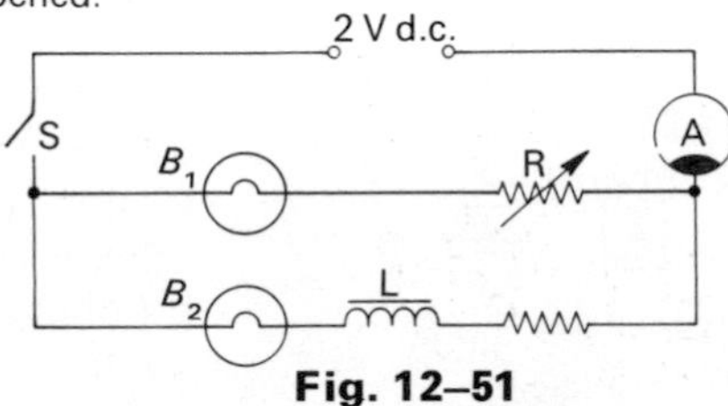

Fig. 12–51

(i) Describe and explain what happens immediately after the switch is closed again.
(ii) If the d.c. supply were replaced by a 2 V a.c. supply, how would the brightness of the bulbs B_1 and B_2 compare with their previous values?
(iii) The variable resistor is now replaced by a suitable capacitor, and the frequency of the supply is increased from a value below the resonant frequency of the circuit to a value above. Sketch a graph of meter reading against frequency for this new circuit, marking clearly the resonant frequency.
(iv) State with reasons how the brightnesses of the bulbs B_1 and B_2 are most likely to change over this frequency range. [S.C.E. (H) 1971]

12.23 The output of an audio-frequency amplifier is connected across XY in the circuit shown below. This is designed to direct the low frequency signals to one loudspeaker and the higher frequency signals to the other.
(i) Suggest which of the loudspeakers P or Q is intended to reproduce the lower frequency signals.
(ii) Explain how the high and low frequencies are separated in this circuit. [S.C.E (H) 1972]

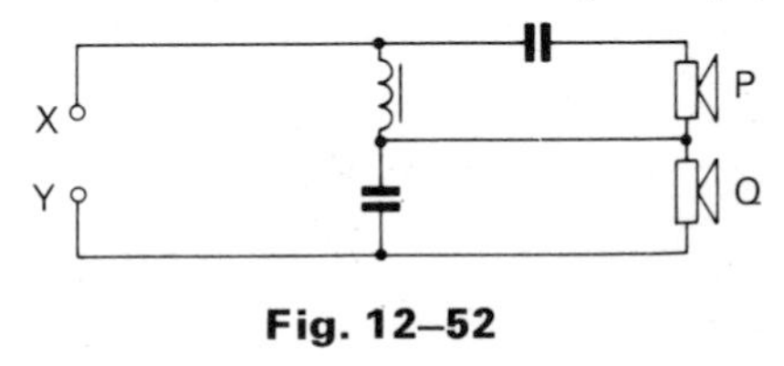

Fig. 12–52

13 Radio Communication

Fig. 13.1

JAMES CLERK MAXWELL

Today we take for granted that waves can travel through space. We no longer wonder at radio and television, radar and radio stars. We have learned to live with electromagnetic radiation of all kinds: VHF, UHF, microwaves, infrared, ultraviolet, X-rays, and gamma rays.

A hundred years ago this amazing spectrum of waves was unknown, and it took the mathematical genius of James Clerk Maxwell to bring it to light. Maxwell was born in Edinburgh in 1831, and from the age of ten to sixteen he attended The Edinburgh Academy. When he was only fourteen Maxwell won the Mathematical Medal in the Academy and wrote a paper on 'Oval Curves', which was published in the Proceedings of the Royal Society of Edinburgh. Ironically his nickname at school was 'Daftie'.

After attending Edinburgh and Cambridge Universities, Maxwell, at the age of twenty-four, was elected to the Chair of Natural Philosophy in Marischal College, Aberdeen. He later became the first Professor of Experimental Physics at Cambridge and directed the plans for the Cavendish Laboratory.

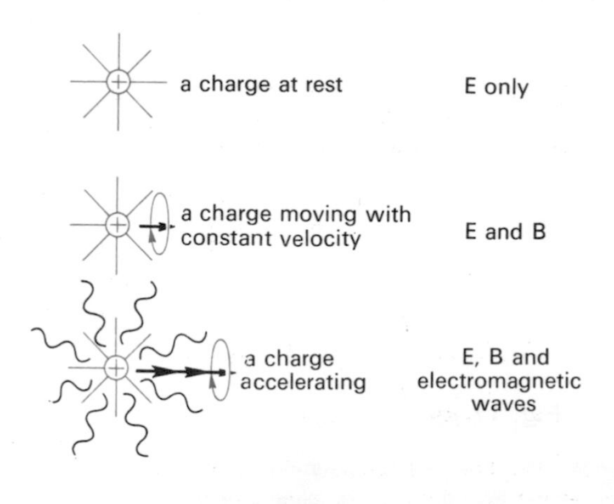

Fig. 13–2

Maxwell's Equations and Electromagnetic Theory

Hans Oersted had shown that a current in a conductor produced a magnetic field and Faraday had discovered that a changing magnetic field could produce a current in a conductor, but there was no unified theory of electric and magnetic behaviour. This was provided by Maxwell's equations. They stated, mathematically, that a changing magnetic field B always produces an electric field E and that a changing electric field always produces a magnetic field. In addition, Maxwell used his equations to predict an entirely new phenomenon—electromagnetic waves. He claimed that oscillating electromagnetic fields were propagated through space in the form of waves which carried energy. Such waves were emitted by any *accelerating* electric charge. These results are summarized in Fig. 13–2. Maxwell's equations also predicted that the speed of propagation of electromagnetic waves was almost exactly equal to the speed of light as measured by Armand Fizeau. Maxwell concluded that light is a kind of electromagnetic radiation.

Still there was no direct experimental evidence for the existence of electromagnetic waves. No one had been able to produce such waves from electromagnetic equipment. Then in 1888, a few years after Maxwell's death, Heinrich Hertz investigated the properties of electric oscillations in a circuit connected to an induction coil. The circuit consisted of two metal plates and two spheres (Fig. 13–3). Imagine it as a spark gap in parallel with a capacitor (Fig. 13–4) in which the plates have been opened up. Each pulse from the induction coil charged up the capacitor until the insulation between the spheres broke down (Fig. 13–5). The

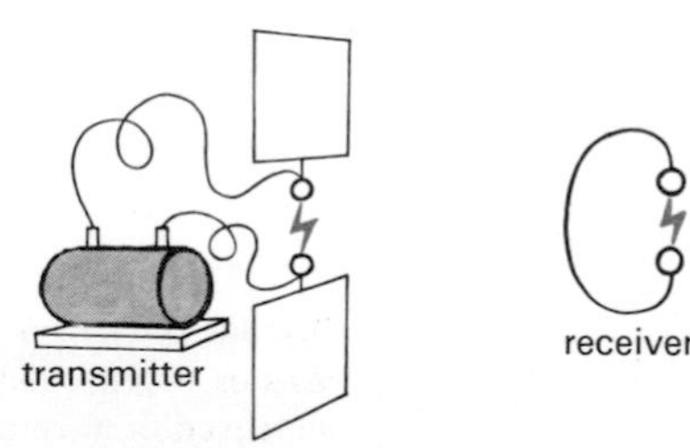

Fig. 13–3

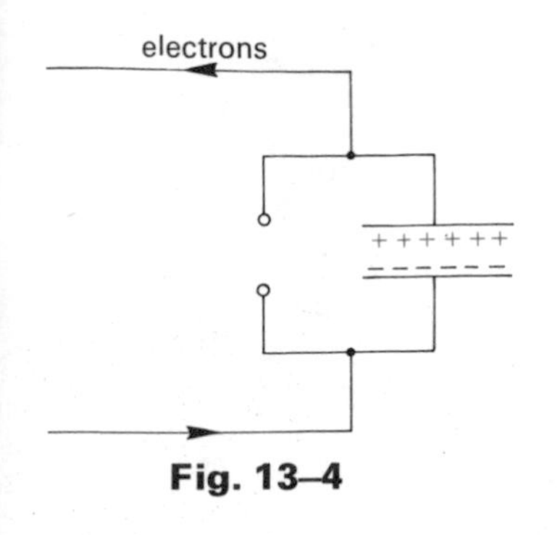

Fig. 13–4

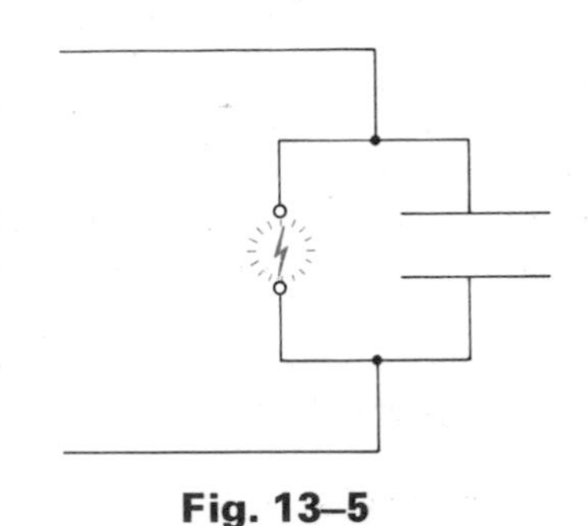

Fig. 13–5

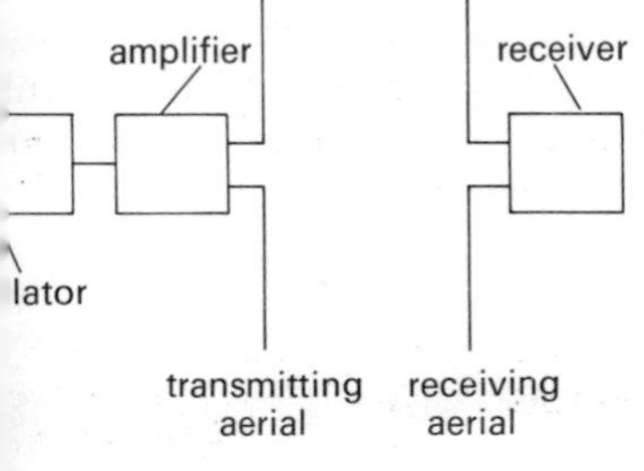

Fig. 13–6

capacitor was then discharged through the ionized air. In this way a damped oscillatory current of about 50 MHz was produced. If Maxwell's predictions were correct, this apparatus should act as a *transmitter* of electromagnetic waves. Hertz used a loop of wire with a small gap in it as his *receiver*. When he placed it several metres from the transmitter, he discovered that a small spark jumped across the gap in the ring whenever the induction coil was switched on.

Hertz showed that these 'rays of electric force' or *radio waves*, as he later named them, could be reflected by metal plates and refracted by a giant prism made from half a tonne of asphalt. He thus confirmed Maxwell's prediction of electromagnetic waves and showed that they behave like light waves.

Spark transmitters are now illegal, as they radiate energy over a wide band of frequencies and interfere with radio and TV reception, as do unsuppressed sparking plugs in car engines. Thermionic valves and transistors are now used in electronic oscillators, which are capable of maintaining a constant frequency. The two plates in Hertz's transmitter have now become the aerial and a similar aerial is used for the receiver (Fig. 13–6).

ELECTROMAGNETIC RADIATION

Electromagnetic waves can best be described by Maxwell's equations. These equations are, however, quite meaningless to anyone who does not have a good grasp of advanced mathematics.

Never mind! When Michael Faraday was con-

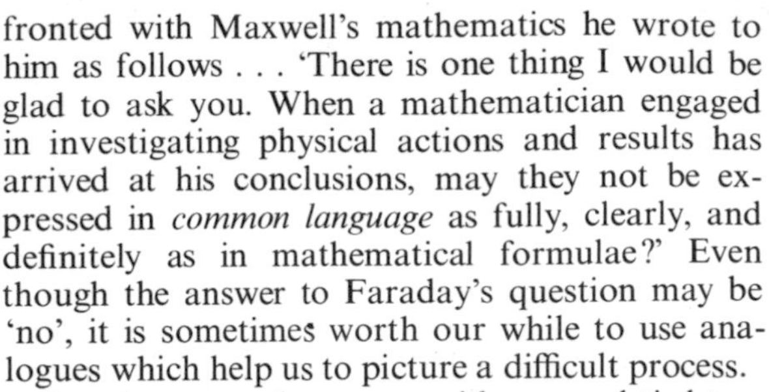

fronted with Maxwell's mathematics he wrote to him as follows . . . 'There is one thing I would be glad to ask you. When a mathematician engaged in investigating physical actions and results has arrived at his conclusions, may they not be expressed in *common language* as fully, clearly, and definitely as in mathematical formulae?' Even though the answer to Faraday's question may be 'no', it is sometimes worth our while to use analogues which help us to picture a difficult process.

Imagine a very long rope with one end tied to a tree. If the free end is waggled up and down waves travel out along the rope towards the tree. The waggler moves the rope vertically and the waves are propagated horizontally. Energy is transmitted along the rope as a wave motion. In a transmitting aerial electric charges are oscillating up and down. These in turn produce electromagnetic waves which carry energy from the aerial. The direction of propagation is perpendicular to the movement of the charges in the aerial.

A more elaborate analogue is illustrated in Fig. 13–7. A positively charged and a negatively charged cylinder oscillate up and down in simple harmonic motion. The negatively charged cylinder can pass through the other. In position (a) the cylinders are furthest apart and starting to move together. At (b) they are closer together and the lines of force still terminate on them. At (c) one cylinder is inside the other so that they produce no external electric field. The original field lines do not disappear instantly but form complete loops and move outwards. At (d) charges are moving apart

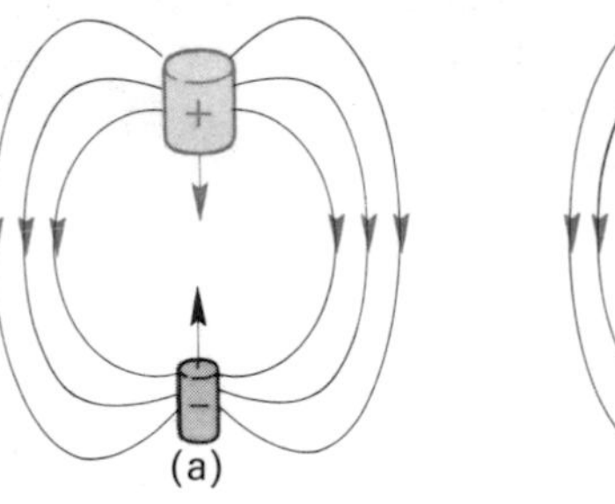

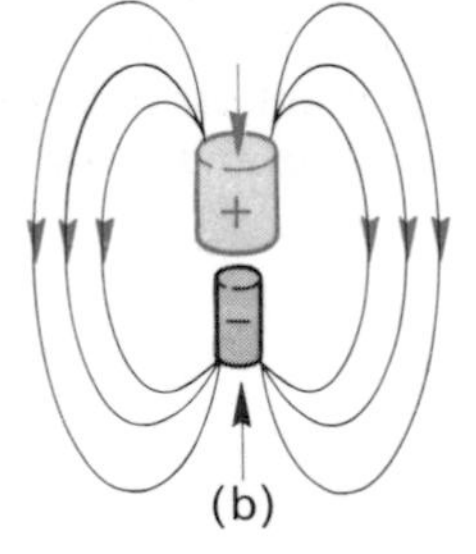

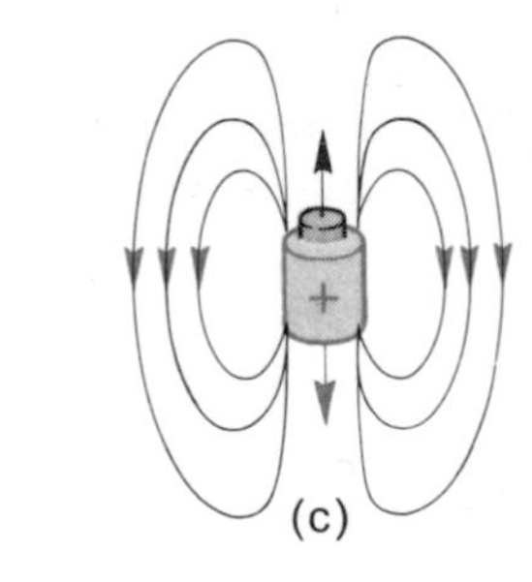

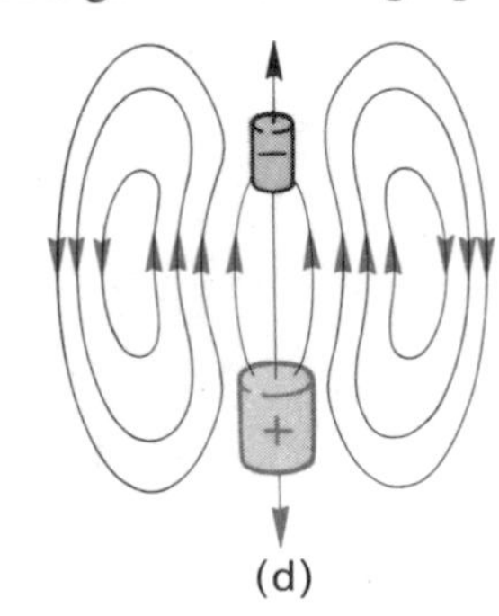

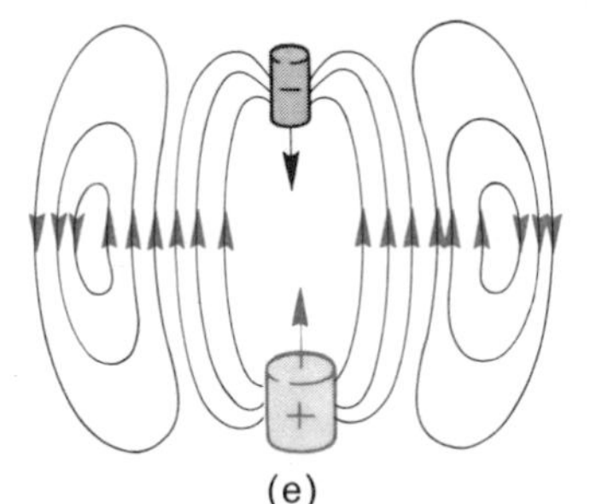

Fig. 13–7

Fig. 13–11

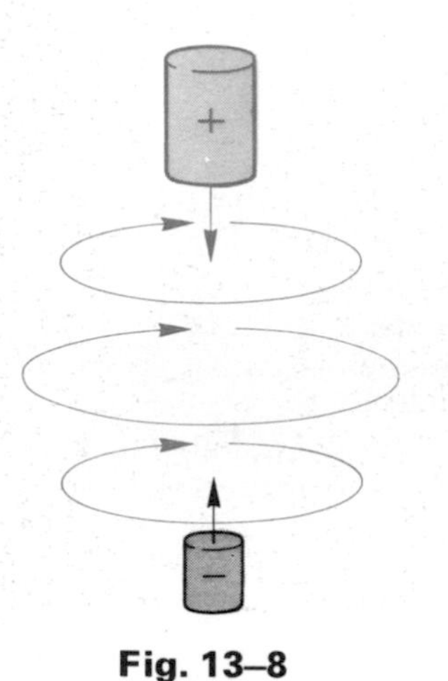
Fig. 13–8

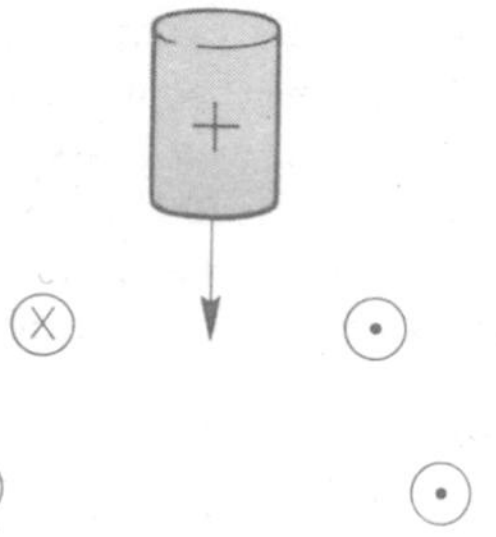

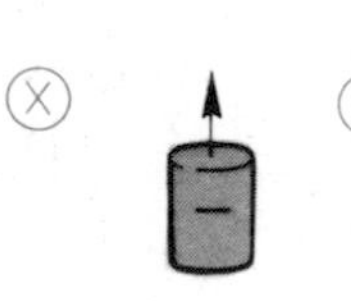
Fig. 13–9

and field lines of opposite sign emerge. The charges are again furthest apart at (e). As the charges continue to oscillate up and down an expanding pattern of closed loops is formed. These loops are radiated out into space at 3×10^8 m/s.

But this is only half the story. We have not yet mentioned the fact that moving charges produce magnetic fields round them. A positive charge moving downwards and a negative charge moving upwards both produce the same magnetic field around them (Fig. 13–8). Unlike the electric field, the magnetic field will be in the same direction when the charges are in positions (a)–(d) in Fig. 13–7, and will change direction only when the charges change direction (Fig. 13–7).

The magnetic field lines are perpendicular to the electric field lines and both are perpendicular to the direction in which the energy is being propagated. To simplify the combined diagram we will indicate magnetic field lines by red crosses and dots as shown in Fig. 13–9. Our model bears some resemblance to the state of affairs in a transmitting aerial in which charges run up and down. Fig. 13–10 shows the pattern produced by such a system. I should, of course, be a three-dimensional pictur and so you must imagine the diagram rotated abou an axis AB to form a laminated doughnut hel together by red rings and expanding at the spee of light!

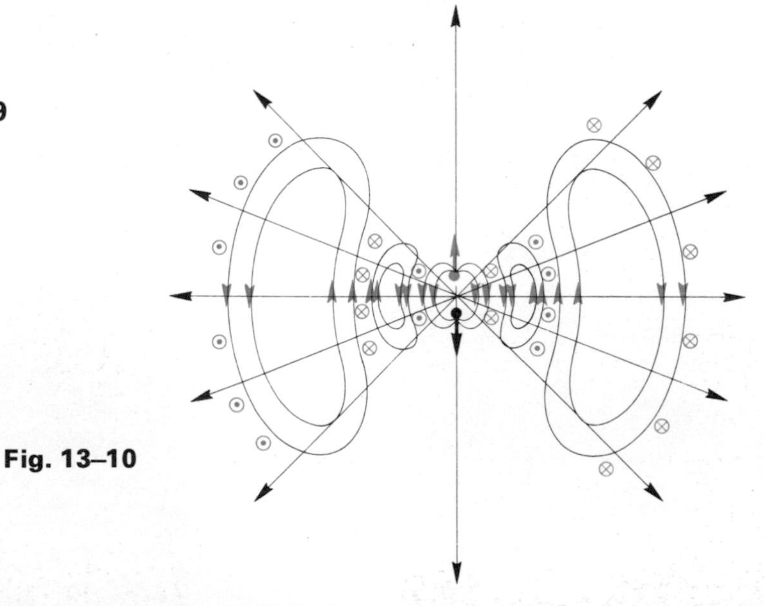
Fig. 13–10

Range of Electromagnetic Waves

In 1901 Guglielmo Marconi (Fig. 13–11) succeede in sending a wireless message across the Atlantic A ring of high aerial masts in Cornwall was use to transmit the signal and a kite carried the receiv ing aerial in Newfoundland. The first trans Atlantic message was 's' in Morse Code. Man scientists thought that the radio waves would shoo

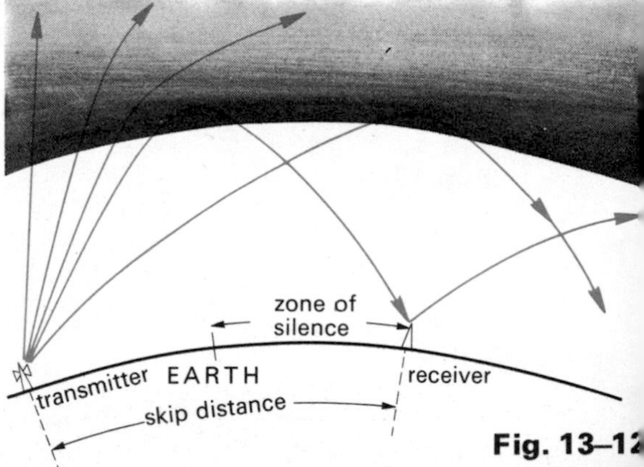

Fig. 13–12

straight out into space and could not possibly benc round the curved earth. In their last assumptio they were correct but they had not reckoned on a 'mirror in the sky'. A layer of ionized gas high above the Earth's surface reflects the radio waves so that they can be picked up thousands of miles from the transmitter (Fig. 13–12). Marconi's signal bounced backwards and forwards between this ionized layer and the surface of the Earth until it reached Newfoundland.

Fig. 13–13

Only waves striking the ionosphere above a certain angle of incidence will be reflected: others simply pass through the electrified layer. There is therefore a minimum distance between the transmitter and the first reflected signal. If this is further than the range of the ground wave (direct signal), there is a 'zone of silence' in which reception is impossible.

Low frequency radio transmissions are reflected most readily by the ionosphere and the new global navigation system *Omega* operates between 10 and 40 kHz. Even submerged submarines can pick up signals at such low frequencies. Very long aerials of the order of several kilometres are, however, needed for such transmissions.

At very high frequencies radio waves are not reflected at all by the ionosphere and other devices have to be employed for ultra high frequency transmissions over long distances. Early communication satellites simply reflected electromagnetic waves but modern satellites pick up and re-radiate the signals, which they aim at densely populated areas. Twelve colour TV channels and nine thousand telephone circuits are provided between America and Europe by the Intelsat IV satellite (Fig. 13–13).

Electromagnetic waves transmitted from a distant galaxy thousands of millions of years ago carry enough energy to induce currents in the aerial of a giant radio telescope today (Fig. 13–14). As long wave radio signals from space are absorbed by the atmosphere, satellites such as RAE-1 (Fig. 13–15) are being used as space-based radio telescopes capable of detecting these waves which never reach the Earth's surface.

Fig. 13–15

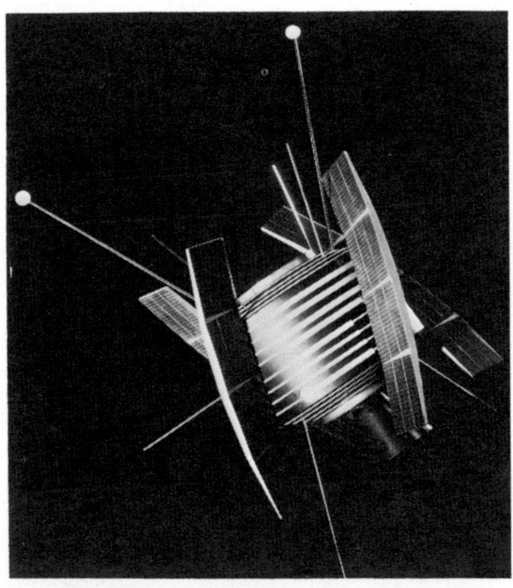

TRANSMITTERS

We have seen that the first radio transmitter was made by Hertz. It was a spark transmitter and radiated energy over a band of frequencies. Spark transmitters were superseded by triode oscillators, in which the frequency could be much more easily controlled by varying the values of capacitance and inductance in the circuit.

Fig. 13–16 shows a triode valve used as such a transmitter. The inductor is a single loop of wire and the capacitor is the capacitance between the wires of the circuit. The transmitter operates at very high frequencies (VHF) of the order of 140 MHz. *If the speed of electromagnetic waves is* $3 \times 10^8 \, m s^{-1}$, *what is the wave length of this radiation ? (13.1)*

Fig. 13–14

Fig. 13–16

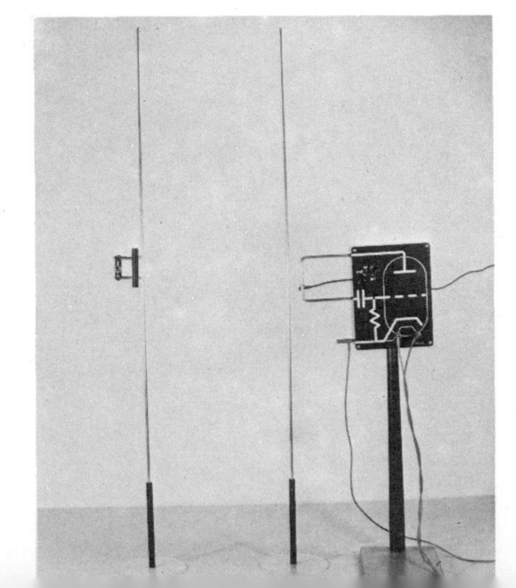

Fig. 13–17

A half wave dipole is coupled to the oscillator to act as the radiating aerial and a second dipole fitted with a bulb at the centre picks up enough energy to light the bulb some distance from the transmitter. Fig. 13–17 shows a small model helicopter driven by the energy carried by electromagnetic waves. The wires merely keep the helicopter centred over the radio beam.

RADAR

The name was coined from *ra*dio *d*etection *a*nd *r*anging. In radar an electromagnetic wave pulse is reflected from a distant object. The time interval between the transmission of the pulse and the reception of the reflected pulse is measured electronically.

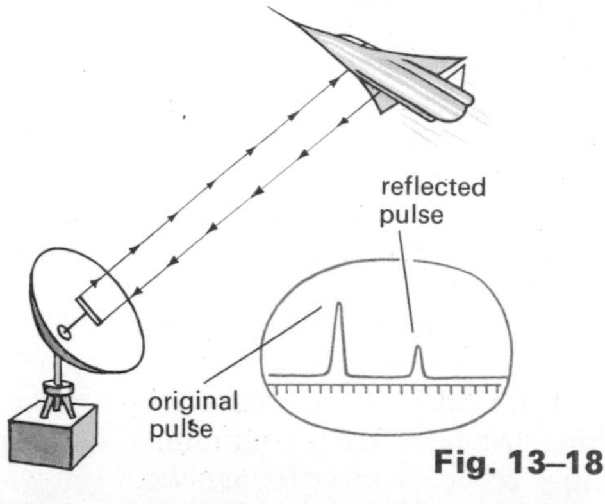

Fig. 13–18

The same aerial system is normally used transmit the pulse and to receive the echo. Betwe pulses the aerial system is switched from the tra mitter to the receiver by an ingenious system cal a duplexer. Both signals are fed to a C.R.O. whi may be calibrated directly in miles or kilomet (Fig. 13–18).

A valley in Puerto Rico has been made int giant 'dish' over three hundred metres in diame (Fig. 13–19). It forms the reflector of a radar te scope used by Cornell University to study planets.

Fig. 13–19

COMMUNICATION

As soon as it was realized that energy could carried by radio waves, attempts were made to them to transmit information. By the use of lo and short bursts of radiation, messages were se by Morse Code. As radio frequencies are all mu higher than the highest audio frequency, we cann use a radio signal to operate a loudspeaker or e phone. How then can we hear such a signal?

Modulation

One method of carrying information on rad waves is to alter the transmitted signal (carr waves) in some way which depends on the inform tion being carried (Fig. 13–20). For example, if

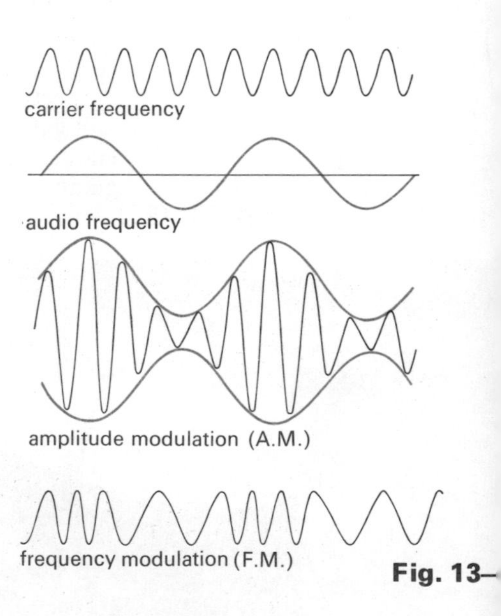

Fig. 13–

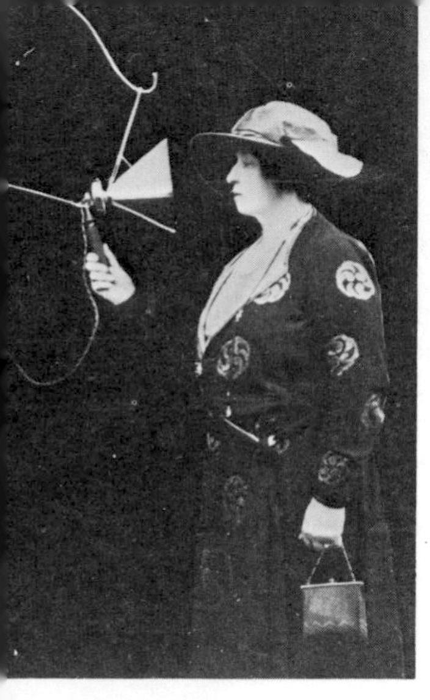

Fig. 13–21

want to send audio frequencies (A.F.), the *amplitude* of the carrier wave can be altered at the audio frequency. This is called *amplitude modulation* (A.M.). This method was used in 1920 when Nellie Melba made an experimental broadcast from the Marconi Company's transmitter at Chelmsford (Fig. 13–21). Amplitude modulation is still used by many radio stations today.

In another type of modulation the *frequency* of the carrier is varied at the audio frequency. This is called *frequency modulation* (F.M.) and is used in the BBC VHF transmitters. We will, however, consider only amplitude modulation in this course.

In a triode oscillator the amplitude of the oscillations depends on the H.T. voltage. If this is varied at an audio frequency, the amplitude of the oscillations varies at the audio frequency. This is called

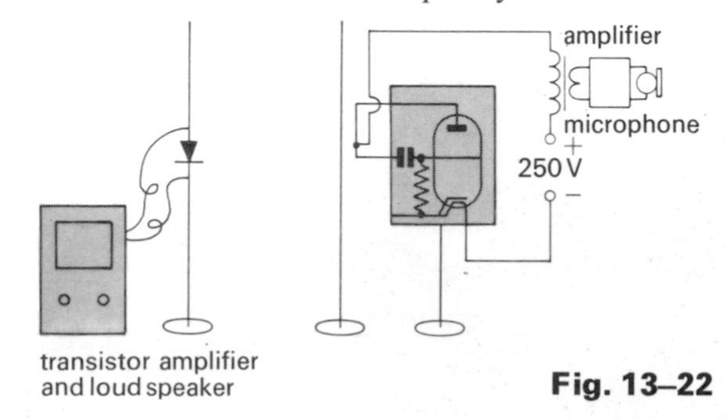

Fig. 13–22

anode modulation. Fig. 13–22 shows how this can be done, using the transmitter illustrated in Fig. 13–16.

The oscillator of the transmitter is not normally modulated directly, as this affects its frequency to some extent. The oscillator output is fed to a power amplifier, the output of which is controlled by an audio signal to produce amplitude modulated waves. (Fig. 13–23.)

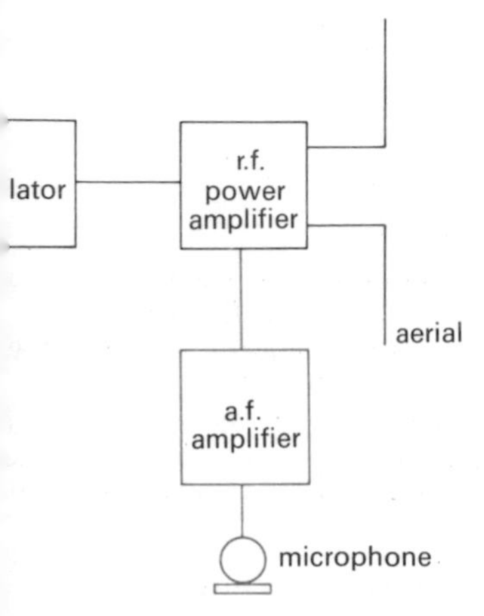

Fig. 13–23

Demodulation

If an amplitude modulated carrier wave such as is shown in Fig. 13–24 were picked up by a receiver, amplified, and passed through a loudspeaker, nothing would be heard. The average value of the current would be zero and it would not vary at the audio frequency. To obtain the audio frequency component, the signal has to be demodulated or detected. This is most easily accomplished by passing the signal through a rectifier, so that only pulses in one direction remain (Fig. 13–25). The average value of the current is now changing at the audio frequency and this will be heard in the loudspeaker.

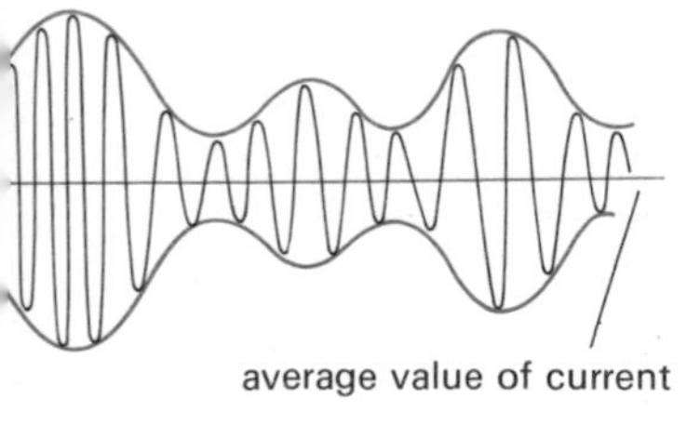

Fig. 13–24

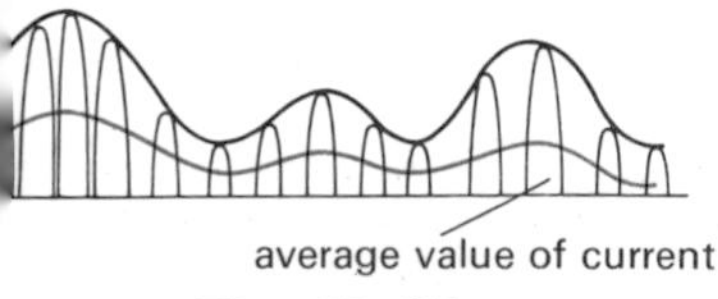

Fig. 13–25

A SIMPLE RECEIVER

In a radio receiver one particular radio frequency must normally be selected from a number of different frequencies present in the aerial. To do this the aerial and earth are connected to a tuned

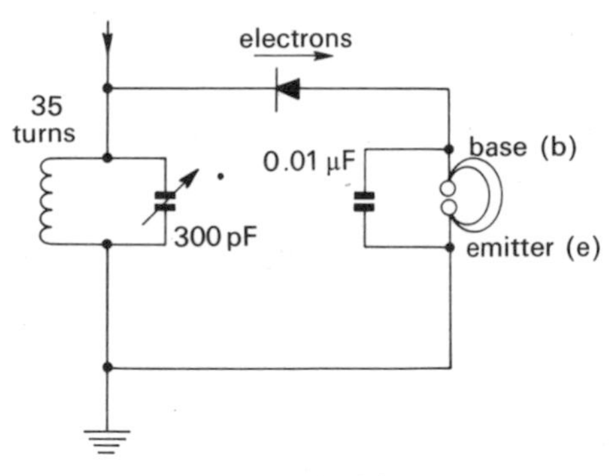

Fig. 13–26

circuit (Fig. 13–26). The variable capacitor is then adjusted until the resonant frequency of the circuit is the same as the frequency of the signal required. As the tuned circuit will then have its maximum impedance at the required frequency, a large voltage at that frequency will then be produced across it. Other voltages at unwanted frequencies will normally be much smaller.

The receiver shown in Fig. 13–26 should enable you to receive programmes from local BBC transmitters. The capacitor across the high impedance earphone acts as a smoothing capacitor as far as the radio frequency component is concerned. It must not, however, be so large that the audio frequency component is also smoothed out! The coil must be wound on a ferrite rod, and a good aerial and earth used. The addition of a transistor amplifier (Fig. 13–27) will greatly improve the performance.

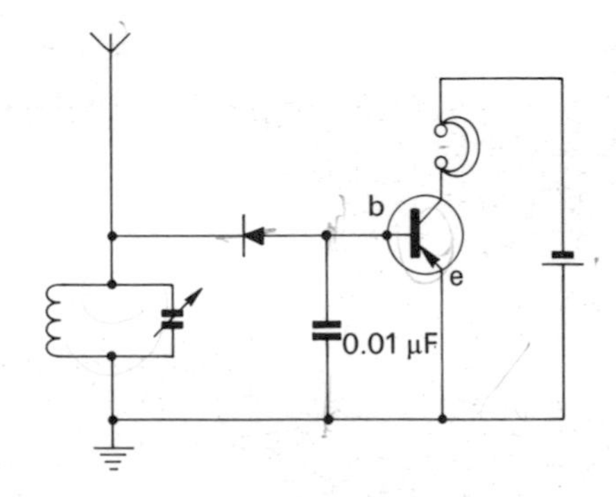

Fig. 13–27

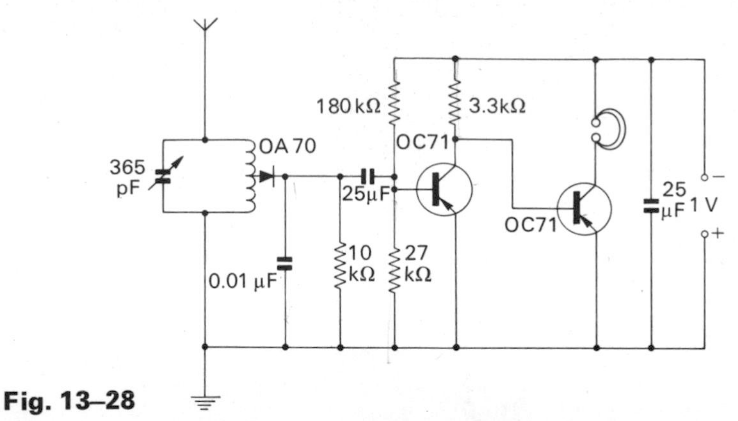

Fig. 13–28

A much more effective radio receiver can be built using the circuit shown in Fig. 13–28. When completed it could look like Fig. 13–29.

Fig. 13–29

One of the most recent developments in two-way communication is the portable radio telephone. Video telephone systems are also being developed (Fig. 13–30).

Fig. 13–30

SUMMARY

Radio waves are radiated when charges a *accelerated* in a transmitter aerial. The electr magnetic radiation has electric and magne components which are perpendicular to ea other and to the direction of propagation.

The range of frequencies runs from abo 10^4 Hz (very long waves) to 10^{11} Hz (mic waves). They all travel in free space at 3×10^8 m

Audio frequency information can be tran mitted by varying the amplitude (A.M.) or t frequency (F.M.) of a radio wave at the aud frequency. A detector in the receiver is th needed to retrieve this information.

PROBLEMS

13.2 What frequency of radio wave has the same wa length as a sound wave of frequency 3300 Hz?

Mention *one* difference between these two types of w other than their respective speeds and frequencies.

Over moderate distances (e.g. in Great Britain) a lo wave radio transmitter has a greater service area tha medium wave transmitter of the same power. Why is th

Speed of sound $= 330\,\text{m s}^{-1}$
Speed of radio waves $= 3\times10^8\,\text{m s}^{-1}$

[S.C.E. (H) 19

13.3 A simple radio receiver may be constructed as sho in Fig. 13–31.

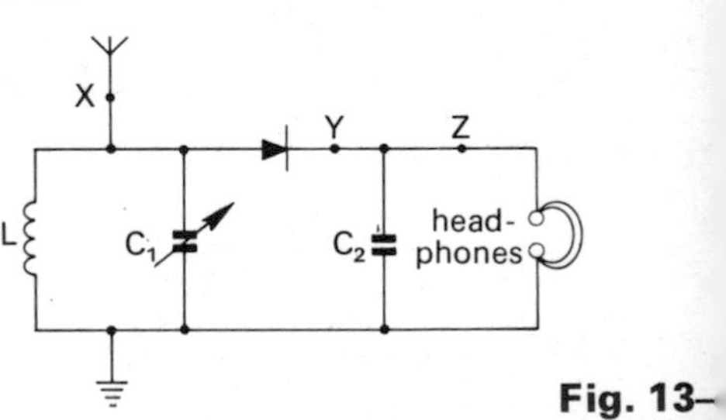

Fig. 13–

(a) Sketch three graphs of current against time to illust the nature of the current at the points X, Y and Z in circuit.
(b) What is the purpose of the network formed by components L and C_1 connected to aerial and earth?
(c) Why is the capacitor C_1 made variable?

[S.C.E. (H) 19

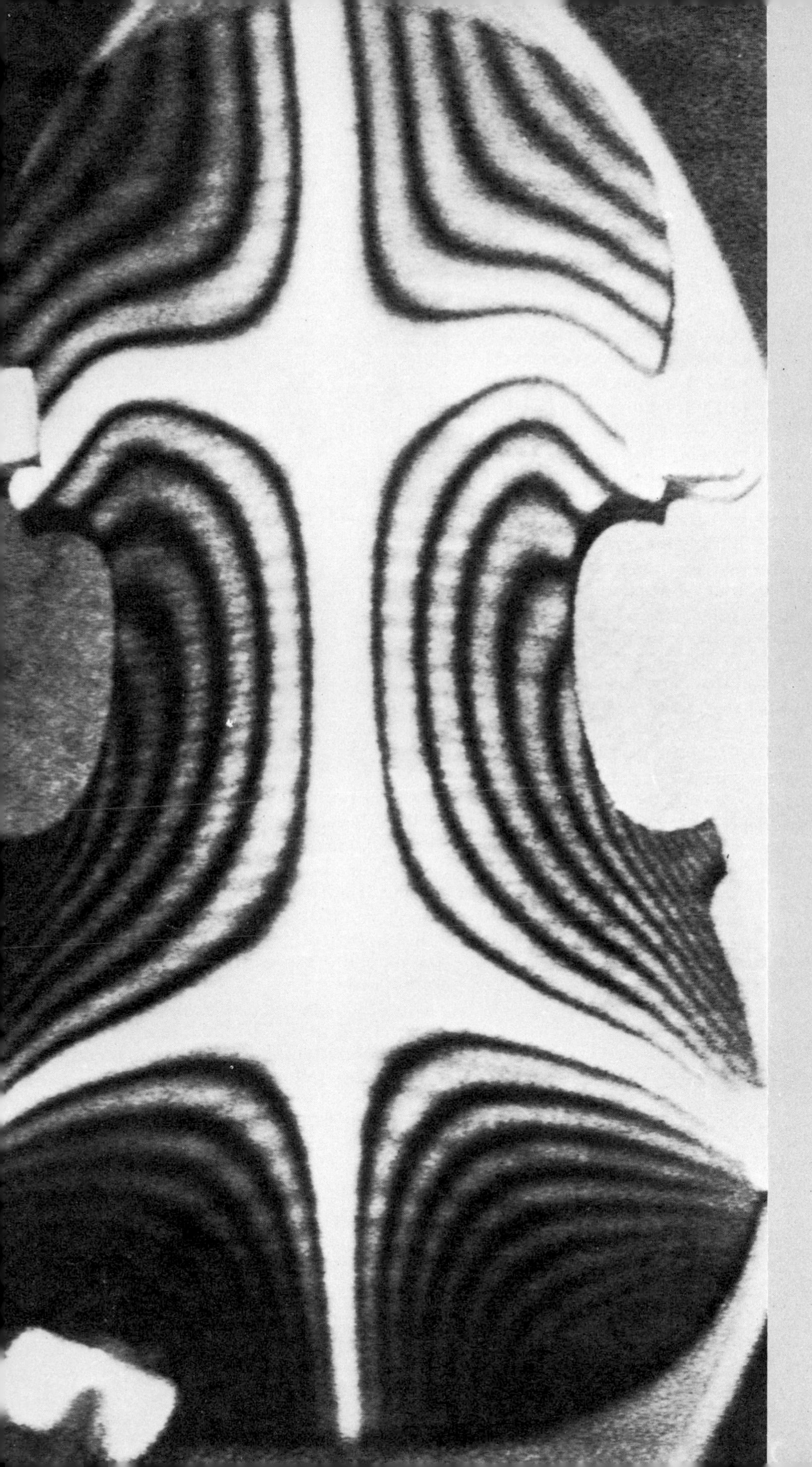

UNIT THREE

Energy in Motion

The Wave-Particle Model

14 Let there be Light

LIGHT AND SIGHT

Fig. 14–1

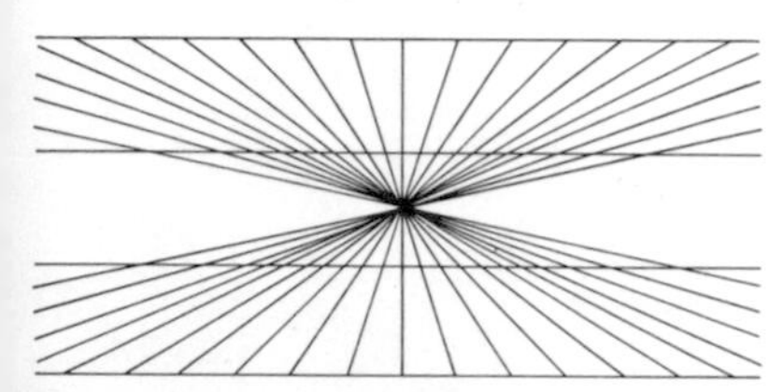

Fig. 14–2

Fig. 14–3

Philosophers, artists, theologians, and scientists have been captivated by the mystery of light since the beginning of recorded history. That light has been thought to carry with it an authority—almost a revelation of truth—is reflected in the analogous use made of the term in our language. We say that a man is enlightened or 'has seen the light' when he agrees with our own views! For most of us the information which light brings is more convincing than any other. We appreciate and often share the sentiments of Doubting Thomas—'Unless I see . . . I will not believe'. But before proclaiming the infallibility of vision we ought to remind ourselves that our eyes can play tricks. For example our home movies never move. Yet a series of stills on the screen produces a convincing enough illusion of motion.

Two popular examples of geometrical optical illusions are shown here. Compare the width and the height of the duck in Fig. 14–1 and see if the horizontal lines in Fig. 14–2 are parallel. Although these can be easily verified by measurement, the illusions remain.

On a warm day we often see what appear to be wet patches on the road. The image produced on the retina of the eye or on the film of a camera (Fig. 14–3) is in this case very similar to that produced by reflections on water (Fig. 14–4). We call this effect a *mirage*. As we are more familiar with reflections on water, we *interpret* the mirage in this way.

Fig. 14–4

Fig. 14–5

How do you interpret the photograph of a[n] etched diamond's surface in Fig. 14–5? Now tur[n] the picture upside down. If you see the structur[e] altering from *blocks standing out* to *hollows in a wall*, it is probably because you are used to seein[g] things in light which comes from above them an[d] you therefore expect the shadows to be in thei[r] normal positions.

Our past experience, then, influences the way w[e] interpret our retinal images. We often see what w[e] believe or unconsciously assume. It would be wise therefore, to test our visual observations and i[n] particular to rely on measurement rather tha[n] subjective judgement whenever possible.

EARLY THEORIES OF LIGHT

By the fifth century B.C. various Greek philosopher[s] had turned their attention to the study of light There were at least two rival theories, each pur porting to explain how we see things. Empedocles said that the eye radiates a kind of beam which when it comes in contact with an object, signals back information. This theory seems to regard sight as an optical form of radar. The other theory suggested by Pythagoras claimed that objects could be seen because they shot out tiny particles which entered the eye. *What evidence would you use to support or refute these theories? (14.1)*

Strange as it may seem, Empedocles' theory persisted in one form or another for over 1000 years. Yet it was strongly opposed by many Greeks, including Aristotle, who laid great stress on the medium through which light travels. He believed that all space was filled with 'something' capable of transmitting the radiation which emanated from the source of light. Two thousand years later, when light was regarded as a wave motion, it was thought necessary to find something to wave and the ether was invented. Both 'fillings' are now thought to be imaginary and there is no evidence of any detectable physical property in empty space . . . unless, of course, you call the transmission of radiation a physical property.

Fig. 14–6

You may not have cooked a meal by using a concave reflector to focus the sun's rays (Fig. 14–6) but you may have set light to a piece of paper by using a shaving mirror. Tradition has it that Archimedes used huge curved mirrors to set fire to the Roman fleet near Syracuse in the second century B.C. The story at least indicates that the principles of reflection ($\angle i = \angle r$) illustrated in Fig. 14–7 were well understood 2000 years ago.

The Greeks also studied the change in direction of a ray of light when it enters water, and in the second century A.D. Ptolemy of Alexandria made some extremely accurate measurements of the angles of refraction in water. It was not, however, until a thousand years later that the next major development in optics—the lens—was recorded. There is little doubt that the lens has been one of the most important of all scientific discoveries. It made possible the development of the microscope, the telescope, the spectroscope, and the camera, each of which revolutionized practically every branch of science in its turn. Yet oddly enough it is impossible to say when the first lenses were made and used. Perhaps the emerald which Nero used to watch gladiatorial combats was ground into a lens. There is evidence which suggests that Ptolemy in the second century and the Arab scientist Alhazen in the eleventh century were familiar with convex lenses. Yet it was to a monk in Oxford, Roger Bacon, that credit must be given for developing practical uses of such lenses. Perhaps we should say blame rather than credit, as his efforts to enable long-sighted elderly monks to continue copying manuscripts landed him in prison for ten years. Surely only a wizard inspired by the devil could cure blurred vision or devise a 'magic' lantern (Fig. 14–8)! But spectacles were here to stay, as the portrait of Cardinal Ugone (Fig. 14–9), painted in 1352, shows.

Fig. 14–7

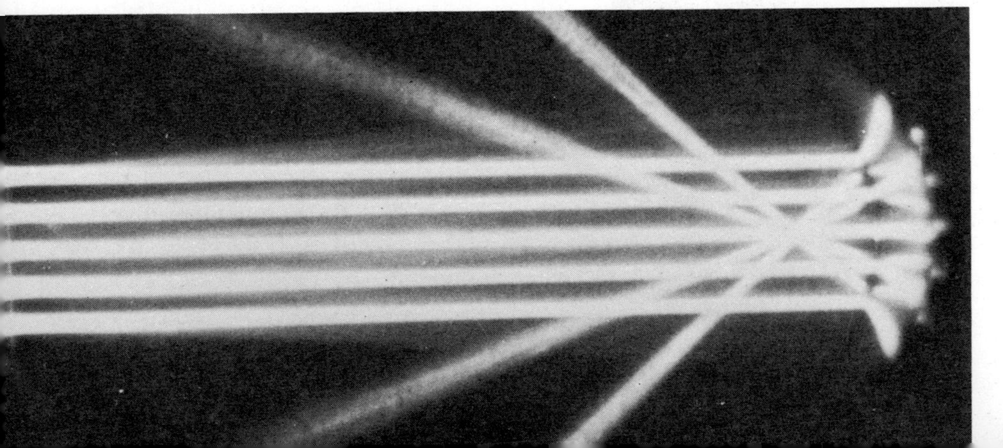

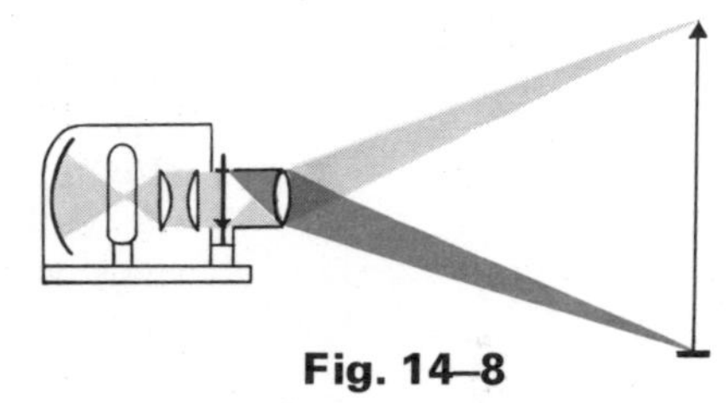

Fig. 14–8

Fig. 14–9

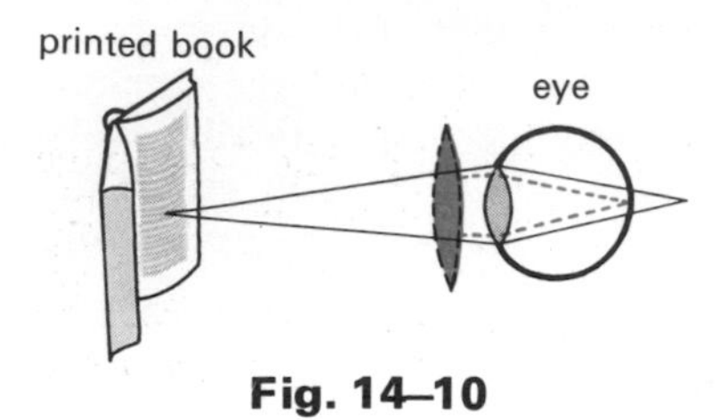

Fig. 14–10

As we grow older most of us find it more and more difficult to focus on objects close to our eyes and eventually we cannot read small type in a book. This defect is called presbyopia (Fig. 14–10). A convex lens placed in front of the eye causes the rays to converge and form an image on the retina.

Sooner or later someone was bound to look through *two* convex lenses and thus invent the telescope. Perhaps Bacon had done this, for he wrote, 'We can give figures to transparent bodies and dispose them . . . that the rays be refracted . . . so that we shall see the object near at hand . . . thus from an incredible distance we may read the smallest letters.' This may, of course, refer to his magic lantern. In any case it is usually to the Dutch spectacle maker Hans Lippershey that the discovery of the telescope in 1608 is attributed. It was from this telescope that Galileo developed his version a year later and thus inaugurated modern astronomy. To obtain a high magnifying power the focal length of the objective should be as great as possible. Fig. 14–11 shows an early seventeenth-century attempt to support such an objective lens in what was called an 'aerial telescope'. The intermediate rings do not hold lenses.

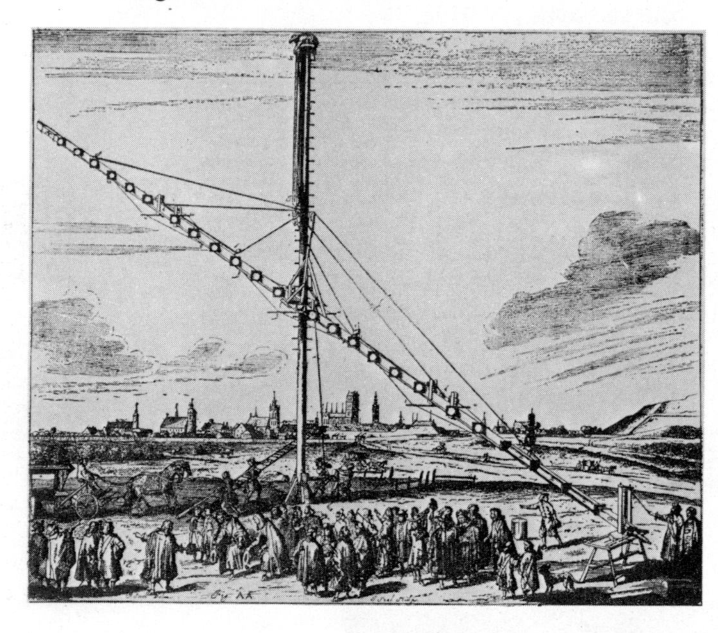

Fig. 14–11

Although refracting telescopes have now been replaced by large reflecting telescopes for most astronomical work, a few refracting telescopes remain. The diameter of the objective lens of the telescope at the Yerkes Observatory in America is 1 metre (Fig. 14–12).

Fig. 14–12

It is interesting to note that the design a
manufacture of lenses used in the early spectacle
telescopes and microscopes were based entirely c
rule-of-thumb methods. For centuries scientis
had looked in vain for a law relating the angle
incidence to the angle of refraction. Even the gre
German astronomer Johannes Kepler, whose le
systems are still used in the eyepieces of telescop
and microscopes, was unable to find a simp
mathematical relationship between these angles.

The breakthrough came in 1621, when anoth
Dutchman, Willebrord Snell, discovered the la
which now bears his name. Fig. 14–13 shows thr
rays of light entering a block of glass. The ang
of incidence is represented by i and the angle
refraction by r. Snell showed that when a ray

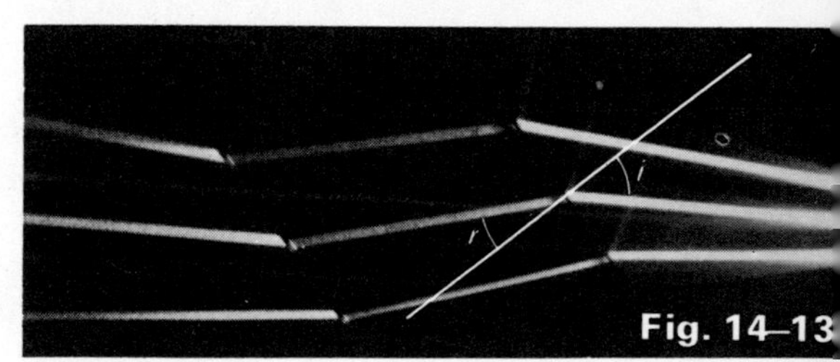

Fig. 14–13

light passes from one medium into another the *si*
of the angle of incidence was proportional to t
sine of the angle of refraction. In other words f
two given media (air and glass in this case),

$$\frac{\sin i}{\sin r} = \text{a constant}$$

The value of this ratio depends on the two med
and on the colour (frequency) of the light. I
Chapter 17 we will be studying the effects of colo
on refraction.

REFRACTIVE INDEX

If we want to use the ratio $\sin i/\sin r$ as a measure of the ability of one medium, e.g. glass, to change the direction of a ray, we will have to state (a) the other medium at the interface and (b) the colour of the light ray. A vacuum is taken as the standard medium, i.e. light is assumed to pass *from* a vacuum *into* the other medium (glass), and yellow light (sodium D line) is taken as the standard colour. Under these conditions, the ratio $\sin i/\sin r$ is a measure of the absolute refractive index of glass n_g. In practice, certainly in school practice, you are likely to meet three modifications to the scheme outlined above.

1. A narrow beam of white light is often used instead of a monochromatic source of yellow light.
2. The absolute refractive index of the substance, e.g. glass, is usually referred to simply as the 'refractive index of glass'.
3. If the ray were passed from a vacuum into a vacuum, clearly there would be no deviation, so that the absolute refractive index of a vacuum is, by definition, *one*. As the absolute refractive index of air is less than 1.0003, it is usual to measure, say, the refractive index of glass at an air/glass boundary.

Although the refractive index of air is very small, it can have quite startling effects under certain conditions. The refractive index varies with density and on a warm day the air in contact with the ground is so hot that there is a rapid change in refractive index immediately above the ground. Some light rays coming from an object A (Fig. 14–14) travel directly to an observer at B, while others are refracted as shown. The observer then sees the object as if it were reflected in a pool of water. From a low level (Fig. 14–3) an observer sees the mirage effect illustrated in Fig. 14–4 (page 106). Of course the density and therefore the refractive index of air normally decrease with height. *What effect would you expect this to have on the light entering the Earth's atmosphere from a star? (14.2)*

Waves and Speed

We have seen that, since the time of the Greek philosophers, light has been thought to consist of tiny particles emitted from the source. Earlier in this course we discussed another theory, which suggested that light is a kind of wave motion. We will return to a discussion of these theories later. For the moment let us look at refraction from the point of view of the wave theory. For simplicity we have on Fig. 14–15 grossly exaggerated the wavelength of yellow light, which is about 590 nm.

Fig. 14–15

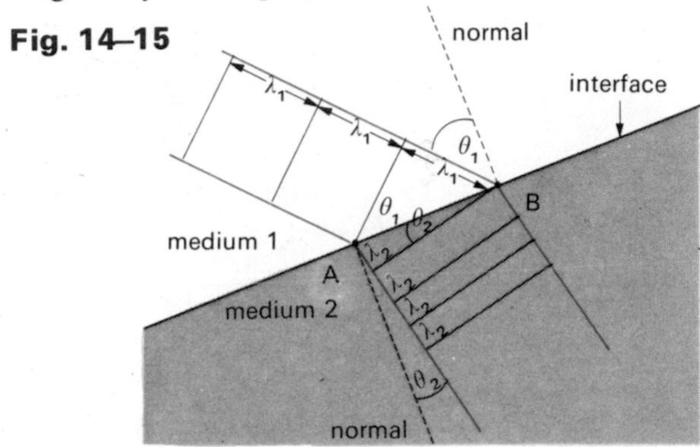

Suppose that a beam of light of wavelength λ_1 is travelling through medium 1 and strikes the plane boundary of another medium at an angle of incidence θ_1. It is bent so that it enters medium 2

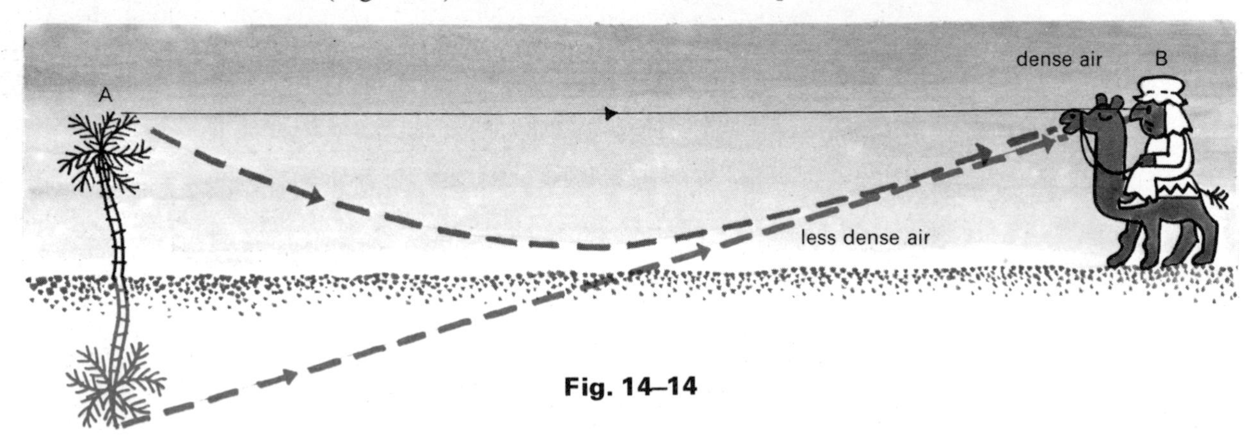

Fig. 14–14

at an angle of refraction θ_2, the new shorter wavelength being λ_2. According to the wave theory the wave front will be perpendicular to the direction of propagation, as were the waves in a ripple tank. From the geometry of the figure you can see that

$$\sin\theta_1 = \frac{\lambda_1}{AB} \quad \text{and} \quad \sin\theta_2 = \frac{\lambda_2}{AB}$$

$$\Rightarrow \frac{\sin\theta_1}{\sin\theta_2} = \frac{\lambda_1}{\lambda_2} \qquad (1)$$

But the frequency (colour) of the light is not altered by refraction and, as the speed of a wave is given by $v = f\lambda$, we have

$$v_1 = f\lambda_1$$

$$v_2 = f\lambda_2$$

$$\Rightarrow \frac{v_1}{v_2} = \frac{\lambda_1}{\lambda_2} \text{ where } f \text{ is constant} \qquad (2)$$

From equations (1) and (2) we see that

$$\frac{\sin\theta_1}{\sin\theta_2} = \frac{\lambda_1}{\lambda_2} = \frac{v_1}{v_2} \qquad (3)$$

If now we consider medium 1 to be a vacuum and medium 2 to be, say, water, equation 3 becomes absolute refractive index of water $n_w = \sin\theta_1/\sin\theta_2$

$$n_w = \frac{\text{speed of light in vacuum } (v_0)}{\text{speed of light in water } (v_w)}$$

$$\Rightarrow n_w = \frac{v_0}{v_w} \qquad (4)$$

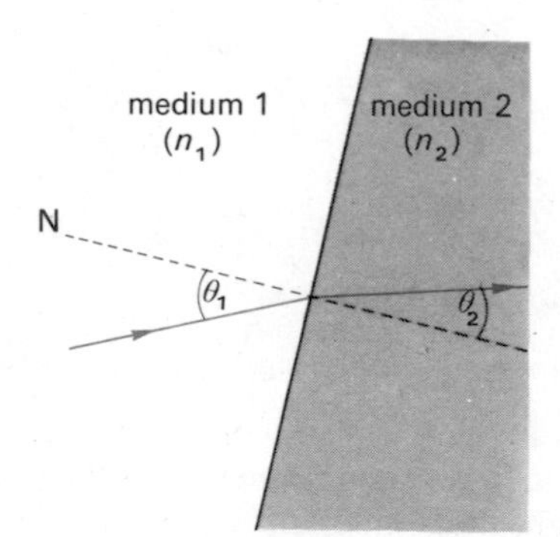

Fig. 14–16

Experimental measurements of the speed of light in a vacuum and in water give this ratio as 1.33. This is also the value of the absolute refractive index of water calculated from $\sin\theta_1/\sin\theta_2$. Results of this kind when first announced were hailed as a great triumph for the wave theory of light, since the particle theory had predicted that light would speed up when it entered a denser medium. The absolute refractive index of a medium n_m is in fact now *defined* by the speed ratio

$$n_m = \frac{\text{speed of light in a vacuum}}{\text{speed of light in the medium}}$$

General Case

Finally, we can consider the general case, in which light travels through any transparent medium 1 of absolute refractive index n_1 and enters another medium 2 of absolute refractive index n_2 (Fig. 14–16). Again taking the speed of light in a vacuu as v_0, we can rearrange equation (3) and multip both sides by v_0 to give

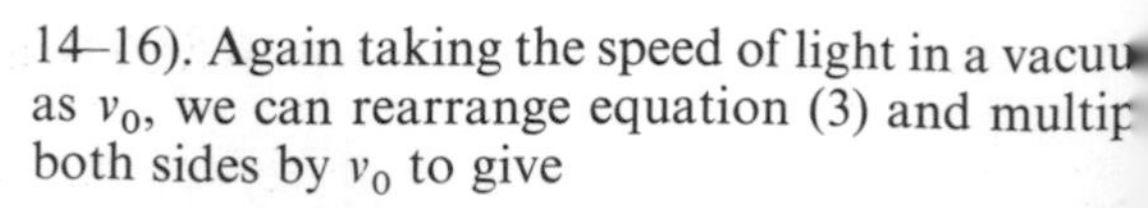

$$\left(\frac{v_0}{v_1}\right)\sin\theta_1 = \left(\frac{v_0}{v_2}\right)\sin\theta_2$$

$$\Rightarrow n_1 \sin\theta_1 = n_2 \sin\theta_2$$

This equation can be used when a ray of light passing from, say, water to glass or from one ty of glass to another. You can see that, in the spec case where medium 1 is a vacuum or air, n_1 = and the equation simplifies to

$$\frac{\sin\theta_1}{\sin\theta_2} = \frac{\sin i}{\sin r} = n_2 \text{ as before.}$$

UNDERWATER VISION

If a ray box at the bottom of a tank of wa projects a ray of light on to the water/air interfa at an angle θ_w (Fig. 14–17), a small amount of lig

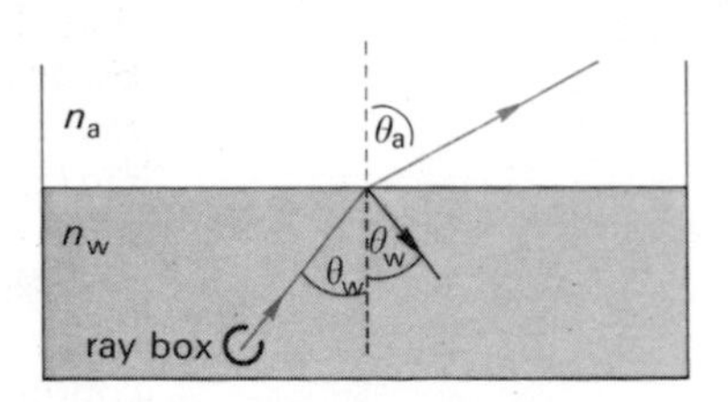

Fig. 14–17

will be reflected at an equal angle, θ_w'. Most of t light will pass through the interface and be fracted at an angle θ_a. If θ_w is increased, θ_a w increase and the amount of light which is reflect internally will also increase. A point will be reach when $\theta_a = 90°$; that is, the emergent ray will along the water surface. If θ_w is increased s further, *all* the light will be reflected internal. This is called *total internal reflection.*

Fig. 14–18

A fish-eye lens produces the kind of picture shown in Fig. 14–18. You have probably seen a similar view of the world from under a water surface. A complete shore to shore picture is then observed within the lightly shaded cone (Fig. 14–19).

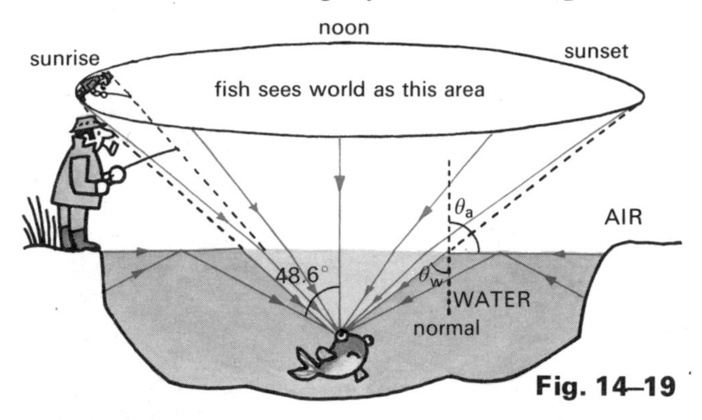

Fig. 14–19

Beyond that cone you see only reflections from under the water surface. Other partial reflections have been omitted for clarity. The changeover to total internal reflection occurs when the incident ray is parallel to the water surface, i.e. $\theta_a = 90°$. The corresponding angle of refraction θ_w is called the critical angle c. In Fig. 14–19 $\theta_w = c$. From equation (5)

$$n_a \sin\theta_a = n_w \sin\theta_w$$

$$n_a \sin 90° = n_w \sin c$$

but $$n_a = 1$$

and $$\sin 90° = 1$$

$$\Rightarrow n_w \sin c = 1$$

$$\Leftrightarrow \quad n_w = \frac{1}{\sin c}$$

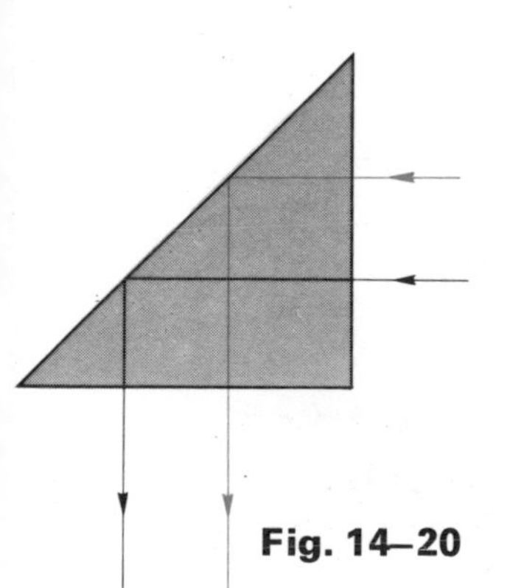

Fig. 14–20

By measuring the critical angle in a medium we can calculate its absolute refractive index. In the case of water the critical angle is 49°, which corresponds to a refractive index of 1.33. At a glass/air interface the critical angle is less than 45°, and total internal refraction can be produced by a 45° prism such as the one illustrated in Fig. 14–20. Prisms can thus be used instead of mirrors in periscopes, binoculars, compasses, and other instruments.

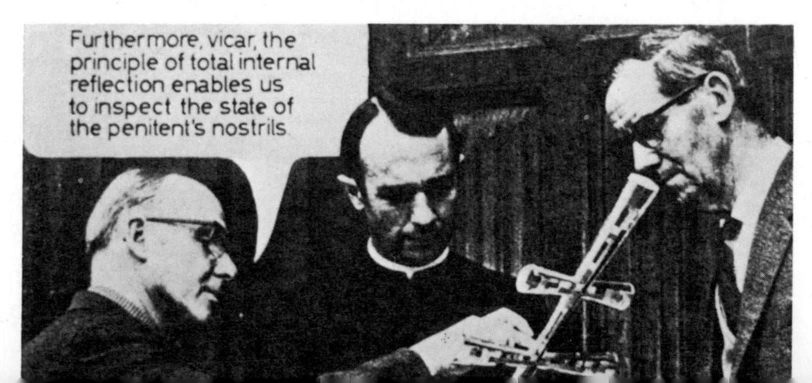

Furthermore, vicar, the principle of total internal reflection enables us to inspect the state of the penitent's nostrils

Fig. 14–21

SUMMARY

Reflection

Angle of incidence (i) = angle of reflection (r)
The incident ray, reflected ray and normal lie in the same plane.

Refraction

$$\frac{\sin\theta_1}{\sin\theta_2} = \frac{\lambda_1}{\lambda_2} = \frac{v_1}{v_2} = \frac{n_2}{n_1}$$

For a ray entering a medium (m) from air

$$\frac{\sin i}{\sin r} = n_m$$

The incident ray, refracted ray and normal lie in the same plane.

Critical angle

$$n_1 \sin\theta_1 = n_2 \sin\theta_2$$

or $$n_a \sin\theta_a = n_w \sin\theta_w$$

and when $\theta_a = 90°$ and $n_a = 1$

$$\sin\theta_w = \sin c = \frac{1}{n_w}$$

where c is the critical angle.

PROBLEMS

14.3 Two parallel solid blocks of glass of different refractive indices are placed together as shown in Fig. 14–22.

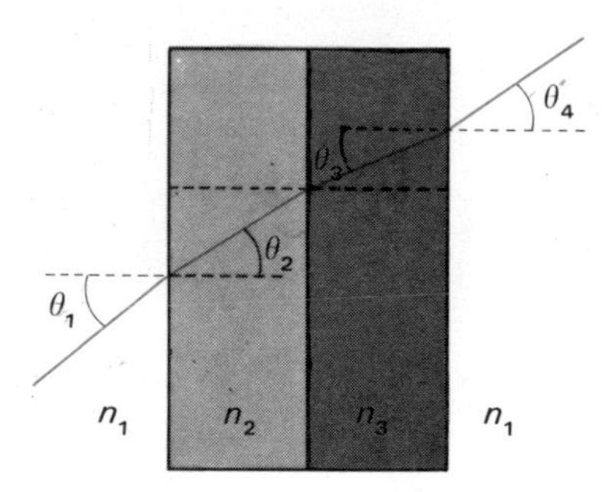

Fig. 14–22

A ray of light is incident on to one block at an angle θ_1. If the ray passes through the blocks as shown,
(a) is n_2 greater than, less than or equal to n_3;
(b) what can you say about the size of the angle θ_4;
(c) how will θ_4 be affected if the first block is replaced by another of the same size and absolute refractive index equal to n_3?

14.4 Explain, using Fig. 14–23, why water appears to be shallower than it is.

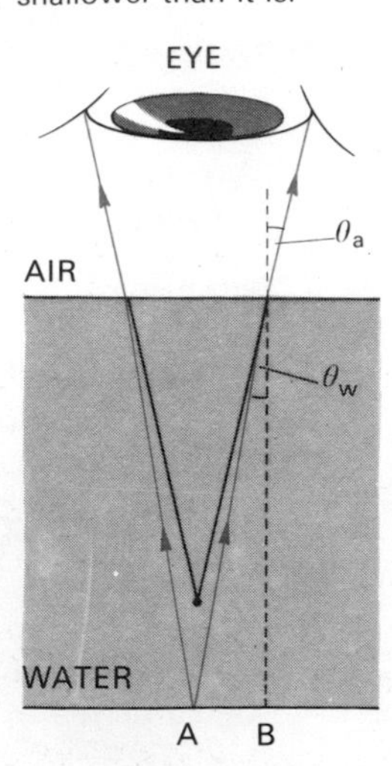

Fig. 14–23

If $\theta_w = 6°$ and $\theta_a = 8°$, find the absolute refractive index of water, the speed of light in water and the critical angle for water/air interface.

14.5 One triangular glass prism has two 45° angles, and another triangular prism has a 30° and a 60° angle. Show the direction through each prism of a ray of light incident normally on the hypotenuse face, if the absolute refractive index is 1.5 in both cases.

14.6 A mixture of 59% carbon tetrachloride and 41% benzene has the same absolute refractive index as Pyrex (1.47). If you were given a number of glass and Pyrex rods, how could you identify the latter?

14.7 Find the absolute refractive index of glass from Fig. 14–13 (page 108).

14.8 If glass prisms each with 45° angles are to be used in prismatic binoculars, what is the least value for the refractive index of glass to give total internal reflection as shown in Fig. 14–24.

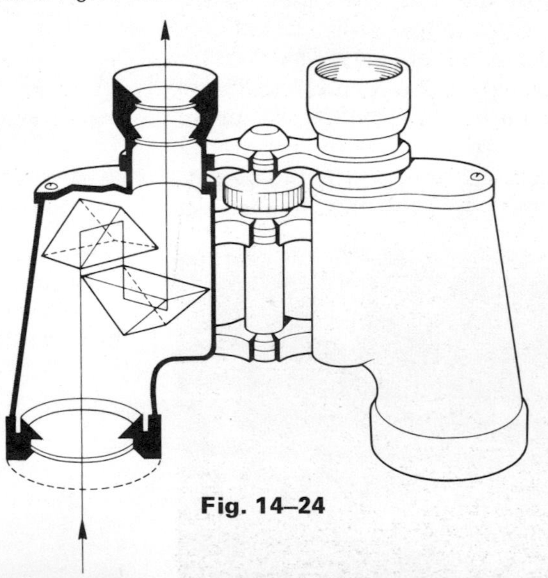

Fig. 14–24

14.9 A small electrolytic cell replaces the slide carrier projector. Unfortunately the image appears upside down the screen. Devise a way of using a glass prism with 45° angles to re-invert the image.

14.10 In fibre optics, light is transmitted through a la number of transparent plastic fibres (Fig. 14–25).

Fig. 14–25

Show by a diagram the part played by total internal reflecti

14.11 Light is incident on the circular end face of cylindrical glass rod (Fig. 14–26).

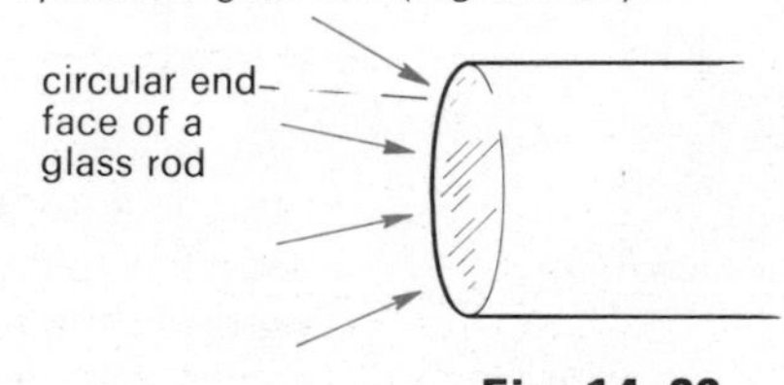

Fig. 14–26

Show that no ray entering through this end face can esca through the sides of the cylinder.

14.12 The critical angle of diamond is 27°. Discuss effect of this small critical angle on the beauty of a well stone. What is the absolute refractive index of diamond?

14.13 Refraction does not affect the colour of a ray light, yet colour affects refraction. Explain carefully w these statements mean and give examples to illustrate y answers.

15 Visual Aids

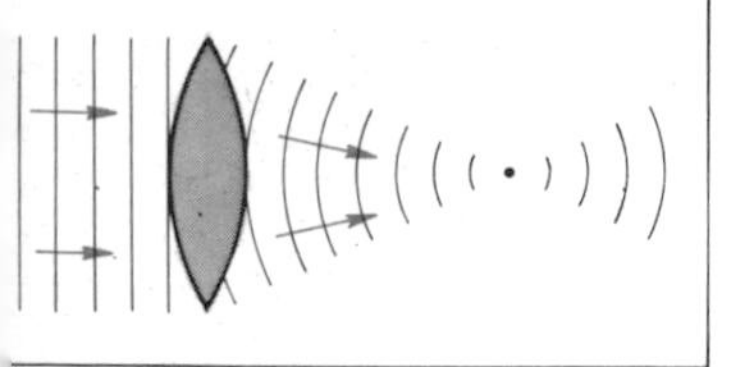

Fig. 15–1

With Snell's discovery a new era of applied optics was born and many new optical instruments were soon developed. In this course we will be concerned with converging and diverging lenses and their use in a few simple cases.

CONVEX LENS

Ripple tank experiments have shown us that, when plane water waves are slowed down over a *shallow* convex-lens-shaped area, they converge to a point beyond the lens (Fig. 15–1). Parallel light rays also converge to a point after passing through a convex lens (Fig. 15–2). The point is called a principal focus F. The word *focus* comes from a Latin word for hearth. If a lens is used as a burning glass, it will form an image of the sun at this point. When parallel rays enter the lens, the distance from the centre of the lens to the focus is called the *focal length f*. As we could obtain the same effect in reverse by passing parallel rays from the left to the right, the lens must have two principal foci. These are at equal distances from the lens. The ray photograph of Fig. 15–3 may be thought of as a two-dimensional representation of the three-dimensional phenomenon associated with a spherical lens (Fig. 15–4).

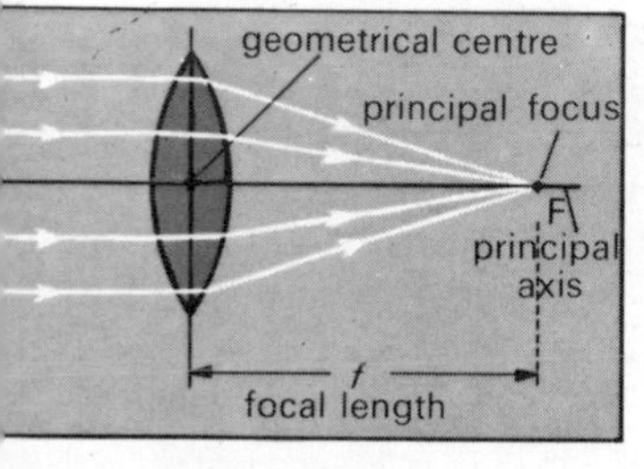

Fig. 15–2

Fig. 15–3

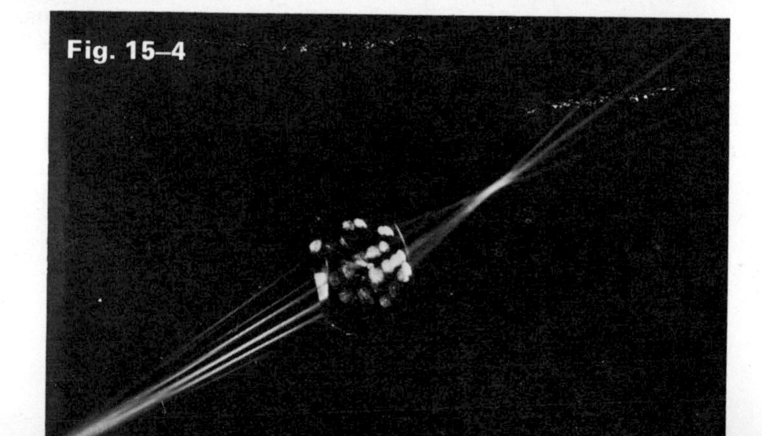

Fig. 15–4

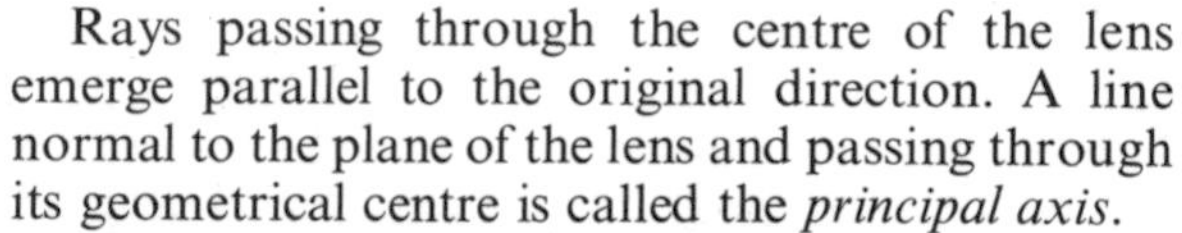

Rays passing through the centre of the lens emerge parallel to the original direction. A line normal to the plane of the lens and passing through its geometrical centre is called the *principal axis*.

When an object is so far away that the rays of light coming from it are (very nearly) parallel, we say that the object is 'at infinity'. For such an object situated on the principal axis of the lens, the image will be formed at the principal focus. Fig. 15–2 represents this situation. The same lens was used in Figs. 15–3 and 15–5.

Fig. 15–5

CONCAVE LENS

When plane water waves are slowed down as they pass over a shallow concave-lens-shaped region, they diverge beyond the lens (Fig. 15–6). Parallel

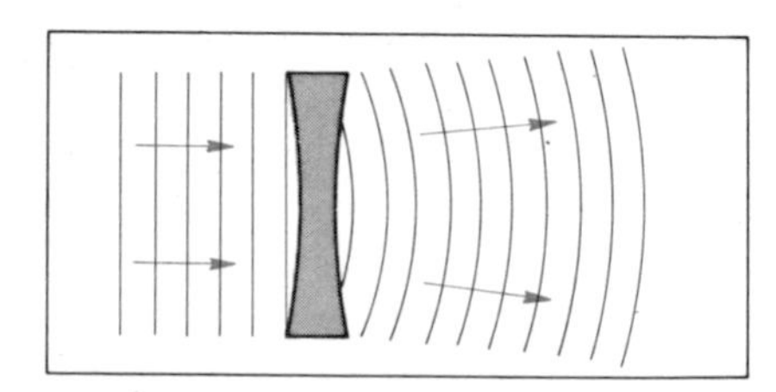

Fig. 15–6

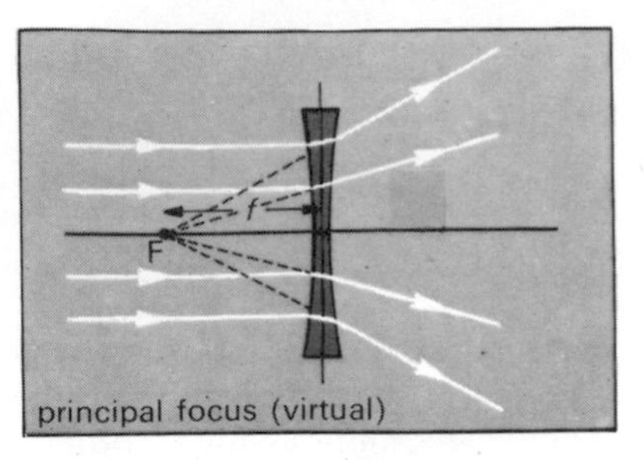

Fig. 15–7

light rays diverge as shown in Fig. 15–7 after passing through a concave lens. The rays now have the same directions as they would have had if they had come from a point F to the left of the lens. This point is therefore called a principal focus. No rays of course actually do come from this point and so it is called a *virtual* principal focus.

POWER OF A LENS

A lens whose opposite faces are almost parallel does not bend light rays very much. Consequently it has a very long focal length, perhaps several metres. Such a lens would be of little use as a magnifying glass. A thick bulging lens on the other hand has a short focal length and makes a good magnifying glass. It is often described as very 'powerful'. In fact we define the *power* of a lens as the reciprocal of its focal length measured in metres. The unit of power is the dioptre (D).

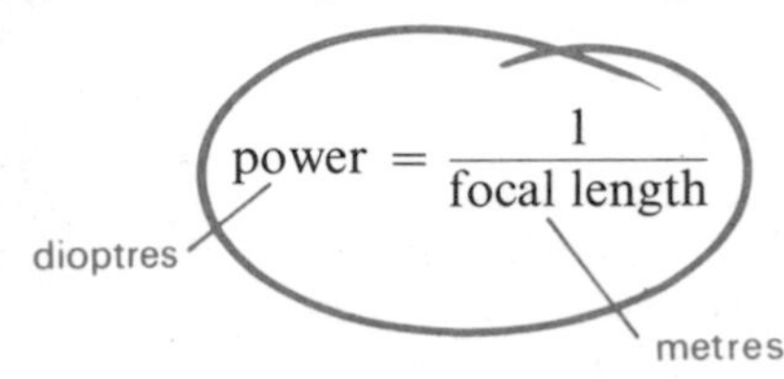

Converging lenses have positive powers (e.g. +7D, +10D, +17D) and diverging lenses have negative powers (e.g. −10D, −17D). *What are the focal lengths of these lenses? (15.2)*

THE CAMERA

A lens camera requires a convex lens to produce a picture. Fig. 15–8 shows the inverted image focused on the translucent screen of such a

Fig. 15–8

Fig. 15–9

camera. We can show on our two-dimensional r diagram how the image of one of the bridge tow is formed on the screen or film. To do this we v imagine a beam of reflected light coming from point A on the top of the bridge, a point B the railway, and a point C on the stone base. ray from each of these points will pass through t optical centre undeviated (but in practice sligh displaced) and so the image must be inverted shown in Fig. 15–9. However, all the other ra which enter the lens from *each of these three poi* on the bridge will be bent so that they arrive on t screen at, very nearly, the same places as the u deviated rays. The shaded areas represent t beams entering the lens from the three points. course there will be a similar beam of light comi from every point on the bridge thus producing complete image. An image such as this which c be projected on to a screen is called a *real image*.

It is usual to label the object-lens distance u a the lens-image distance v. You will have fou from experiment that as u is *decreased* v must *increased* if the image is to stay in focus. For clo up photography the lens has to be moved outwa away from the film.

Depth of Focus

With a camera focused on a bridge in the dista and the iris diaphragm fully open, a flute pla placed close to the camera was photographed (F 15–10). He is out of focus because light from, say point on his nose tip forms a *patch* of light on film (Fig. 15–11).

Without moving the camera or the lens setti the diaphragm was partially closed and anot photograph taken using a longer exposure (F 15–12). As the cone of light entering the camera now much narrower, it forms a smaller patch the film (Fig. 15–13). The image of the tip of

Fig. 15–10

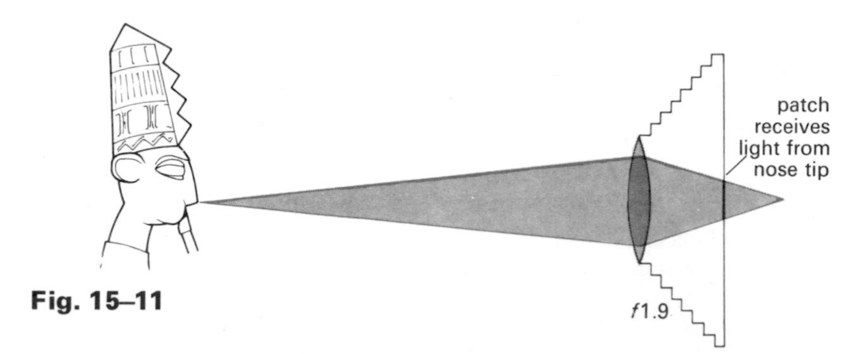

Fig. 15–11

Fig. 15–12

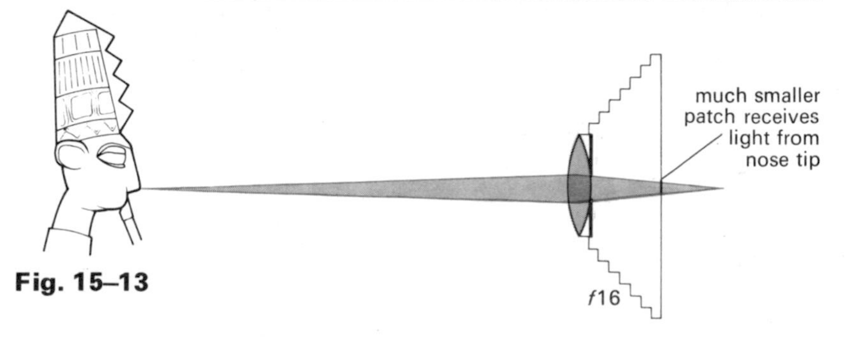

Fig. 15–13

flautist's nose and every other point reflecting light into the camera are therefore much sharper than before. So if we use a small aperture (for those of you who are photographers, this means a large f number) the focusing of the camera is less critical; that is, the *depth of focus* is greater.

You can see a similar effect when the aperture of your eye is reduced. Move a book towards your eye until you can no longer focus clearly on the print and then insert a card with a pin hole in it

Fig. 15–14

close to your eye (Fig. 15–14). All the rays through your eye lens must then pass through the small hole and you should be able to read the book. If you break your glasses, it is worth remembering this method of reading the optician's phone number!

VISUAL PERCEPTION

If a camera with a translucent screen in place of the film is set up as shown in Fig. 15–15, a small pencil of light will enter the lens. A pin held *upright* and close to the lens casts an *upright* shadow (not an image) on the screen. You can try a similar

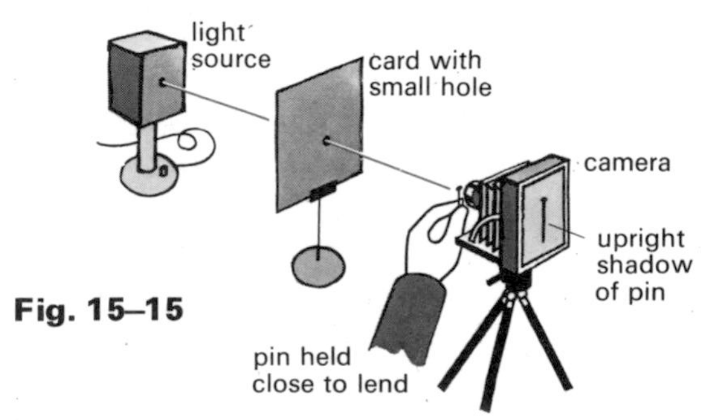

Fig. 15–15

Fig. 15–16

experiment to see if your eye does the same kind of thing (Fig. 15–16). A card with a pin hole in it allows only a narrow pencil of light from, say, the sky to enter your eye. When you hold a pin *upright* and close to your eye it will cast an *upright* shadow on the retina of your eye as in the case of the camera. *How do you see this upright shadow? What does this tell you about visual perception? (15.3)*

Many fascinating experiments have been conducted with special glasses which enables the wearer to see everything upside down. If, when he is awake, he wears these glasses continually for several weeks, he gradually starts to see things the right way up again. There is a transition period and it is thought that by touching things either directly or with a stick the relearning process is speeded up. Once the person wearing the glasses has learned to see things the right way up, he is invited to remove the glasses and for a short time he then sees everything upside down! This lasts for only a few minutes, however, as years of learning to see without the inverted glasses have made a much greater impression on him than a few weeks with the glasses. Normal service is resumed almost immediately.

The Eye

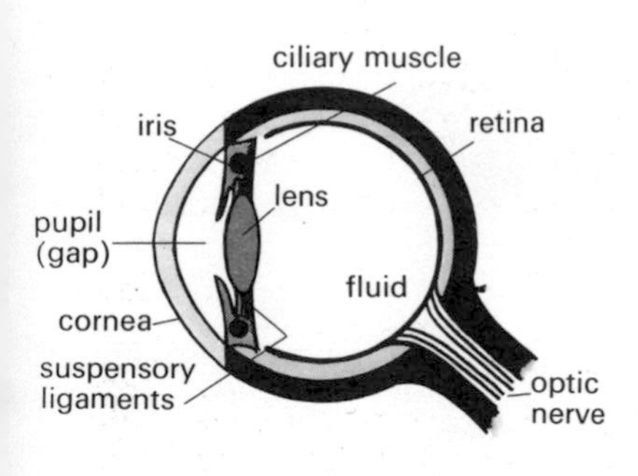

Fig. 15–17

The human eye (Fig. 15–17) is a highly complex form of camera in which a multiple lens system focuses incoming rays on the retina. The cornea and the lens, with fluid either side of it, form the lens system. Unlike the hard glass lenses we normally deal with, the eye's plastic lens tends to bulge—as would a balloon—when it is free to do so. However, it is held round the circumference by *suspensory ligaments,* which in turn are connected to the ring-like *ciliary muscles* (Fig. 15–18). When they are contracting, these muscles act rather like a stretched rubber ring. The ring becomes smaller and the tension on the suspensory ligaments is reduced. This allows the eye to bulge, thus reducing its focal length and enabling the eye to focus on nearby objects. The ability to do this is called *accommodation.* The nearest point on which the eye can focus is called the *near point*. It is usually between 10 and 30 cm from the eye. The accepted 'standard' near point is 25 cm.

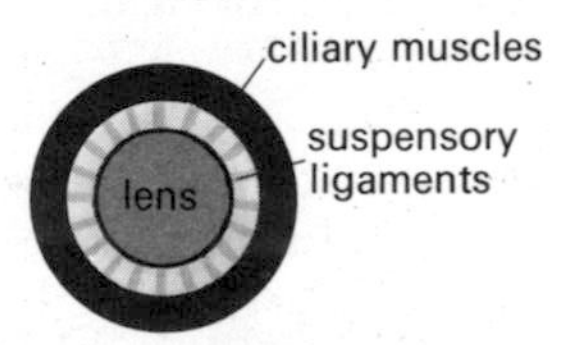

Fig. 15–18

When our eyes are resting, the ciliary muscles are relaxed and the tension in the suspensory ligaments increases, so that the lens becomes thin and our eyes are focused on infinity. So we focus our eyes by changing the shape of the lens as contract or relax the ciliary muscles.

The iris is an adjustable diaphragm with a cent aperture, the pupil, through which light enters t eye. The size of this aperture and the sensitivity the retina change automatically with the amou of light coming into the eye. Images are formed the retina, which then transmits signals to the br via the optic nerve. The sensations persist for least 1/20 of a second after the light stimulus h been removed, which is why movie films a television produce apparently continuous mo ment from a series of still pictures.

The Retinal Image

An image can be formed on the retina of the e if parallel rays from an object at infinity (Fig. 15– or diverging rays from an object close at ha (Fig. 15–20) enter the lens. If converging rays en the lens (Fig. 15–21), no image will be formed the retina.

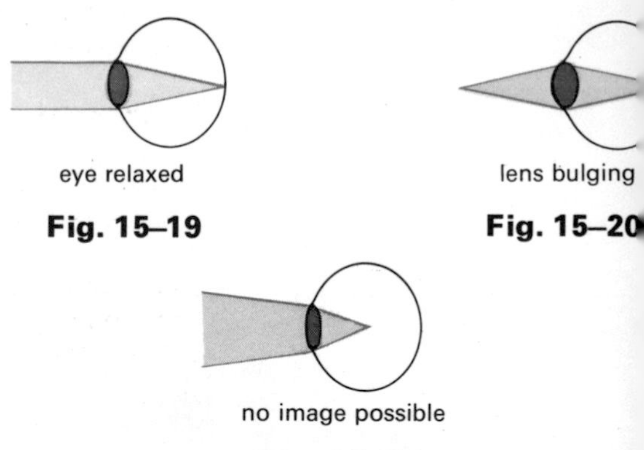

Fig. 15–19

Fig. 15–20

Fig. 15–21

To examine an object in detail we want *largest possible image* to be formed on the ret of the eye. The size of this retinal image depe on the *visual angle*, that is, the angle the obj subtends at the eye. This angle depends on the of the object and on the distance it is away fr the eye. An object AA′ (Fig. 15–22) will produc

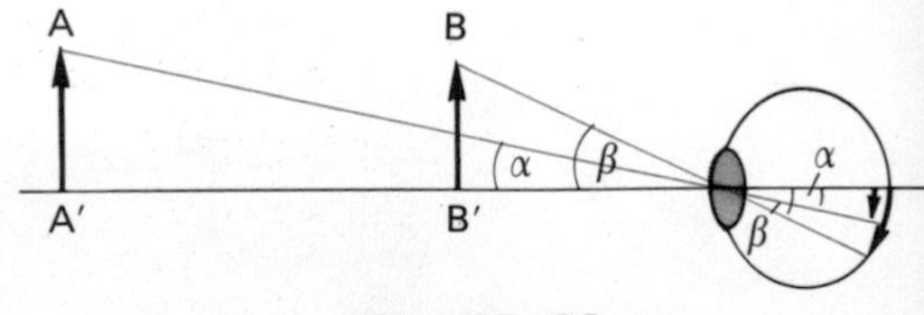

Fig. 15–22

larger retinal image if it is brought closer to the eye, e.g. to position BB′. The visual angle has then increased from α to β. In Fig. 15–23 the same visual angle and therefore the same size of image are produced by a large object at CC′ and a small object at DD′.

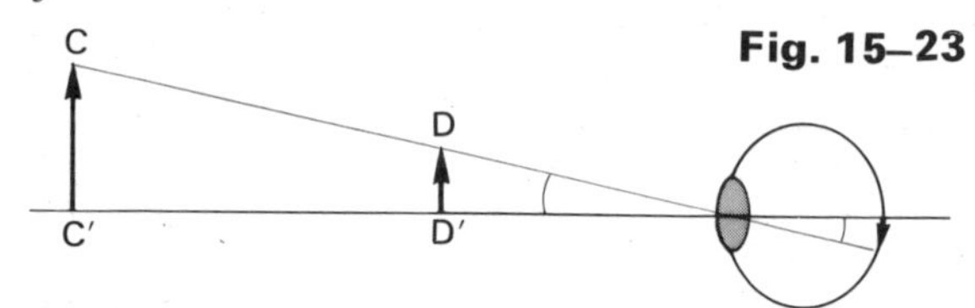

Fig. 15–23

If we look at a distant object, an inverted image will be formed on the retina. This we interpret, by experience, as upright. To construct the ray diagram (Fig. 15–24) we have shown a small cone of rays coming from the top and bottom of the object. The rays passing through the optical centre of the lens are undeviated. The eye is focused on infinity.

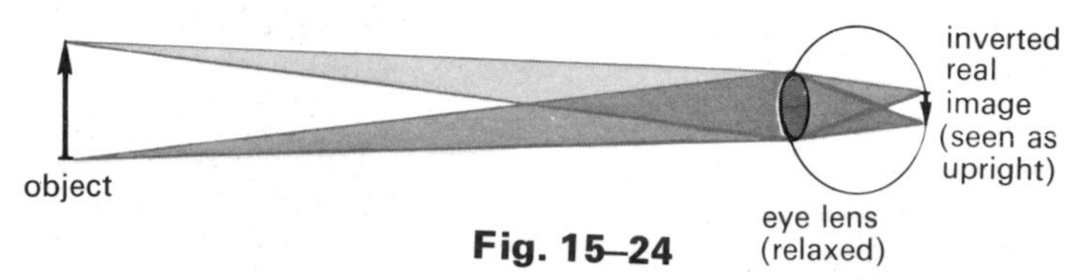

Fig. 15–24

Converging Lens

If now a camera is focused on the image formed by a magnifying glass placed in front of it, the final image will be upright as shown in Fig. 15–25A. If you were to place your eye in the camera position you would see an inverted bridge (Fig. 15–25B), although the retinal image would, of course, still be upright.

Representing this situation on a ray diagram is a little more difficult, and it will help if the object is made to stand on the principal axis. The base of the image must then also be on the principal axis, as rays from the base of the object pass through the optical centre of the lens without deviation. To locate the tip of the image, at least two rays are needed. One can be the undeviated (red) ray from the tip of the object through the optical centre (Fig. 15–26). The other can be the dotted ray drawn parallel to the principal axis so that it passes through the principal focus F after refraction. The tip of the image is located at the point of intersection of these lines.

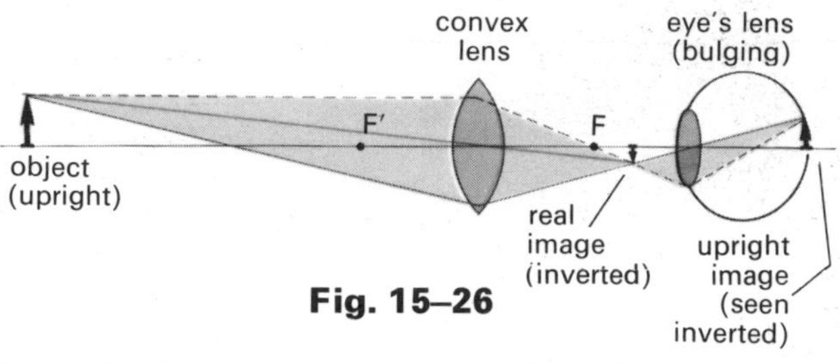

Fig. 15–26

If you are not satisfied, you can check this by drawing a third ray from the tip of the object through the other principal focus F′. After refraction it should be parallel to the principal axis and pass through the tip of the image.

Although the image formed by the magnifying glass is not projected on to a screen, it is nevertheless a *real* image. It is formed by real rays of light and if a sheet of paper is held there an image will be projected on to it. If the eye is placed between this image and the magnifying glass, *converging* rays will enter the eye and no image can be formed on the retina.

Fig. 15–25A

Fig. 15–25B

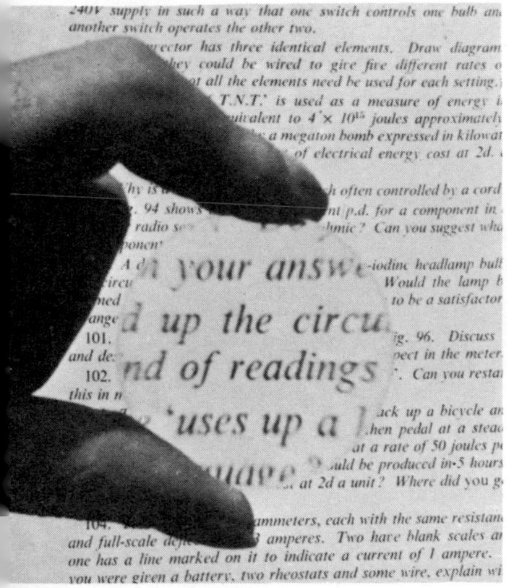

Fig. 15–27

Fig. 15–28

A similar magnifying glass is used in Fig. 15–27 to produce an *upright virtual* image which is bigger than the original object (type face). The image of this virtual image is, of course, inverted on the retina of the eye.

So far we have drawn eye lenses big enough to allow all the rays to pass through them. This can, however, lead to unnecessarily swollen eyes! In a simplified geometrical representation (Fig. 15–28), it is enough to indicate whether the lens is convex or concave and then to show the rays bending somewhere along the vertical *line* through the optical centre of the lens.

Two (real) rays are first drawn from the tip of the object; one parallel to the principal axis and then through F, and one through the optical centre of the lens. To locate the tip of the virtual image these rays are produced back until they meet (light dotted lines). Other rays from the tip of the virtual image can then be drawn, including one to pass undeviated through the optical centre of the eye's lens. This ray will locate the tip of the image on the retina. The shaded areas represent two of the many possible pencils of light from the tip of the object to the retina of the eye.

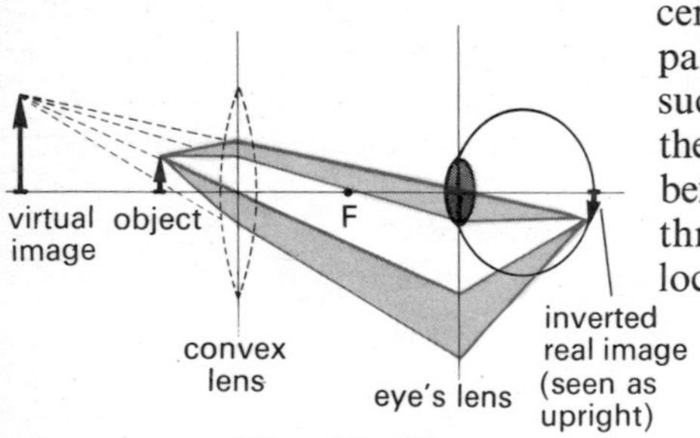

Diverging Lens

The same kind of ray diagram can be used to illustrate the formation of the *virtual* image formed by a concave lens (Fig. 15–29). One (red) ray from the tip of the object passes through the optical centre of the lens, and another (broken red line) parallel to the principal axis diverges at an angle such that when produced back (light dotted line) the line passes through the virtual focus F. As before, a ray drawn from the tip of the image through the optical centre of the eye's lens will locate the tip of the inverted image on the retina.

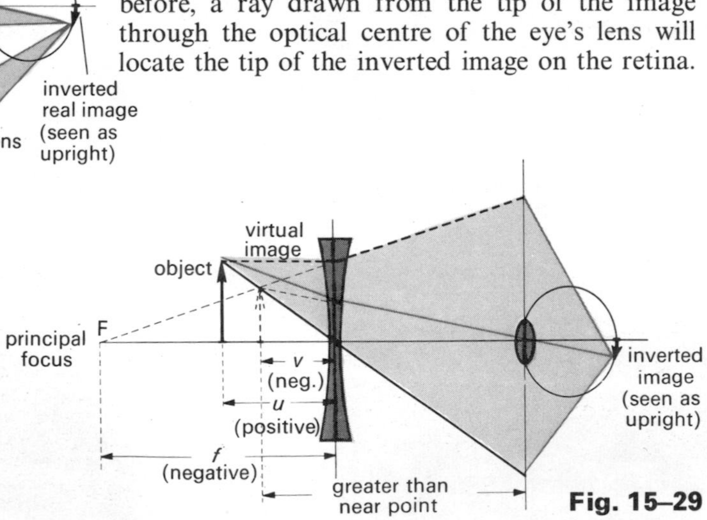

Fig. 15–29

Although all the light represented by the shad area cannot pass through the eye's lens shown, t geometrical construction is simplified by using t large scale. With the sign convention we are usi (see page 120), distance from the lens to *virtu* images, objects and foci will be given *negat* values. The shrunken image formed by a divergi lens is illustrated in Fig. 15–30.

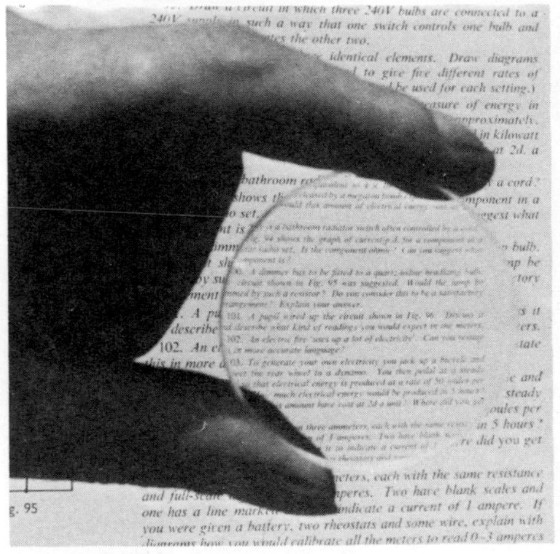

Fig. 15–30

ASTRONOMICAL TELESCOPE

If you look back to Fig. 15.25B (page 117) you see that a weak lens (low power—large fo length) produces a relatively large *real* image o distant object. A strong lens (e.g. the lens in 8 mm cine camera) would make the refracted r bend more steeply and so produce a much sma real image. On the other hand a strong lens needed to produce a large *virtual* image of a clo up object (Fig. 15–28). A weak lens would prod little bending and therefore a much smaller virt image.

Suppose then that we use a weak lens to prod a large real image and then look at this im through a strong magnifying glass. In other wo we combine the arrangements of Fig. 15.25B a 15–28 to make a telescope. First consider the im formation when the telescope is used to look a not-too-distant object. The same construction before is used in Fig. 15–31 but the eye has b omitted to avoid, or at least reduce, confusion. the image is inverted, this telescope has limi uses as a terrestrial instrument. When study heavenly bodies, however, this is less of a advantage.

We have already seen that the eye is in its m relaxed condition when focused on infinity and

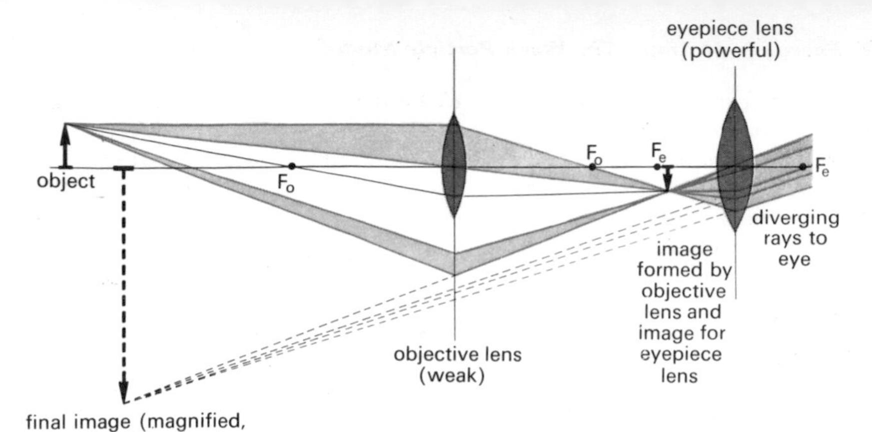

Fig. 15–31

we might expect a telescope to be used most comfortably—particularly for sustained viewing—if the final image was also at infinity; that is, the rays entering the eye from a point on the object should be parallel. The telescope is then said to be in *normal adjustment*. This state of affairs can be accomplished by making one principal focus of the objective lens coincide with a principal focus of the eyepiece lens. The first image is then focused at this common point and the rays entering the eye are parallel (Fig. 15–32).

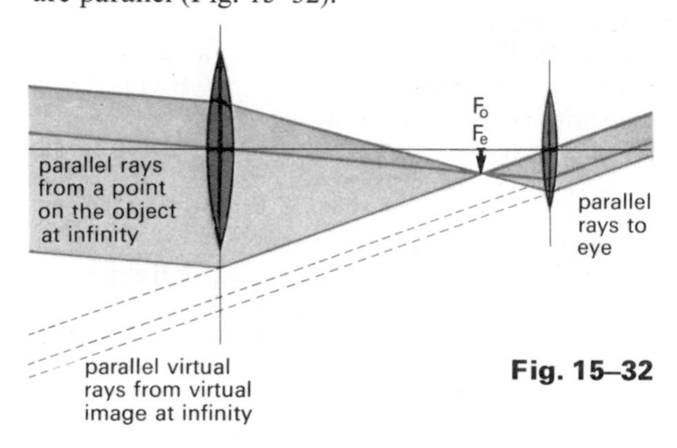

Fig. 15–32

The magnifying power of a telescope is defined as the ratio of the angle subtended by the image α to the angle subtended by the object β. Fig. 15–33 shows the relevant parts of Fig. 15–31. If we consider α and β $(=\beta')$ to be very small angles, each angle, in radians, is approximately equal to the tangent of the angle.

Then $\alpha = \dfrac{\text{image size AB}}{f_e}$ and $\beta = \dfrac{\text{image size AB}}{f_0}$

thus the magnifying power $M = \dfrac{\alpha}{\beta} = \dfrac{f_0}{f_e}$

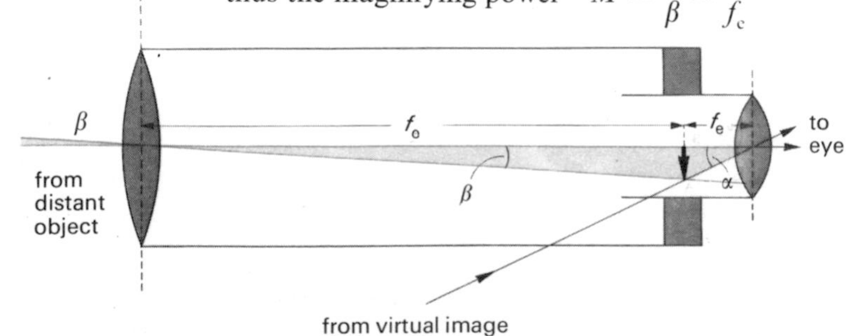

Fig. 15–33

Fig. 15–34

This shows that, if you want to build such a telescope with a large magnifying power, you should use an objective lens with a long focal length (see Figs. 14–11 and 14–12) and an eyepiece lens with a short focal length.

The telescopic lens of a camera (Fig. 15–34) is a particularly useful adaptation of the astronomical telescope. Fig. 15–35 shows a photograph taken with an ordinary lens and Fig. 15–36 is taken from the same place using a powerful telescopic lens.

Fig. 15–35

Fig. 15–36

Can you suggest a slight modification to the telescope illustrated in Fig. 15–33 to enable it to be used as a telescopic lens in place of the ordinary lens in a camera? Remember the real image will have to be formed on the photographic film. (15.4) Would you have to hold the camera upside down? (15.5)

Fig. 15–37

PHYSICS AND FORMULAE

Most of the problems you are likely to meet in this course can be solved by scale drawings (ray diagrams) such as those we have been using. You might in fact be forgiven for thinking that there is more geometry than physics involved which explains why the subject is called geometrical optics! As a check on your diagrams you might like to use some algebra as well! This necessitates adopting a sign convention. We are going to use the 'Real Is Positive' (R.I.P.) convention, in which all distances are measured from the *optical centre* of the lens and in which we attach positive signs to the numbers (but not the letters) representing distances to real objects and real images. In this convention the focal lengths and powers of converging lenses are also positive. Consequently the distances from the optical centre to a virtual image or virtual object are negative, and the focal length and power of a diverging lens are also negative.

The 'lens formula' can be derived by simple geometry from Fig. 15–38. The linear magnification m of a lens is the magnitude of the ratio image size/object size.

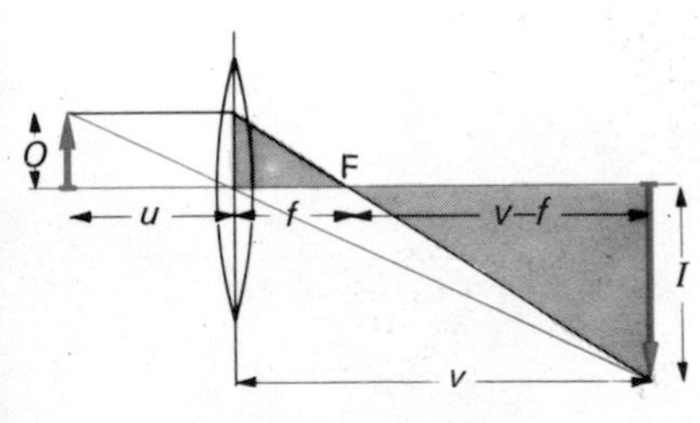

Fig. 15–38

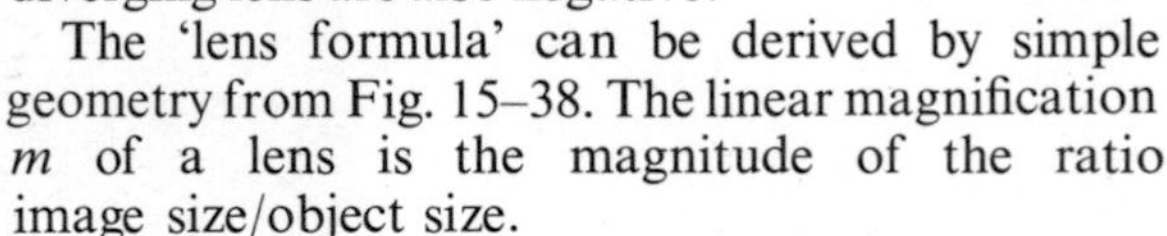

$$m = \frac{I}{O} = \frac{v}{u} \quad \text{(by similar red triangles)}$$

also

$$\frac{I}{O} = \frac{v-f}{f} \quad \text{(by similar shaded triangles)}$$

$$\Rightarrow vf = uv - uf$$

and, dividing through by uvf, we have

$$\frac{1}{u} = \frac{1}{f} - \frac{1}{v}$$

$$\Leftrightarrow \frac{1}{f} = \frac{1}{u} + \frac{1}{v}$$

This formula can be applied to a convex or a concave lens, provided we apply the sign convention given above.

Example

Find the image position and the magnification of a -10D lens when an object is placed 15 cm from it.

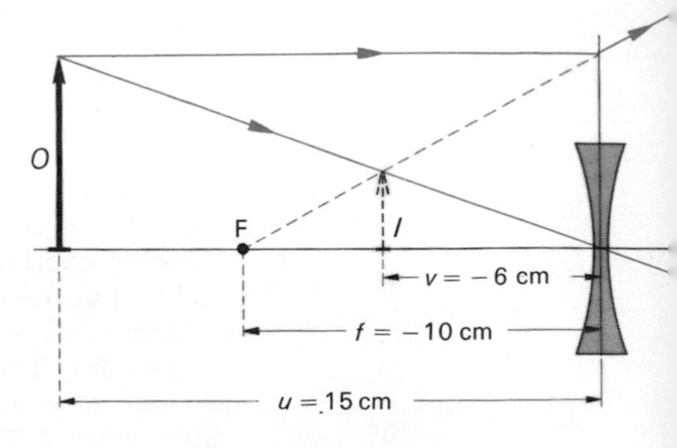

Fig. 15–39

(a) As power $= 1/f$, the focal length of the lens is -10 cm. By scale drawing (Fig. 15–39) the virtual image is 6 cm from the lens on the same side as the object. The object size on the ray diagram is 2 cm and the image size 0.8 cm. The magnification is therefore $0.8/2 = 0.4$.

(b) By the lens formula

$$\frac{1}{u} + \frac{1}{v} = \frac{1}{f}$$

$$\frac{1}{0.15} + \frac{1}{v} = \frac{-10}{1}$$

$$\Leftrightarrow \frac{1}{v} = \frac{-10}{1} - \frac{1}{0.15}$$

$$\Leftrightarrow \frac{1}{v} = \frac{-150-100}{15}$$

$$\Leftrightarrow v = \frac{-15}{250} = -0.06 \text{ m}$$

The image is therefore 0.06 m from the lens. As v is negative the image must be virtual.

$$m = \frac{v}{u} \text{ numerically}$$

$$= \frac{0.06}{0.15}$$

$$= 0.4$$

SUMMARY

Power of lens $= \dfrac{1}{\text{focal length in metres}}$

Power of converging lens is *positive*.
Power of diverging lens is *negative*.

Lens formula: $\dfrac{1}{f} = \dfrac{1}{u} + \dfrac{1}{v}$

Linear magnification $= \dfrac{v}{u}$ numerically

The image position can be located using a ray diagram as follows.

(a) Draw a ray, parallel to the principal axis, from the tip of the object to the lens. This ray should then be bent to pass through the appropriate principal focus.
(b) Draw another ray from the tip of the object straight through the optical centre of the lens.
(c) To confirm the point of intersection of these rays, a third ray can be drawn from the tip of the object through the *other* principal focus. When it passes through the lens this ray turns parallel to the principal axis.

For a telescope in normal adjustment (i.e. focused on infinity)

$$\text{magnifying power, } m = \frac{f_o}{f_e}$$

PROBLEMS

15.6 An astronomical telescope is focused on infinity (normal adjustment). Describe as fully as possible the nature and position of the image formed by the objective lens. What is the object for the eyepiece lens? Describe the final image.

15.7 A −10 D lens is placed 0.3 metres from an object 4 cm tall. Find by scale drawing or calculation where the image will be formed. Describe its nature and size.

15.8 A converging lens of focal length 20 cm produces a virtual image 30 cm from the lens. If the image is 4 cm high, find graphically the height of the object and its distance from the lens.

15.9 A slide projector is switched on and focused correctly but the picture does not fill the screen. Explain what you would do (a) to increase the picture size and then (b) to refocus it.

15.10 A converging lens produces a 4 cm image of a lamp post 4 metres high situated 50 metres from the lens. Find the power of the lens.

The same lens is now used as the objective of a telescope to view the lamp post from 50 metres. How far from the lens would a 10D eyepiece lens have to be situated to produce the final image at infinity?

15.11 Why is it possible for a cheap camera to have a 'fixed focus'?

15.12 A man has to hold a newspaper well away from his eyes to focus on it. What kind of spectacle lenses does he need?

15.13 Explain the use of the condenser lenses C_1 and C_2 in the cine projector illustrated in Fig. 15–40.

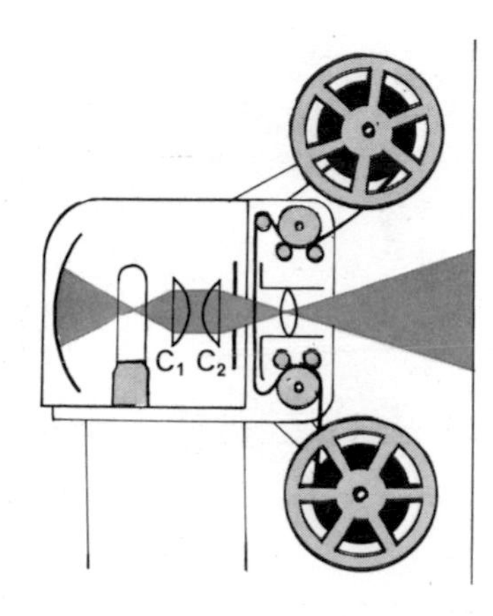

Fig. 15–40

15.14 A lens is used to project a colour transparency on to a screen. If the slide is 35 mm wide, what width of picture will be produced on a screen 4 metres from the 10.25D lens? How far is the slide from the lens?

15.15 A convex lens of focal length 6 cm held close to the eye is used to inspect a small coin 1 cm in diameter. If the coin is placed 4.8 cm from the lens, find *by scale drawing* the diameter and position of the image. Check your results algebraically.

15.16 How many lenses of one kind or another can you find at home? List the apparatus which uses them.

15.17 (a) Draw a ray-diagram to show how the final image of a distant object is formed in an astronomical telescope. Explain why it would be inconvenient to use an astronomical telescope as a terrestrial one.
(b) When an object is viewed through a 'magnifying glass' it may appear upright and enlarged *or* inverted and smaller. Draw sketches to show the positions of the object with respect to the lens, and the formation of the image in each case.
(c) A home-made photographic slide viewer (Fig. 15–41) produces an upright image of the slide with a linear magnification of 4, when the slide is placed 15 cm from the lens.

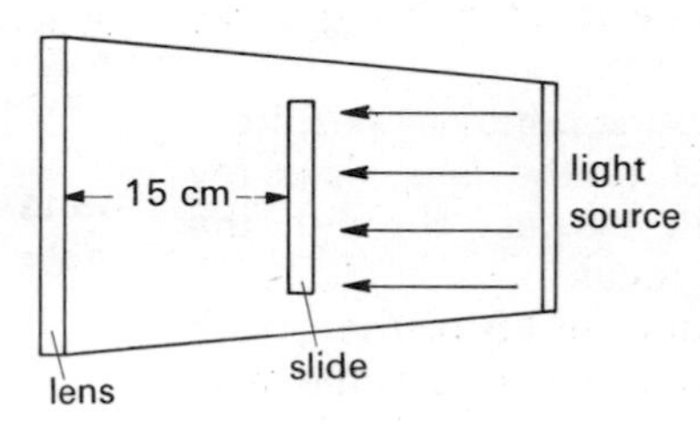

Fig. 15–41

What type of lens (converging or diverging) is used and what is its focal length? [S.C.E. (H) 1970]

15.18 Fig. 15–42 is a ray diagram for an astronomical telescope focused on a distant star.

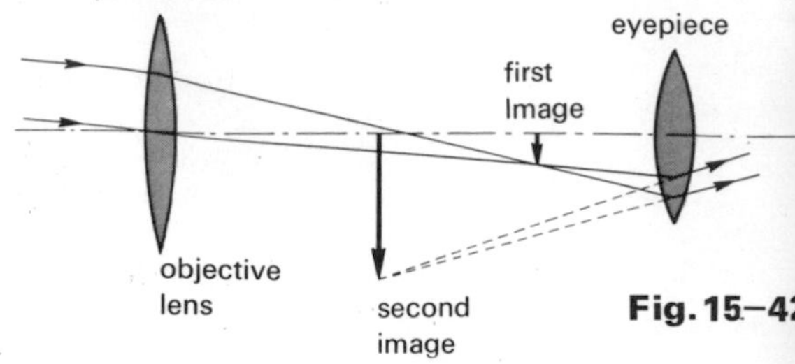

Fig. 15–42

(a) What is the location of the first image?
(b) Describe fully the nature of the second image.
(c) How far apart would the lenses need to be for the second image to be seen at infinity?
(d) Why should the objective lens have a large aperture?
(e) Suggest two modifications which could be made to the telescope to increase the size of the second image.
[S.C.E. (H) 1971]

15.19 The lens of a particular camera has a focal length of 50 mm and the film to lens distance can be adjusted to any value between 50 mm and 60 mm.
(a) Where should the lens be positioned relative to the film in order to produce a well-defined photograph of a distant object?
(b) By drawing or by calculation, determine the shortest distance an object can be placed from the lens in order to obtain a sharp image on the film. [S.C.E. (H) 1971]

16 Graven Images

Natural philosophy thrives on controversy. In their attempts to make sense of the physical world, the early philosophers suggested models and explanations based on their everyday experience. Often, however, their followers judged these explanations in terms of the authority of the philosopher rather than on the merit of the theory. The views of Aristotle in particular seemed to have carried great weight for nearly two thousand years. But by about the seventeenth century there was a growing awareness that theories were of value only if they explained all or at least most aspects of a particular phenomenon. Moreover, assessment of a theory in terms of the results of experiment rather than the authority of a particular philsopher was gaining popularity.

We cannot and should not avoid models in science, yet the very language we use can often mislead us. If we speak of heat or electricity *flowing* along a copper bar, we seem to imply that some kind of fluid is moving through it. If we can find other theories which more usefully describe and more accurately predict the behaviour of heat and electricity, we should discard the fluid model. Of course, we may still find it helpful to *compare* some aspects of these phenomena with water flowing in pipes (analogy), but this is quite different from thinking of heat or electricity *as if they were* fluids (model).

It is perhaps worth spending some time thinking more deeply about this distinction. Earlier we described the flow of electric charge in a metal conductor in terms of a gradual drift of electrons superposed on a much faster random movement (page 60). This could be called the particle model of electric current. The water circuit board illustrated in Fig. 16–1 is an analogue. We compare, for example, rate of flow of water (in litres per second) with rate of flow of charge (in coulombs per second) and opposition to the flow of water in narrow tubes with resistance.

Now let us consider an example in which the analogue and the model are less distinct. The bouncing ball analogue (Fig. 5–15, page 43) helped us to think of the way in which molecules in a gas behave. We compared the behaviour of the balls with the behaviour of the molecules. The kinetic *model* of the gas is, however, quite different. In it we make several assumptions which do not apply to the analogue. For example, real molecules fly around for ever without energy being continually fed into the system; that is, the collisions are elastic. We use a *model* to think about the actual physical phenomenon. An *analogue* on the other hand can, at least in principle, be built from nuts and bolts or string and sealing wax!

We are not likely to get very far in our study of light if we refuse to think of it in terms of some analogue or other. A bullet from a gun provides a pretty crude analogue for the particle model of light. Do the light particles have mass like bullets? Do they speed up from rest as they leave the source and slow down afterwards? What particular characteristics of a particle do we want to use to describe light?

The other analogue often used is the ripple tank. The water waves provide an analogy for the wave model of light. Again we have to ask the same kind of questions. What can possibly wave in a vacuum? What particular characteristics of a water wave are relevant to light?

Living as we do hundreds of years from the start of the 'nature of light' controversy, it is easy to be wise after the event and to describe the particular properties that, for us now, define the sense in which we want to use these terms. However, to be fair we ought to look first at the situation in the seventeenth century.

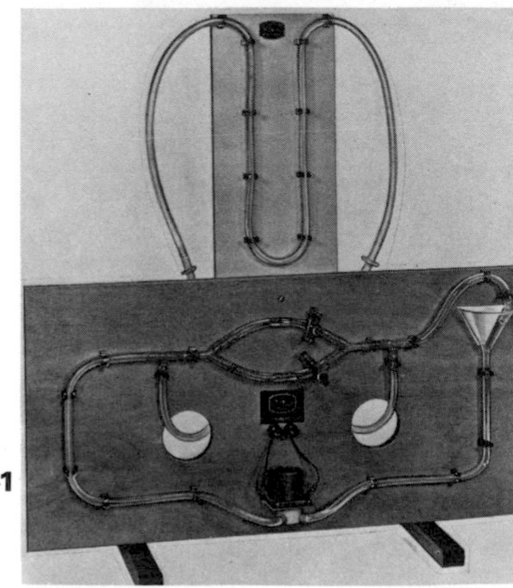

Fig. 16–1

WAVES IN—PARTICLES OUT

When water waves pass a sharp edge such as the breakwater in the model harbour shown in Fig. 16–2, they are slightly bent. This diffraction effect with water waves and sound waves was well known by 1665, the year in which Francesco Grimaldi, an Italian Jesuit, observed a similar effect with

Fig. 16–2

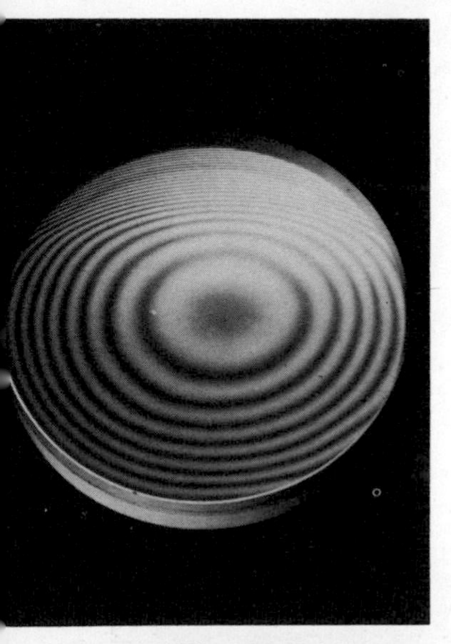

Fig. 16–3

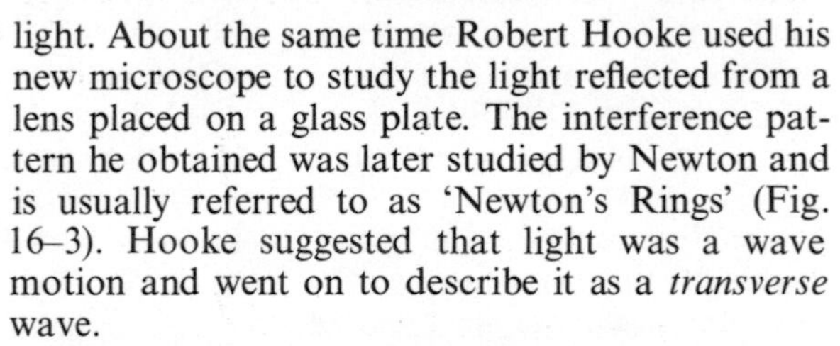

light. About the same time Robert Hooke used his new microscope to study the light reflected from a lens placed on a glass plate. The interference pattern he obtained was later studied by Newton and is usually referred to as 'Newton's Rings' (Fig. 16–3). Hooke suggested that light was a wave motion and went on to describe it as a *transverse* wave.

The next piece of evidence in favour of the wave theory came from Denmark. Erasmus Bartholinus discovered that, when light passed through a calcite crystal, it emerged as two separate beams. A few years later the Dutch scientist Constantijn Huygens showed that these two beams were *polarized.* You can see this effect with two pairs of Polaroid sunglasses. When two of the lenses are held in their normal position and overlapped as shown in Fig. 16–4, light passes through the two lenses. If, however, one lens is rotated through 90°, practically no light gets through (Fig. 16–5).

It is difficult to see how this effect could be produced by particles or by longitudinal waves such as sound waves. However, if we imagine light as

Fig. 16–4

Fig. 16–5

transverse waves a reasonable analogue can b constructed. Vertical transverse waves on a rop pass easily through a fence (Fig. 16–6) but ar stopped by a gate. If we imagine light as transvers waves in all directions, then only waves in a parti

Fig. 16–6

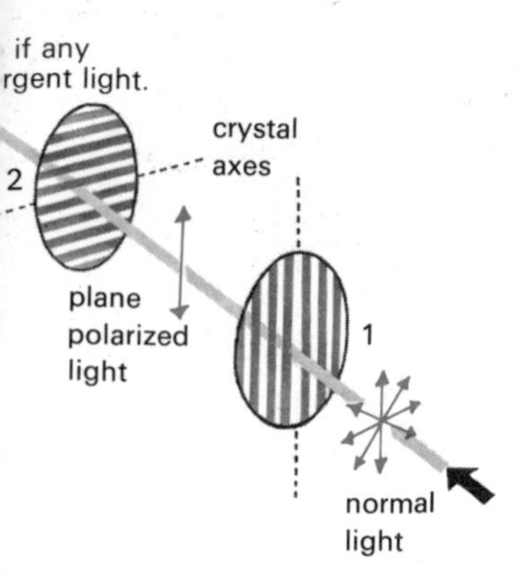

Fig. 16–7

cular direction can pass through Polaroid No. 1 (Fig. 16–7). A second Polaroid could be placed in the light beam and rotated until it would not allow these waves to pass. Huygens certainly supported the wave theory but he thought of light not as continuous waves but rather as a succession of pulses in a medium made up of elastic spheres in contact with each other.

Newton is sometimes regarded as the main opponent of the wave theory of light. His corpuscular theory certainly seemed to favour the idea of particles moving at 3×10^8 m/s, a speed calculated earlier from Ole Roemer's measurements of the eclipse times of Jupiter's moons. But Newton was very much aware of the properties of waves and had already devised a method for calculating the speed of sound waves. In fact, it may have been his familiarity with *longitudinal* sound waves which prevented his appreciating Hooke's suggestion of *transverse* light waves. Certainly Newton refused to accept a wave theory to explain the polarization effects demonstrated by Huygens.

Newton's corpuscles were not, however, little bullets. Polarization and interference patterns could not be produced by a shower of tiny particles. So Newton had to endow his corpuscles with various wave characteristics!

Whenever the corpuscles arrived at a boundary between two media, a kind of oscillation was set up so that there was a certain *frequency* associated with the particles. These oscillations consisted of alternate 'fits' of easy reflection or easy transmission. From his interference experiments Newton actually calculated the distance travelled by the corpuscles between these 'fits' and obtained a value which is in fact very close to the distance we now call the *wavelength*. It seems surprising then that Newton, having attributed speed, frequency, and wavelength to his corpuscles, should still feel that the wave model was inappropriate. He clearly did not see enough common ground between the ripple tank effects (analogue) and the concept of light waves (model) to justify it.

Young's Slits

For more than a hundred years the rival theories survived but Newton's reputation and authority were such that his corpuscular theory was generally accepted. In 1801 the crunch came. Thomas Young showed by simple experiment that light could produce interference patterns similar to those in a ripple tank. To understand his argument imagine that in a round of golf your partner and you both drive off at the same instant and in the same direction. If the balls both land on the green, what do you think are the chances that you will find not two golf balls but *either* one large ball of twice the size or none at all? It is not the nineteenth hole!

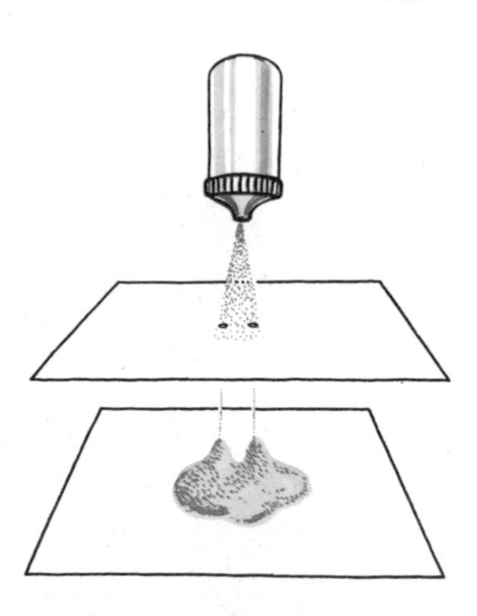

Fig. 16–8

Fig. 16–8 shows another analogue. Sand is poured from a salt cellar on to a board in which two small holes are drilled. The sand forms two piles as shown. Even if the two holes were very close together we would obtain simply the sum of the two piles. If light behaves like particles, we might expect a similar effect when light passes through two tiny holes or slits. Allowing for some bending at the edges of the slits we should obtain a light distribution not unlike that illustrated in Fig. 16–8. Young found something quite different. When light from a small source was passed through two narrow slits, he obtained a pattern similar to that shown in Fig. 16–9. Moreover, when one slit was closed the pattern disappeared. Apparently the two light beams were adding to produce bright bands (constructive interference) and dark bands (destructive interference).

Fig. 16–9

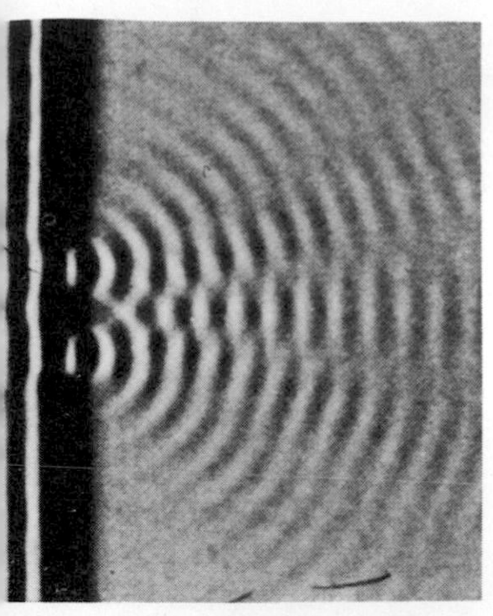
Fig. 16–10

Fig. 16–11

Fig. 16–13

In the ripple tank, interference patterns can be produced by plane waves passing through two slits thus producing two sets of semi-circular waves which are always in phase with each other (Fig. 16–10). Two dippers produce similar waves and thus similar interference patterns (Fig. 16–11). The pattern along a cross-section AB of the ripple tank consists of alternate areas of reinforcement (constructive interference) and cancellation (destructive interference) and bears a strong resemblance to the interference pattern projected on to the screen in the two-slit light experiment. Furthermore, if one of the slits in the ripple tank is closed, the interference pattern disappears.

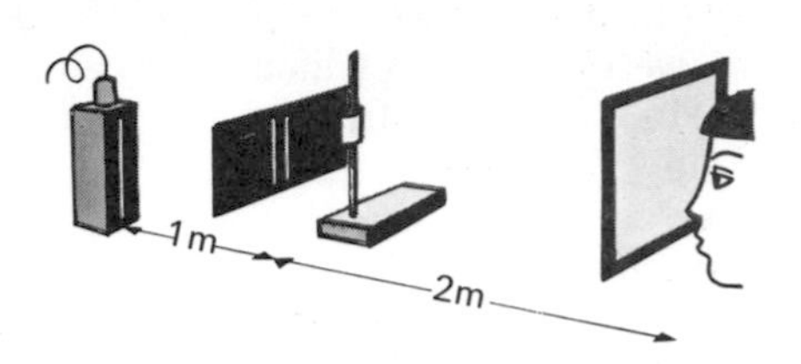

Fig. 16–12

Young's double-slit experiment at last convinced the world's scientists that light was a form of wave motion. No particle theory could explain the interference so clearly demonstrated by this experiment. A practical arrangement using a lamp with a very small filament is shown in Fig. 16–12, and the theory is illustrated in Figs. 16–13 and 16–14. When the path difference from the two slits is the same a bright central line will be produced. When the path length differs by one wavelength λ, the next (first order) bright line is formed. If the distance between these lines D is measured and the separation of the slits d known, we can calculate the wavelength of monochromatic light.

The two shaded triangles in Fig. 16–14 can be taken as similar for all practical purposes since l is very much larger than d or D. Thus

$$\frac{\lambda}{d} = \frac{D}{l}$$

$$\Leftrightarrow \lambda = d \times \frac{D}{l}$$

or

$$\lambda = d \sin\theta$$

The distance between the central bright fringe and the first *dark* band is $D/2$.

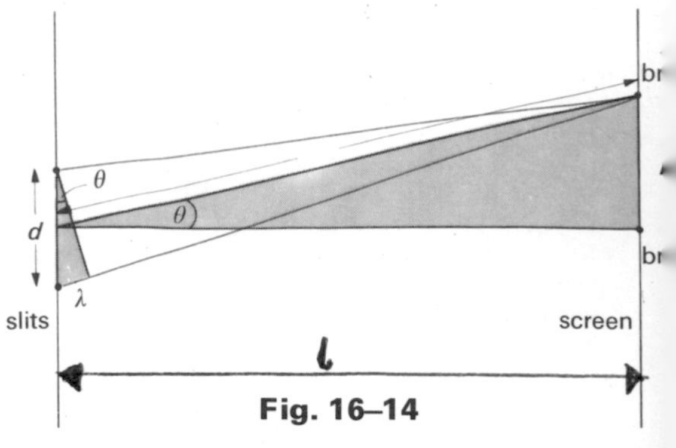

Fig. 16–14

Constructive interference
Bright bands are formed where the path differen from the slits is an integral number of wavelengt (i.e. $n\lambda$, where $n = 0, 1, 2, 3 \ldots$).

Destructive interference
Dark bands are formed where the path differen is an odd number of half wavelengths [i.e. $\left(\frac{2n+1}{2}\right.$ where $n = 0, 1, 2, 3 \ldots$].

DIFFRACTION GRATING

From Young's double slit, the diffraction grati has evolved. By using a large number of slits can pass much more light through and the lin produced *for each wavelength* are much mo

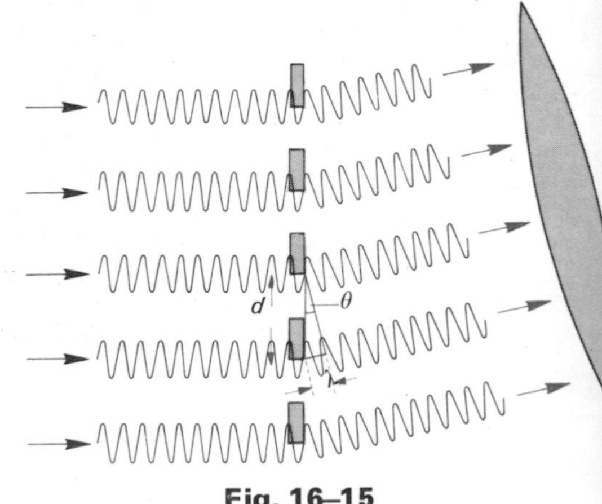

Fig. 16–15

clearly defined. Fig. 16–15 shows how one line one wavelength might be produced. As each wa train differs from the adjacent wave-train by o wavelength, the waves will produce constructi interference when they are brought together by

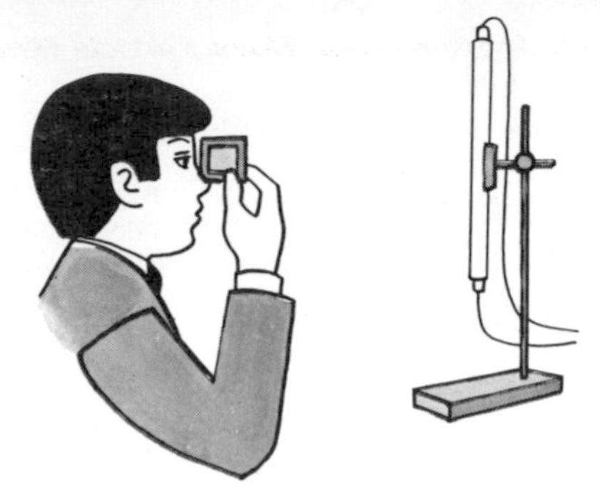

Fig. 16–16

lens system. A spectroscope or the human eye (Fig. 16–16) can be used to bring these waves to a focus, thus producing a single line.

If white light is passed through a diffraction grating continuous spectra are formed.

For light of one particular wavelength λ, there will be certain directions in which the waves will reinforce each other. As with the double slit, the first order line is given by $\sin\theta_1 = \lambda/d$, where d is the distance between adjacent lines of the grating and θ_1 is the deviation. As $\sin\theta_1$ is proportional to λ, the deviation of the longer wavelengths is greater. If λ is to be measured accurately, d should not be very much greater than λ, otherwise $\sin\theta_1$ will be extremely small.

The diffraction grating also produces *second order spectra* when the path difference is 2λ, *third order spectra* when the difference is 3λ, and so on. For second order spectra $\sin\theta_2 = 2\lambda/d$, for third order spectra $\sin\theta_3 = 3\lambda/d$, etc.

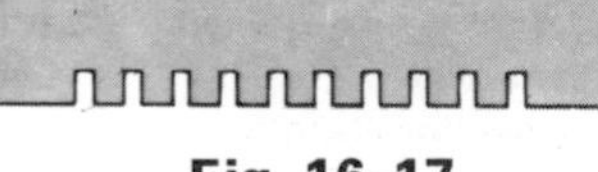

Fig. 16–17

An interesting analogue of the diffraction grating can be produced in a ripple tank using a multi-slit barrier (Fig. 16–17). The resulting pattern shows the central reinforced wave and first order deviations (Fig. 16–18).

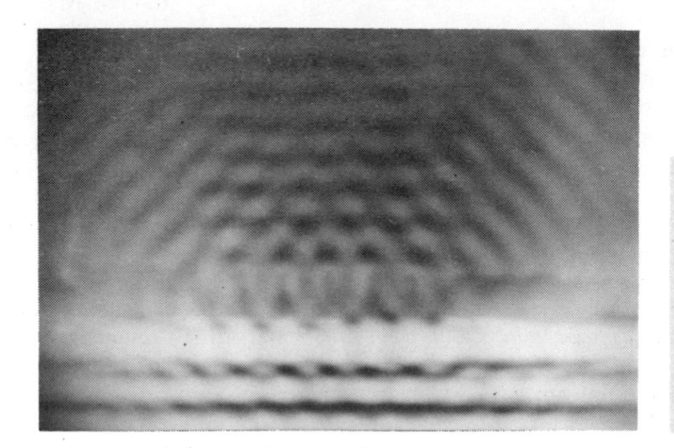

Fig. 16–18

BLOOMED LENSES

As some light is always reflected from the surface of a lens, there is a reduction in the light transmitted and in the intensity of the image formed. The amount of light reflected can, however, be reduced by coating the surface of the lens with a layer of magnesium fluoride. Lenses in modern cameras, telescopes, and binoculars are normally *bloomed* in this way. By making the thickness of the coating $\lambda'/4$ (where λ' is the wavelength in the coating), light rays reflected from its outer surface and from the surface of the lens are 180° out of phase and so cancel. As, in practice, this only happens perfectly for one wavelength, an average value of about 600 nm in air is chosen. The reflected

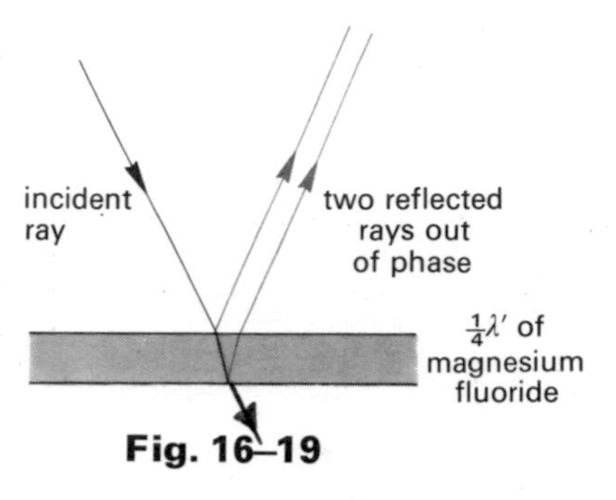

Fig. 16–19

rays shown in red in Fig. 16–19 do not, of course, exist. As energy is conserved, the light which would have been reflected had the lens not been bloomed is now transmitted through the lens. This is the point of the exercise.

If the absolute refractive index of magnesium fluoride is 1.38, find the thickness of the coating needed to prevent the reflection of the above wavelength of light, i.e. 600 nm in air. (16.1)

SUMMARY

1801 Young produced an interference pattern from the light passing through two narrow slits. The wave nature of light was thus confirmed.

For Young's slits or a diffraction grating $\lambda = d\sin\theta_1$ for the first order spectrum.

PROBLEMS

16.2 How do you reconcile the formation of sharp shadows with the diffraction of light?

16.3 A car bulb has two parallel filaments close together. If they are both switched on, will they produce an interference pattern similar to Young's slits? Explain.

16.4 A monochromatic source is used to illuminate a pair of Young's slits which are separated by 0.2 mm. Bright lines 5.9 mm apart are produced on a screen 2 metres from the slits. Find the wavelength of the light. What might the source be?

If the slits were separated by 0.4 mm, would the bright lines on the screen be further apart?

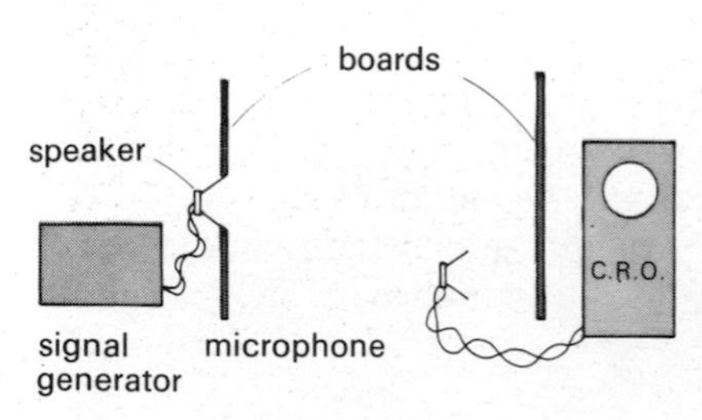

Fig. 16–20

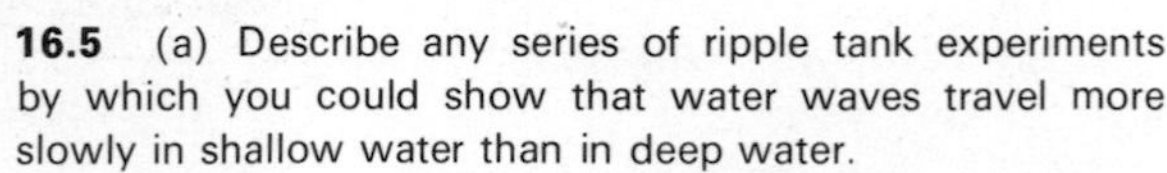

16.5 (a) Describe any series of ripple tank experiments by which you could show that water waves travel more slowly in shallow water than in deep water.
(b) With the arrangement shown in Fig. 16–20, the frequency of the sound emitted by the loudspeaker is 3300 Hz. The microphone is moved about in the space between the boards and detects points of minimum intensity at various positions. Explain why these minima occur. If the speed of sound is 330 m s^{-1}, how far apart will the positions of minimum intensity be?
(c) The two loudspeakers A and B are connected to a signal generator as shown in Fig. 16–21, and give a maximum of intensity at points on the perpendicular bisector of the line joining them.

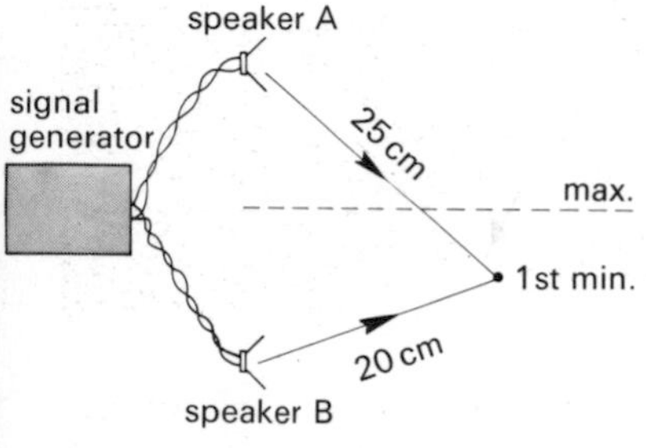

Fig. 16–21

The detector microphone is moved to one side and detects a first minimum position when the distances from the loudspeakers are 25 cm and 20 cm respectively. What is the frequency of the sound used?

If the microphone is moved further in the arc of a circle at a distance of 25 cm from loudspeaker A, at what distance will it be from loudspeaker B at the next minimum position?
(d) What would be the effect on the intensity of the signal picked up at one of these minimum positions if
(i) *one* of the loudspeakers were covered by a sound absorbing cloth;
(ii) the wires to *one* of the loudspeakers were connected the other way round?
Which of the alterations (i) or (ii) results in the greater strength of signal being detected? Explain why.

[S.C.E. (H) 1969]

16.6 (a) Two fine slits, very close together, are ruled on a black-painted microscope slide. When an electric lamp, several metres away, is viewed through the slits, a band of light and dark fringes is seen. Away from the centre of the band, the fringes appear tinged with colour, mainly red and violet.
(i) Explain why the fringes occur and say what they suggest about the nature of light.
(ii) Suggest an explanation for the coloured edges to the fringes.

(b) A microphone, connected to a cathode ray oscillosco is set up about 1.0 m from a loudspeaker which is emitt a note of frequency about 2 kHz. The apparatus is set o bench fairly close to the wall of the room. The wave tr on the oscilloscope is as shown in Fig. 16–22.
As the microphone is moved slowly away from the lou speaker, the wave trace on the oscilloscope gradu becomes smaller, as expected, but then, on further mo ment of the microphone, it increases in size before ag diminishing. The intensity of the sound emitted has changed nor have the controls of the oscilloscope be altered.
(i) Suggest an explanation for this behaviour.
(ii) Show how a single point source in a ripple tank can made to produce water waves which would illustr the behaviour of the sound waves in the above exp ment.
Draw a diagram of the resulting wave pattern.

[S.C.E. (H) 19

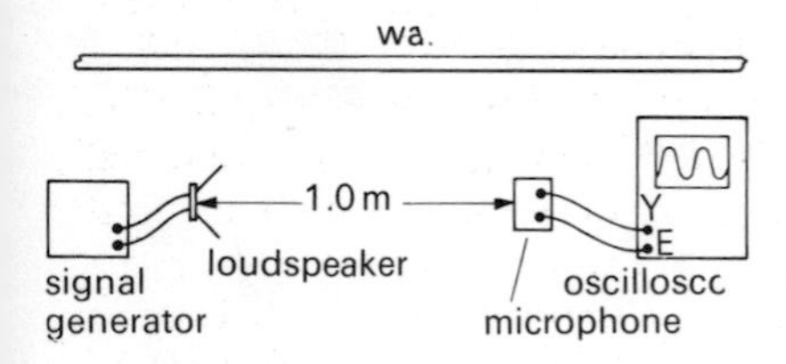

Fig. 16–22

16.7 Describe how you would demonstrate the follow phenomena in a ripple tank: (a) refraction and (b) diffract In each case show an *accurate* drawing of the result wave pattern.

[S.C.E. (H) 19

16.8 (a) Use the values given in the data book for refractive index of water and the velocity of light in va to calculate the velocity of light in water.
(b) In a Young's slits experiment designed to demonstr the interference of light, two parallel slits scratched o blackened microscope slide are illuminated by an inte beam of monochromatic light (Fig. 16–23).
Bright fringes with an average separation Δx given by formula $\Delta x = \lambda D/d$ are observed on a distant screen.
(i) State the effect of
A. bringing the screen closer to the slits;
B. reducing the separation of the slits.
(ii) Explain the effect on the Young's interference patter
A. covering one of the slits;
B. using light of longer wavelength;
C. using white light.
(iii) Two parallel slits 0.5 mm apart are found to prod fringes with an average separation of 1.0 cm o screen placed at a distance of 8 m. What do th figures give for the wavelength of the incident light
(iv) In the practical determination of this wavelength th distances have to be measured. By considering e measurement in turn, explain which one would be most critical in obtaining a reasonably accurate res
(c) Describe with the aid of a sketch how you wo demonstrate the interference of sound waves.

[S.C.E. (H) 19

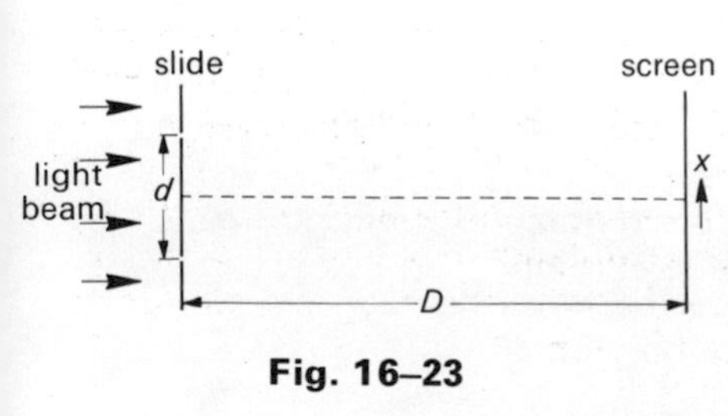

Fig. 16–23

16.9 What are the main characteristics of wave motion
(b) Outline how you would show experimentally that
(i) sound and
(ii) light
are transmitted by waves.
(c) State *two* important differences between light sound waves.

[S.C.E (H) 19

17 Spectra

HISTORY OF SPECTROSCOPY

For thousands of years man watched and wondered about Nature's giant spectrum—the rainbow. A similar array of colour was mysteriously produced when sunlight passed through colourless fragments of precious stones.

Then in 1666 Newton discovered that when a shaft of sunlight passed through a prism it produced the same range of colours—red, orange, yellow, green, blue, indigo, violet. Newton called this the *spectrum*, and spoke of 'the most surprising and wonderful composition . . . of whiteness'.

Little if any use seems to have been made of Newton's discovery during the next century, and it was not until 1752 that Thomas Melville published in Edinburgh the first description of an *emission spectrum*. He used a prism to study a sodium flame.

In 1801, as we have seen, Thomas Young passed a beam of white light through two fine slits and used his interference theory to calculate the wavelength of the colours in Newton's spectrum. Later, Joseph Fraunhofer increased the number of slits to produce the first *diffraction grating*. This device has largely replaced the prism in modern spectroscopy.

Fraunhofer built a refined form of spectrometer to study the sun's spectrum, and discovered that it was crossed with 'an almost countless number of strong and weak vertical lines'. He found that planetary spectra were similar but stellar spectra differed from that of the sun.

During the first half of the nineteenth century, flame and arc spectra were studied by many scientists, but it was not until 1859 that a general pattern emerged. This was largely due to the work of the German physicist Gustav Robert Kirchhoff. He pointed out that each element *emits* its own unique set of colours (line spectra). He also showed that the element in its gaseous state will *absorb* these same colours of light from white light (absorption spectra).

By stating this general law relating emission and absorption spectra, Kirchhoff was able to explain the dark lines across the sun's spectrum, which Fraunhofer discovered. The Fraunhofer lines in-

Fig. 17–1

dicated the presence of elements in the sun's atmosphere. They included many elements already known, such as sodium and calcium, but one set of lines could not be explained. Apparently there was an unknown element in the sun's atmosphere. It was labelled *helium*, after the Greek word for the sun—helios. The chemists had already predicted the possible existence of such an element, but another quarter of a century elapsed before Sir William Ramsay first identified traces of helium in the Earth's atmosphere. Later, small amounts of helium were also found in natural gas from the gas wells of Texas.

The work of Robert Bunsen and Kirchhoff laid the foundation of modern spectroscopy. In the course of their studies they discovered two new alkali metals, caesium and rubidium.

The wavelengths of spectral lines may now be determined so accurately by spectroscopic methods that the international standard of length, the metre, is now based on the wavelength of the red line in the spectrum of krypton 86.

THE SPECTROMETER

The process of splitting up light into coloured beams is called *dispersion*. The amount by which a beam of light is bent when it enters a block of glass depends on the colour of the light. By using a triangular prism, the emergent rays at the violet (short wave) end of the spectrum are bent more than those at the red (long wave) end. This produces dispersion.

Fig. 17–2

One of the simplest ways of using a prism to observe a spectrum is to hold a bright needle in the sun's rays and then view the needle through the prism (Fig. 17–2). Another method is shown in Fig. 17–3. Coloured images of the slit or filament are formed on the card where they merge to form a *continuous spectrum*. *D* represents the angle of deviation.

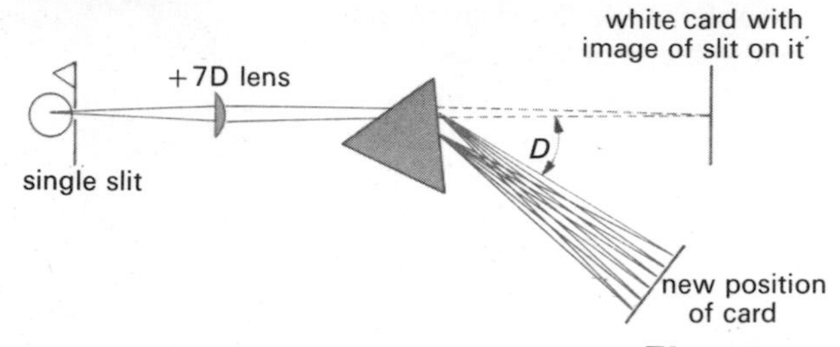

Fig. 17–3

If, however, we wish to make precise measurements of the angle of deviation for different colours, we need an accurately calibrated instrument such as the *spectrometer* illustrated in Fig. 17–4. It consists of three main parts—a tube called the collimator (to collimate means to 'make parallel'), a prism or diffraction grating standing on a turntable, and a telescope through which the spectrum is viewed. Scales on the turntable enable us to measure the appropriate angles. As the lenses used in the collimator and telescope must not produce dispersion, they are made of two types of glass which have different refractive indices. This produces a converging lens which is *colour corrected* (Fig. 17–5).

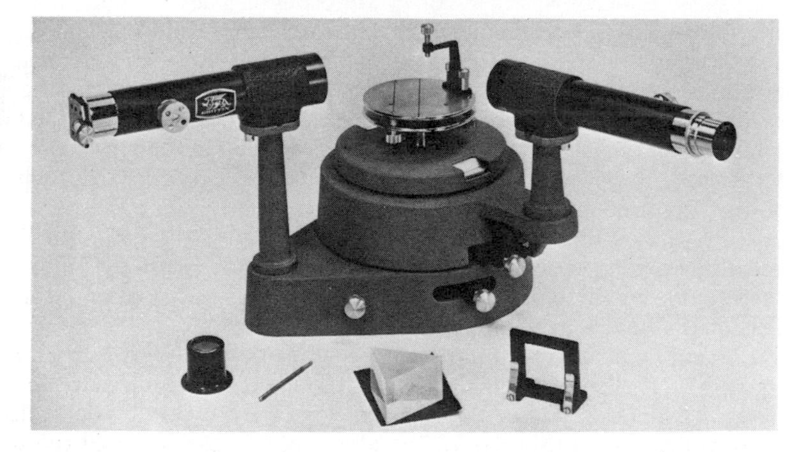

Fig. 17–4

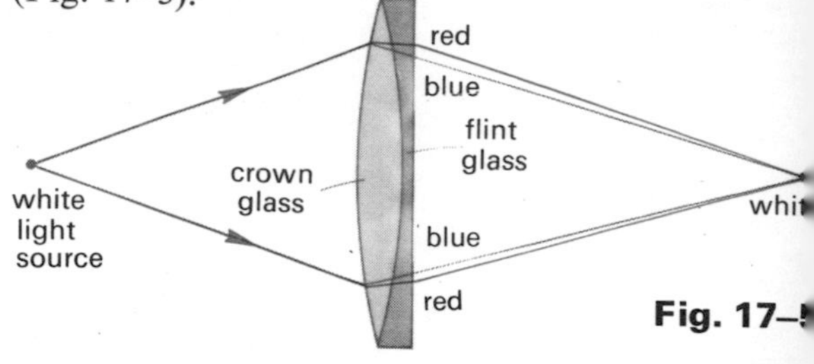

Fig. 17–5

Light from the source being studied first passes through a narrow slit. It is then bent to form a parallel beam by the collimator lens which is fixed one focal length away from the slit. The beam enters the prism or grating and emerges as a set of coloured beams. Each beam produces a coloured image of the slit at the principal focus of the telescope objective lens. The image which may be formed on a small screen or cross wires is then viewed through the eyepiece lens (Fig. 17–6).

Fig. 17–6

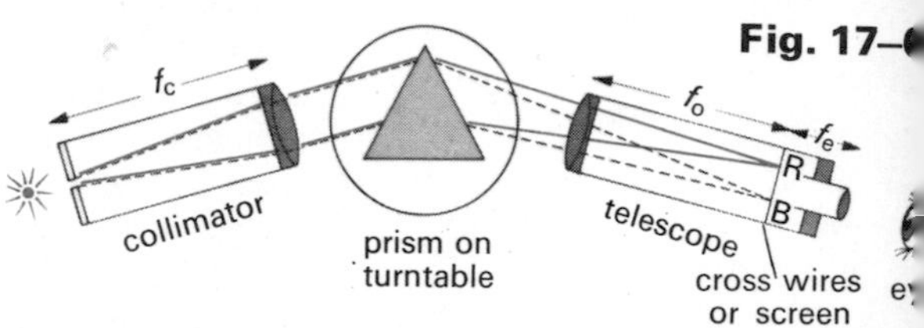

A diffraction grating placed on the spectrometer table can be used to measure the wavelength of the spectral lines. The grating is set at right angles to the beam and the telescope adjusted until the direct image of the slit is formed on the cross wires of the telescope (Fig. 17–7). The telescope is then rotated until the appropriate spectral line appears on the cross wires (Fig. 17–8). The angle of rotation is noted

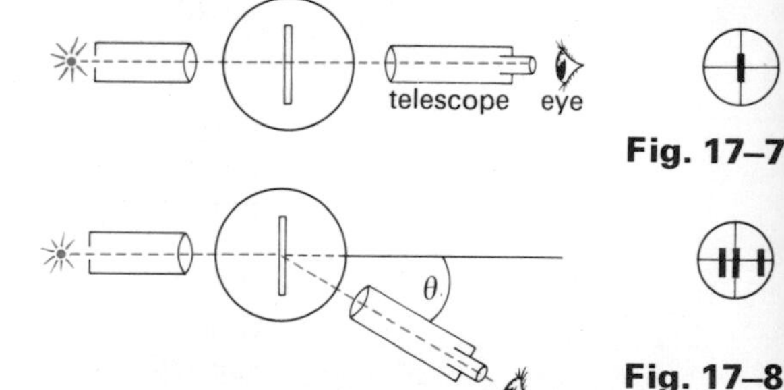

Fig. 17–7

Fig. 17–8

and the wavelength of light calculated from it using $\lambda = d \sin\theta$, where d is the distance between two adjacent lines on the grating.

The Spectroscope

When a spectrometer is used merely to examine spectra rather than to take measurements it is called a spectroscope. Other simpler types of spectroscope can, of course, be used for this purpose. A cardboard tube with a slit fixed at one end and a diffraction grating at the other is perhaps the simplest (Fig. 17–9).

Fig. 17–9

Three prisms made of crown and flint glass are cemented together and used in another form of direct vision spectroscope illustrated in Fig. 17–10.

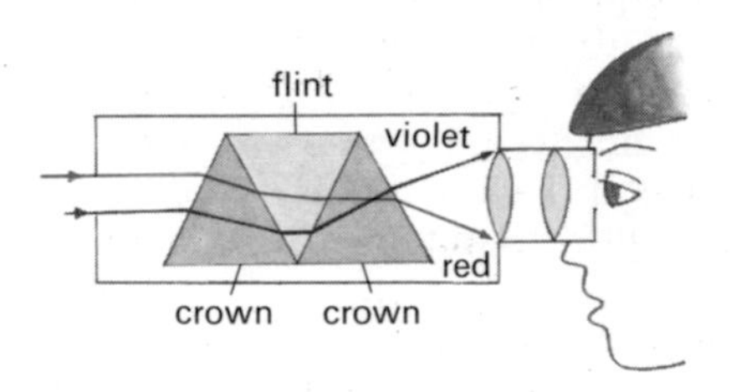

Fig. 17–10

These spectroscopes are particularly suitable for qualitative studies of emission and absorption spectra.

EMISSION SPECTRA

Every source of light, whether it be in a galaxy or a glow worm, emits a particular pattern of colours which gives a clue to the nature of the source.

Continuous Spectra

Our most familiar source, the sun, produces a spectrum ranging from red (long wave) to violet (short wave). Continuous spectra are emitted from glowing solids, liquids, or gases under high pressure. As the temperature of the source rises, the most intense part of the spectrum shifts towards the violet (high frequency) end. Examining the spectrum of a star will tell us something about its temperature.

Line Spectra

A line spectrum is produced by excited atoms of a glowing low-pressure gas, for example, the atoms of an element vaporized in a flame (Fig. 17–9). When a salt is heated its atoms are *excited*; that is, some of their electrons are moved further away from the nucleus, so that they gain electrical potential energy. This is not unlike the gain in gravitational potential energy when a body is moved away from the Earth. In the atom, however, only certain energy levels are 'permitted', and these we represent in Fig. 17–11 by various horizontal

Fig. 17–11

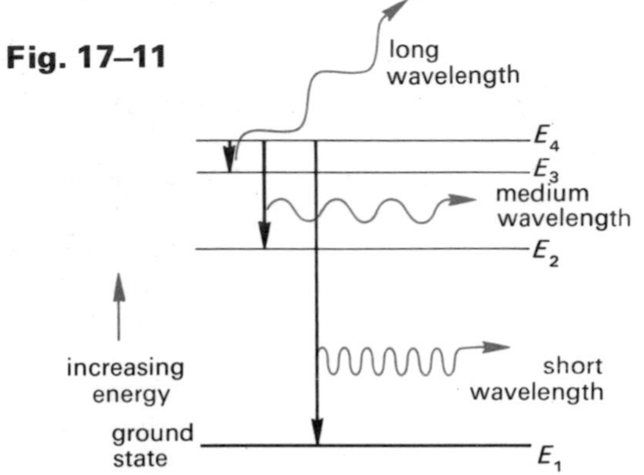

lines. As electrons fall back to a lower state, light of various wavelengths is emitted. The wavelengths depend on the difference between the energy levels. As the atoms of every element produce a different characteristic spectrum, a chemist can, by studying the spectrum produced by a substance, identify the elements in it. The spectrum has been called the 'fingerprint' or the 'signature' of an element. Various line spectra are illustrated inside the front cover.

To examine the spectrum of a gas, a low pressure discharge tube can be used (Fig. 17–12). When a

Fig. 17–12

high voltage is applied to the electrodes, some of the gas atoms are excited in a series of collisions. They then emit light as the electrons return to their original energy level. The spectrum of a metal can be studied by using it to make the electrodes of a spark gap. If a spark is then produced by applying a high voltage across the gap, the electrodes reach a very high temperature and the metal vaporizes. It then emits the spectrum characteristic of that metal.

Band Spectra

Compounds and polyatomic gases produce groups of lines which appear compressed together at the red end of each group. They thus form a band of colour called a band spectrum. As the atoms rotate and vibrate within the molecules, they produce a wide range of different energy jumps.

ABSORPTION SPECTRA

When white light is passed through a colour filter, a dye in solution or a glowing vapour, some of the light is absorbed. The spectrum of the light that gets through is called an *absorption spectrum*. In general there are three types of absorption spectra corresponding to the three emission spectra.

About the middle of the nineteenth century Kirchhoff found that, when he placed a sodium flame in the path of white light, he obtained a continuous spectrum which had dark lines across it in the positions of the sodium emission spectrum. The two spectra are reproduced inside the front cover. Fig. 17–13 shows a similar experiment using a quartz-halogen lamp as a source of white light. An image of the filament is formed in the sodium flame by a large convex lens, and a direct vision spectroscope is used to study the absorption spectrum. From his experiment Kirchhoff concluded that the flame was *absorbing* light of the same wavelength

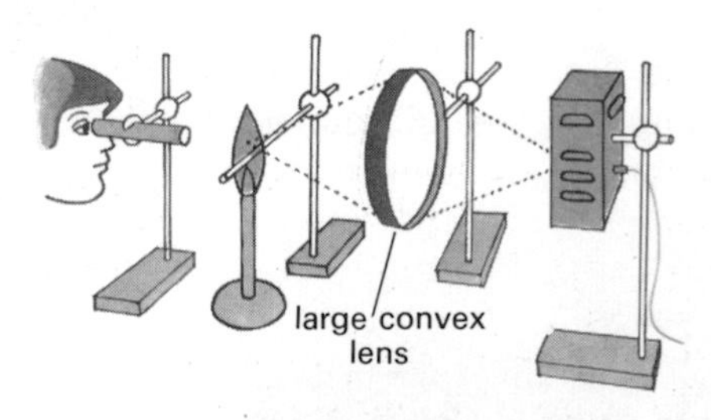

Fig. 17–13

as that which it usually *emits*. The energy absorbe in this way is re-emitted in all directions and so th amount re-emitted in the viewing direction is sma and consequently the line looks dark against th other colours. Kirchhoff's interpretation of th Fraunhofer lines was confirmed when, during total eclipse of the sun, bright *emission* lines wer seen in exactly the same positions as the Fraunhofe absorption lines. The bright lines were emissio spectra of the vapours round the sun.

A much simpler way of observing the absorptio of yellow light by a sodium flame is illustrated i

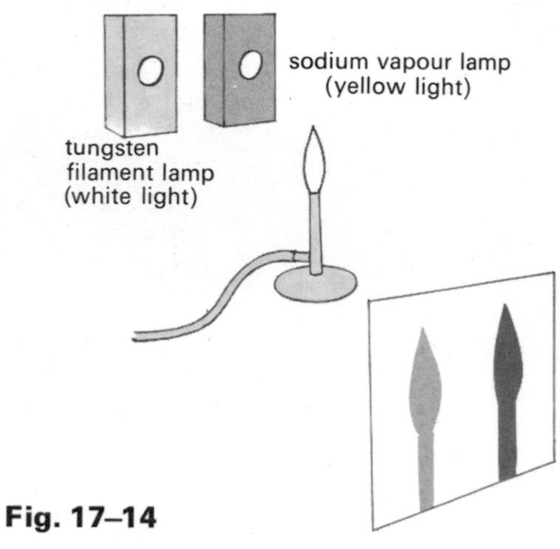

Fig. 17–14

Fig. 17–14. A tungsten filament lamp (white) and a sodium vapour lamp (yellow) cast shadows of a bunsen flame on to a translucent screen. When a sodium pencil is held in the flame, the shadow shown in Fig. 17–15 are produced. *Which shadow i cast by the sodium vapour lamp?* (17.1)

A piano subjected to a very loud note at th natural frequency of one of its strings provides a analogous situation. The string will start to vibrat by absorbing sound energy at the frequency a which it normally radiates.

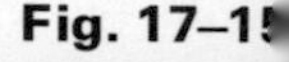

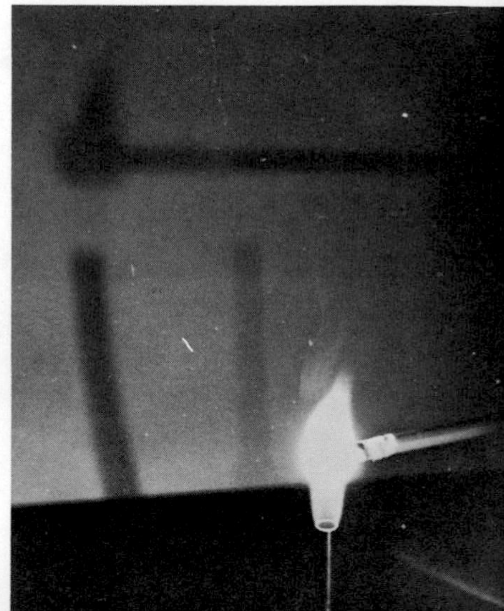

STELLAR SPECTRA

The absorption spectra of some stars have been found to contain sets of lines which are similar to the spectra of elements on the Earth. Other stars have the same spectral lines but they are slightly displaced towards the violet end of the spectrum. Others, again, have their lines displaced towards the red end of the spectrum—the so-called 'red shift'.

When a racing car roars towards you, passes, and then tears away from you, the note of the engine appears to change. In the days of steam trains, the note of the whistle appeared to alter as the train approached, passed, and moved from you. As the train moved towards you, the waves were 'compressed', their wavelength reduced and their frequency increased. As the train left, the waves were 'stretched', their wavelength increased, and the frequency reduced. This is called the *Doppler effect*. The red shift in the spectrum of a star is another example of the Doppler effect. The star is moving rapidly away from the Earth so that the frequency of the light emitted by it appears to be reduced; that is, the spectral lines are shifted a little towards the red end of the spectrum.

As the spectra of the sun, stars, and planets all contain the same spectral lines as those found in the spectra of elements on the Earth, it looks as though all the known universe may be made of the same kind of stuff.

SUMMARY

Emission Spectra

1. *A continuous spectrum* is produced by a glowing solid, liquid or high pressure gas.
2. *A line spectrum* is produced by excited atoms of a glowing low pressure gas.
3. *A band spectrum* is produced by the vibrating molecules in a glowing gas at low pressure.

Absorption Spectra

The glowing vapour of an element emits a bright line spectrum. If, however, white light shines through it, the atoms of the vapour will *absorb* the wavelength it normally emits. The bright lines become dark lines against a continuous spectrum.

In general, there are three types of absorption spectra, corresponding to the three types of emission spectra.

PROBLEMS

17.2 Draw a sketch to represent a glass prism with three 60° angles. Indicate the path of red and blue rays of light incident on one face at 50°. (Refractive index for glass is 1.51 for red and 1.53 for blue light.)

17.3 Give examples of three sources of line spectra.

17.4 A diffraction grating with 600 lines per millimetre is used in a spectrometer. The *first order* diffracted image of the slit is obtained when the telescope is rotated through 20.7° from the position where the direct image of the slit was observed. What is the wavelength of the light used to illuminate the slit?

What would be the value of θ if a 1200 lines per mm grating were used?

17.5 The diagram (Fig. 17–16) is that of a bright-line

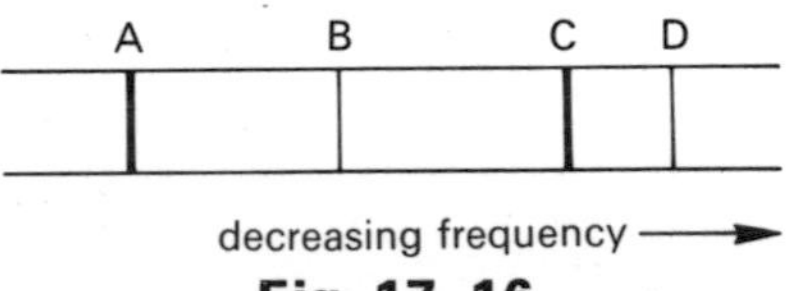

Fig. 17–16

spectrum associated with one kind of atom.
(a) How do you account for there being several lines, A, B, C, D?
(b) Explain why two of the lines (A and C) are more intense than the others. [S.C.E. (H) 1970]

17.6 The sketch (Fig. 17–17) illustrates the main parts of a spectroscope.

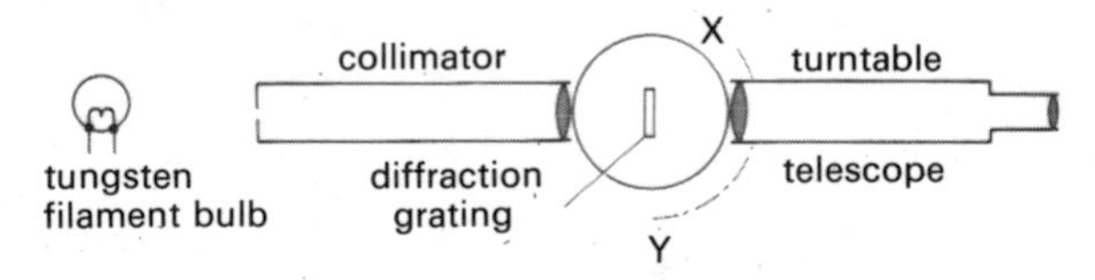

Fig. 17–17

(a) What is the function of (i) the collimator, (ii) the diffraction grating, and (iii) the telescope?
(b) What would be observed (i) when the telescope is in the position shown and (ii) while the telescope is being rotated from X to Y?
(c) State the additional apparatus required to produce an absorption spectrum, and indicate how this apparatus would be used. Describe the appearance of such a spectrum.
(d) Explain why a chemist might find a spectroscope useful in his work. [S.C.E. (H) 1971]

18 The Fickle Photon

A PARTICLE IS A WAVE IS A PARTICLE

By the end of the nineteenth century the scientific world was convinced that light was a form of wave motion. Refraction, as we have seen in Chapter 14, was explained in terms of a change in the speed of light waves when they enter another medium. The refractive index is merely the ratio of the two speeds. Diffraction of light was seen to resemble the diffraction of water waves in a ripple tank and finally the interference experiments of Thomas Young showed, beyond a shadow of doubt, that two whites *could* make a black . . . or an even brighter white. Interference dealt the final blow to the particle theory. It was impossible to imagine two particles cancelling each other.

As so often happens in the history of science, a problem no sooner appears to have been satisfactorily solved than suddenly some new evidence appears which throws doubt on the validity of the neat solution.

Oddly enough it was the experiment by which Hertz confirmed Maxwell's electromagnetic *wave* theory which, by an unexpected and incidental result, reopened the question of the *particle* nature of light! Hertz showed that a spark was produced more easily if the electrodes were illuminated by ultraviolet light. This *photoelectric* effect was investigated by Hallwachs, a student of Hertz. He discovered that, when certain metals were illuminated by ultraviolet light, they became charged. Moreover, the charge was always *positive*. This effect can easily be demonstrated by shining ultraviolet radiation on to a zinc plate attached to a d.c. amplifier as shown in Fig. 18–1.

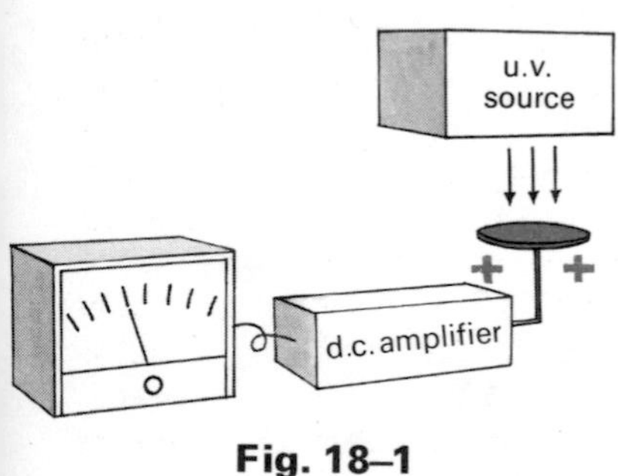

Fig. 18–1

Experiments with a charged electroscope can also help us to investigate the photoelectric effect. For example, if an electroscope fitted with a normal brass plate is charged positively and visible light then made to shine on the plate—nothing happens! Similar unexciting results are obtained when the electroscope is charged negatively and also when we shine ultraviolet light instead of visible light on the plate. However, if the electroscope plate is made of clean zinc, a different result is obtained when it is negatively charged (Fig. 18–2). This last result could be explained by asserting that the ultraviolet radiation has made the zinc lose electrons. Or perhaps the radiation is ionizing the air round the plate, as would the radiation from a radioactive source or a stream of X-rays. Or again the ultraviolet radiation could be a stream of positive charges. *Which of these explanations is most probable—and why? (18.1)*

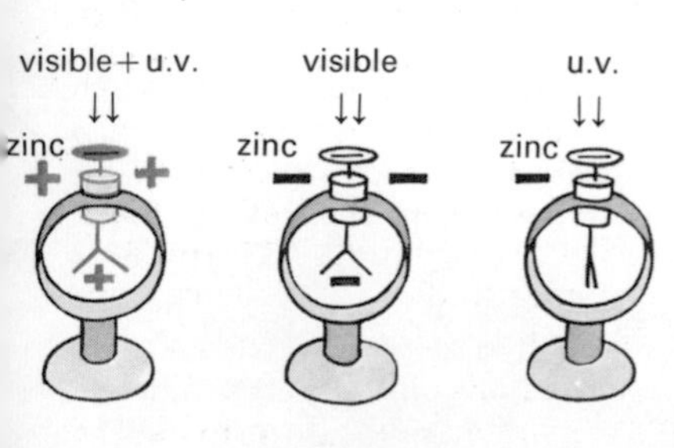

Fig. 18–2

If we assume that the ultraviolet radiation has enough energy to release electrons from the zinc we can see that a positively charged electroscope will attract these electrons back again. If it is negatively charged, the released electrons will be repelled and the electroscope will discharge. But a powerful source of visible light shining for a *long time* carries far more energy than a brief burst of ultraviolet, yet such a beam of light does not discharge the electroscope. There must then be some property of zinc which enables it to lose electrons when ultraviolet radiation falls on it, yet no amount of visible light releases these electrons.

In a series of experiments conducted around the turn of the century, the effects of various radiations on different metals were studied. Fig. 18–3 represents the evacuated tube in which ultraviolet radiation entered through a quartz window and fell on a zinc electrode (photo-cathode). Some of the rays from the cathode passed through a hole in the anode and produced a spot on the fluorescent screen. These rays were subjected to electric and magnetic fields and were identified as *electrons*.

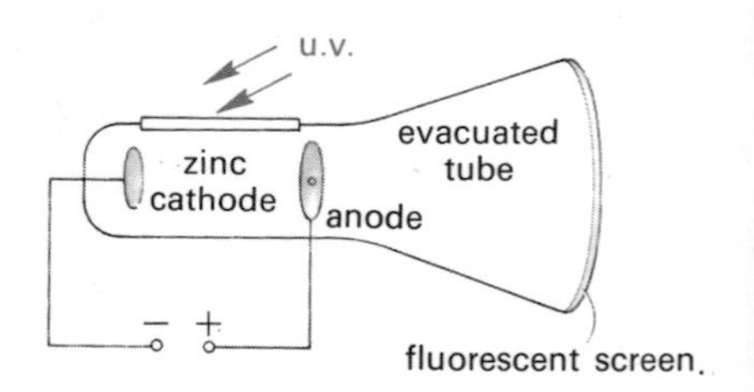

Fig. 18–3

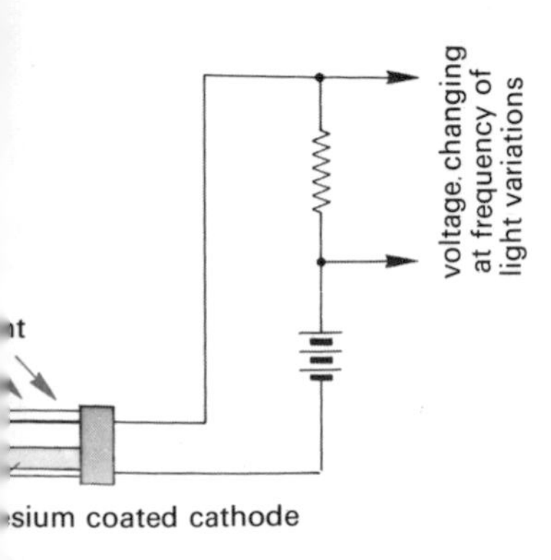

Fig. 18–4

PHOTO CELLS

Electrons are ejected from zinc by ultraviolet radiation but not by visible light. A few metals, including caesium, do, however, emit electrons when light falls on them. In the type of photo-cell used in a cine sound projector, light falls on a caesium coated cathode which then emits electrons. These electrons are attracted to a positively charged wire placed in front of the cathode so that a current, varying with the light intensity, flows through the circuit in Fig. 18–4. If the light first passes through the sound track of a film it will vary in intensity at the sound frequency, thus causing the photo-cell current and the voltage across the resistor to vary at the same frequency. This voltage is then amplified and the output fed to a loudspeaker (Fig. 18–5).

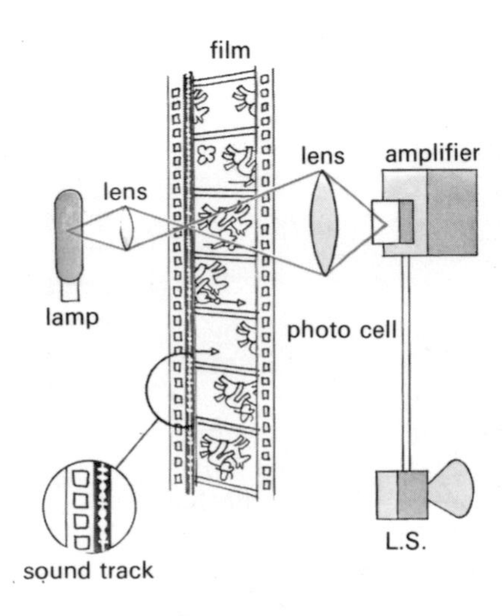

Fig. 18–5

Photo-electric Effect

To study these photo-electric currents, radiations of various frequencies are allowed to fall on a photo-cathode in an evacuated tube and the current is measured by a sensitive microammeter in the anode circuit (Fig. 18–6). The following factors have been found to affect the current:

(a) the intensity of the ultraviolet radiation,
(b) the frequency of the ultraviolet radiation, and
(c) the anode-cathode p.d.

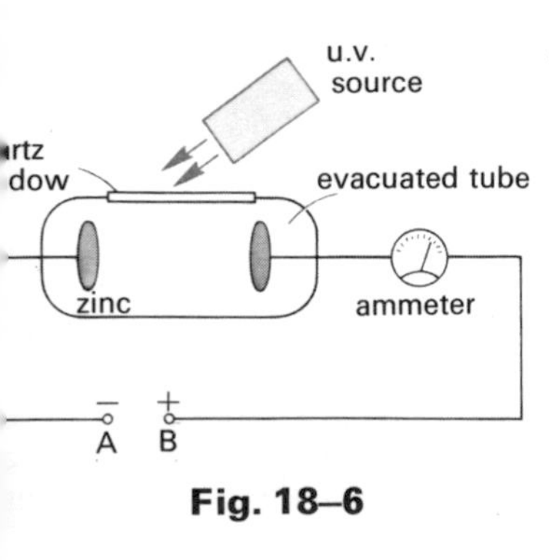

Fig. 18–6

Graphs showing how each of these variables in turn affects the current are shown here. As we might expect, the current varies directly with the

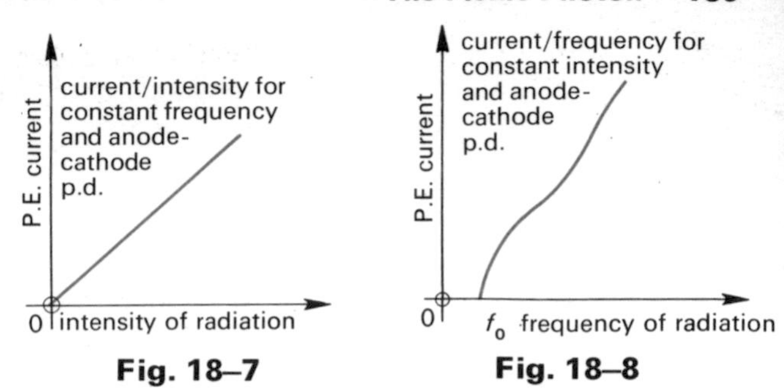

Fig. 18–7 Fig. 18–8

intensity of radiation (Fig. 18–7). The next graph (Fig. 18–8) indicates a less obvious result. There is a threshold frequency (f_0) below which no current is produced. Other experiments show that increases in the intensity or the time of illumination still produce no current below this threshold frequency.

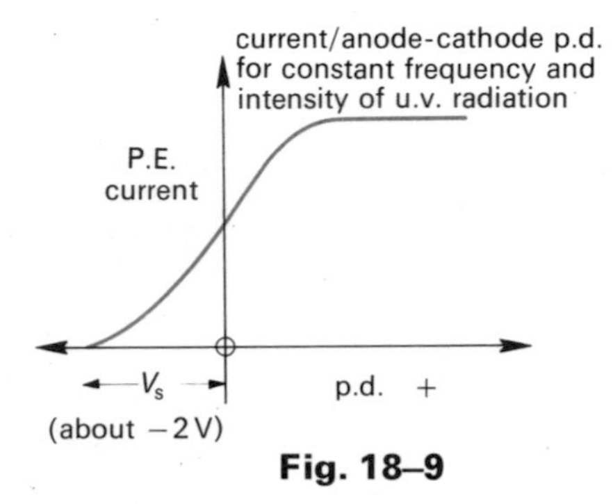

Fig. 18–9

The third graph (Fig. 18–9) is perhaps the most interesting of all. It shows that there is a current even with *no* p.d. applied between the cathode and anode. The energy to make this current flow must come from the ultraviolet radiation. Moreover, once all the photo-electrons are being attracted to the anode, increasing the p.d. makes no difference to the current. This is not unlike the saturation current in a thermionic diode. Finally, a reverse p.d. of about 2 V is needed to stop the current altogether. This stopping voltage V_s can tell us something about the energy of the photo-electrons Remembering that a volt is a 'joule per coulomb' we see that the stopping voltage will indicate the kinetic energy (in joules) per electronic charge (in coulombs) for the most energetic electrons. As Millikan was able to measure the electronic charge ($e = 1.6 \times 10^{-19}$ C), and as the stopping voltage ($V_s = 2$ V, say) can be measured with a voltmeter, the maximum kinetic energy of a photo-electron ($V_s \times e$) must be of the order of 3×10^{-19} joules.

In addition to giving the electron this amount of kinetic energy, the ultraviolet radiation must supply enough energy to overcome the *binding energy*

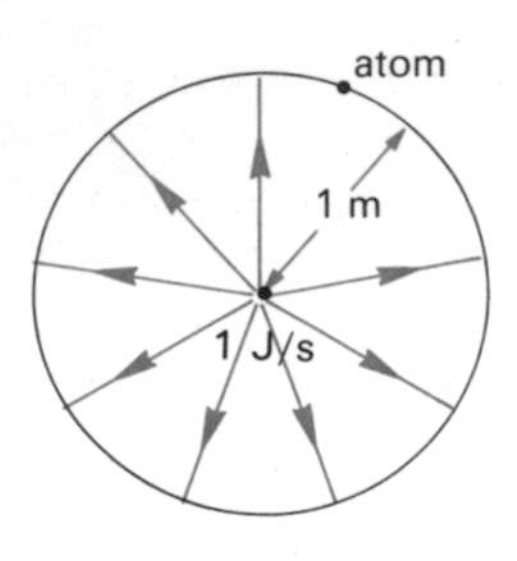

Fig. 18–10

W which keeps the electron in the metal. Therefore, to eject a photo-electron, the ultraviolet radiation must supply

$$\text{energy} = \text{K.E.}_{\text{max}} + W$$

Now suppose that a source emits ultraviolet radiation as a continuous wave at a rate of 1 watt. We can calculate the average energy per second which should pass through an atom in a metal 1 metre away (Fig. 18–10). The oil film experiment suggested that an atomic diameter might be of the order of 0.1 nm and so the energy per second passing through the atom should be a tiny fraction of 1 joule. The fraction is the atomic cross-sectional area divided by the area of a sphere of 1 metre radius. The energy passing through the atom every second is therefore

$$\frac{\pi(0.05 \times 10^{-9})^2}{4\pi(1)^2} \times 1 = \frac{10^{-20}}{16} \text{ joules}$$

If we consider only the kinetic energy of the photo-electron, the time needed for the atom to absorb this amount of energy is

$$\frac{3 \times 10^{-19} \times 16}{10^{-20}} = 480 \text{ seconds}$$

Even if our estimate of atomic area is out by an order of magnitude, the time needed for the atom to absorb the necessary energy could easily be measured. Yet experiments indicate that photo-electrons are emitted immediately ultraviolet radiation falls on the metal. There is no measurable time lag!

In 1905 Albert Einstein proposed a daring solution to this problem. Light is both a wave *and* a particle! It depends on what aspect of light you are considering whether you use the wave model (spread out) or the particle model (not spread out).

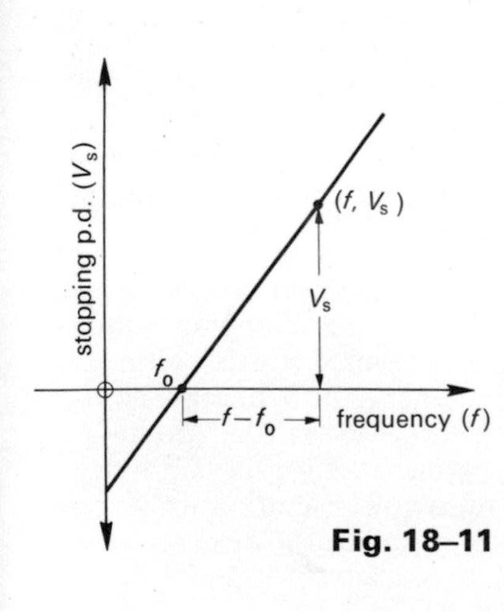

Fig. 18–11

QUANTUM THEORY

Towards the end of the nineteenth century it was found that the electromagnetic theory did not predict correctly the energy radiated at different frequencies from a hot body. In attempting to produce a theory to fit the experimental results, Max Planck suggested his revolutionary *quantum theory*.

If 'atomic oscillators' emit radiation at a particular frequency f, the energy is proportional to the frequency. Moreover, Planck claimed that there is a *minimum quantity* of radiation which he called the *quantum*. As this quantum is proportional the radiated frequency, we can write

$$\text{energy quantum} = hf$$

where h is a constant of proportionality—nc called Planck's constant.

So Einstein argued that perhaps light and ultr violet radiation were not only radiated but al *absorbed* in distinct pulses or 'packets of energ These discrete packets or quanta of radiation a labelled *photons*.

Let us return to the problem of the 1 watt sour situated 1 metre from a piece of metal. If, inste of considering the radiation as spreading out ur formly through a sphere of radius 1 metre, we thi of a large number of high energy photons ea carrying hf units of energy, then *one* of the photons could provide energy to release *one* ele tron from the metal, provided $hf > W$ where W the binding energy, sometimes called the *wo function* of the metal.

Einstein's solution would also explain why t frequency of the radiation determined whether not electrons would be released. If the frequen were so low that $hf < W$, no electrons could emitted no matter how intense the illuminatic was. If we call the lowest frequency which caus electrons to be emitted the threshold frequency ar denote it by f_0, then $hf_0 = W$. We can now wri the equation

$$\text{energy of ultraviolet photon} = \text{K.E.}_{\text{max}} + W$$

as

$$hf = V_s e + hf_0$$

when V_s is the stopping voltage in the phot electric experiment. The equation suggests that, h, e, and f_0 are all constant, a graph of V_s against should be a straight line. Experiments in which t stopping voltages required for various frequenci of radiation were measured produced the resul shown in Fig. 18–11.

The gradient of this graph is $V_s/(f - f_0)$.
From the photo-electric equation

$$hf = V_s e + hf_0$$

$$\frac{V_s}{f - f_0} = \frac{h}{e} = \text{gradient}$$

$$\Rightarrow h = e \times \text{gradient}$$

After measuring the charge on the electron wi his oil drop experiment, Millikan conducted sever photo-electric experiments, and from graphs suc as Fig. 18–11 determined the gradient h/e. Whe he calculated Planck's constant h from these result he found that it agreed to within 0.5% with Planck's own value obtained by a quite different method. Millikan's result confirmed Einstein's quantum theory of light and provided a new accurate method of measuring Planck's constant (6.6×10^{-34} J s).

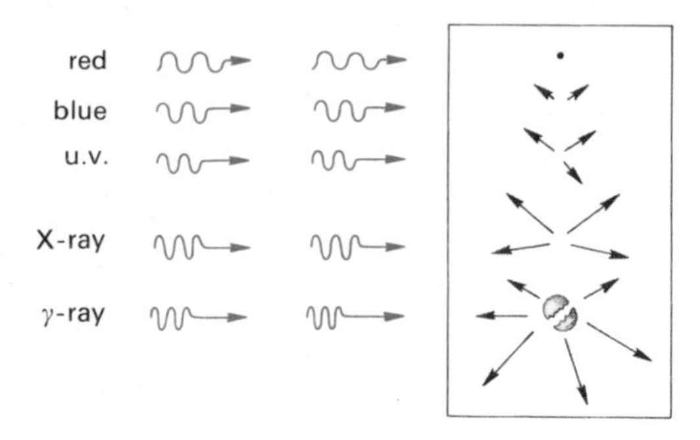

Fig. 18–12

The different energies of light photons are represented in Fig. 18–12. Photons of red light (low frequency) do not have enough energy to eject any electrons from most metals. As the frequency increases, the energy per photon increases. Yet even blue light will release electrons from only a few metals. Ultraviolet photons produce electron emission in many metals and the kinetic energy of the ejected electrons increases as the frequency increases. X-ray photons have even more energy and eject very energetic electrons from any material. A cloud chamber photograph showing the paths of electrons released by X-rays passing through damp air is shown in Fig. 18–13.

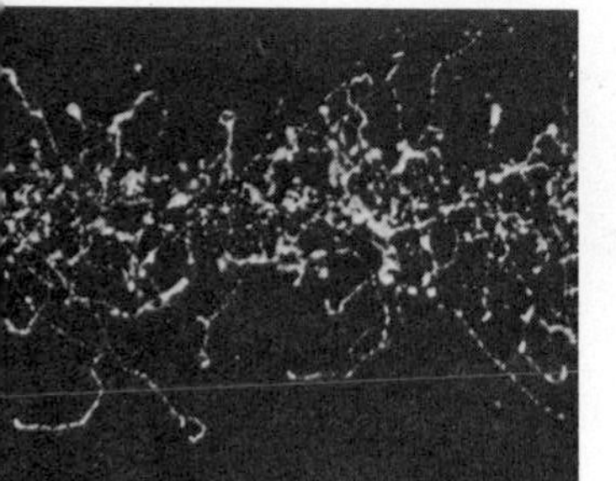

Fig. 18–13

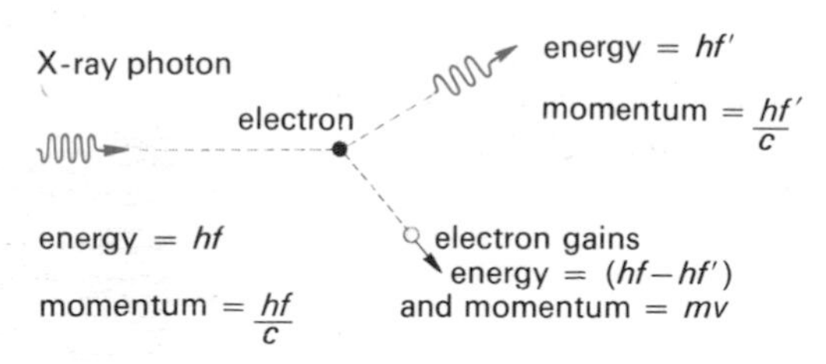

Fig. 18–14

The Compton Effect

In 1923 Arthur Compton produced startling evidence for the particle nature of X-rays. He showed that, when a photon strikes an electron, the wavelength of the photon *increases*; that is, its frequency and thus its energy are *reduced* (Fig. 18–14). Moreover, the electron gains the same amount of energy as the photon loses. Calculations showed that in such a collision both energy and momentum are conserved, as they are in an elastic collision between two solid particles. Later experiments showed that a light photon produces similar results.

The *Compton effect*, as it is now called, provided new evidence for the particle-like properties of radiation. It shows that the photon not only carries energy but also exhibits the perfectly respectable mechanical property of momentum! For his discovery Compton received the Nobel prize for physics in 1927.

Planck's equation gives the energy of a photon as hf. Einstein's famous equation expresses it as mc^2, where c is the speed of light. Thus

$$hf = mc^2$$

$$\Rightarrow \frac{hf}{c} = \text{the momentum of a photon.}$$

The momentum of photons was dramatically demonstrated recently in the Bell laboratories. A small 20 μm glass sphere was supported in mid air for several hours in a vertical beam of laser light.

THE LASER

The laser illustrates a spectacular and fascinating application of quantum theory. The first instrument was made in 1960 by Theodore Maiman in California.

An electron in an atom can be raised from its ground state to a higher energy level by the absorption of a photon. But, as there are only certain permissible energy states for the atom, the energy of the photon (hf) has to be exactly equal to the difference between the original and the final energy levels, i.e. $hf = E_2 - E_1$ (Fig. 18–15). Of its own accord the excited atom can then release its stored energy by emitting a photon identical to the photon absorbed earlier (Fig. 18–16). This process is called *spontaneous emission*. The red parts of each diagram indicate the action we are considering.

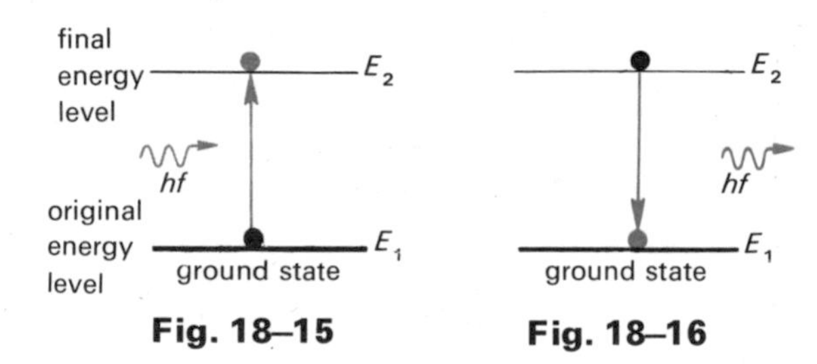

Fig. 18–15 **Fig. 18–16**

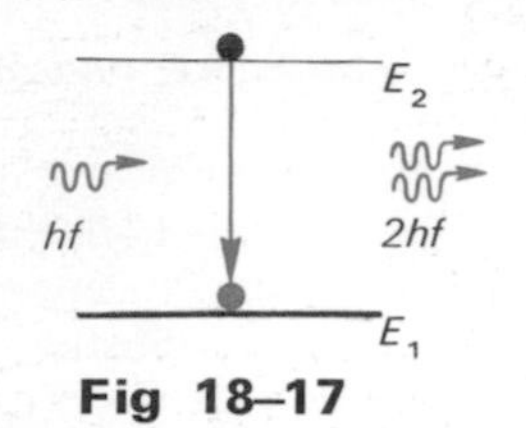

Fig 18–17

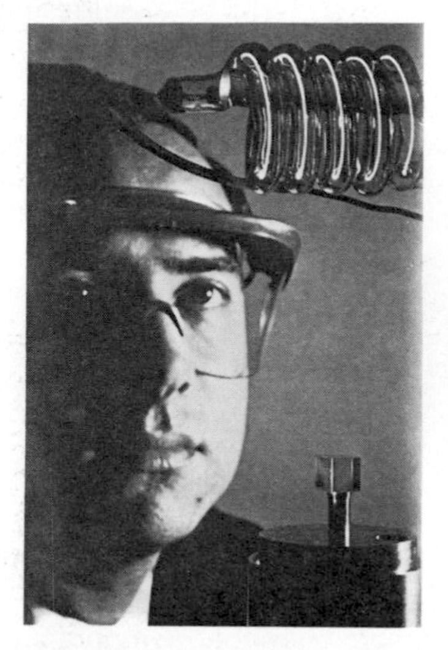

Fig. 18–18

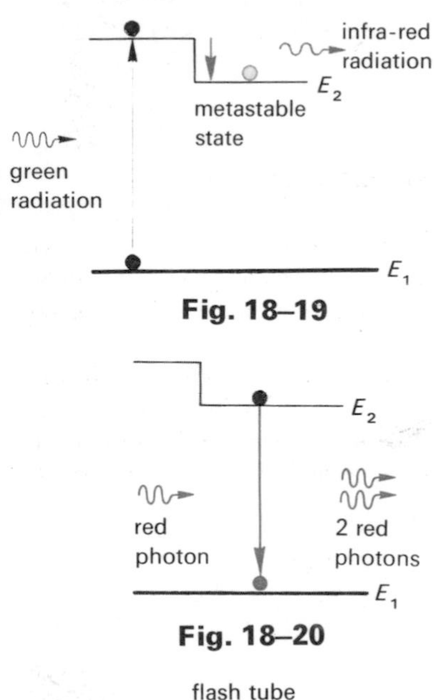

Fig. 18–19

Fig. 18–20

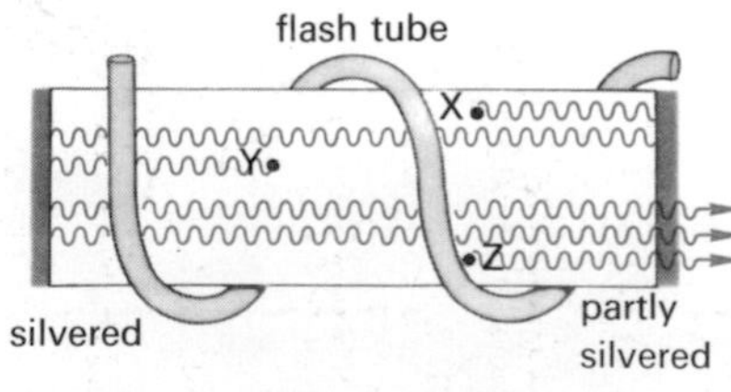

Fig. 18–21

In 1917 Einstein predicted that an excited atom could be 'triggered' to release a photon if another photon *of the same frequency* passed close to it. Moreover, the two photons would then move off together *in the same direction* and exactly *in phase* with each other (Fig. 18–17). This is called *stimulated emission.*

If these two photons then pass close to another two excited atoms, four identical photons will move off in phase and in the same direction. If this avalanche process continues, a powerful beam of light will soon be produced.

However, this *light amplification* can take place only if the photons are *more likely* to meet excited rather than normal atoms. If the photons hit normal atoms and are absorbed by them, they merely raise these atoms to an excited state. Each of these excited atoms then radiates *one photon* at random, as far as time, direction, and phase are concerned.

The problem then was to produce a population of atoms in which there were more excited than normal atoms present. Maiman accomplished this by subjecting the chromium atoms in a ruby cylinder to bright green flashes from a helical flash tube (Fig. 18–18).

Ruby is a form of aluminium oxide which contains a few chromium atoms (about 1 in 5000). The energy levels of the aluminium and oxygen atoms are such that the green light does not excite them but most of the chromium atoms present are raised to a high energy level. They then return to their ground state in two stages. During the first spontaneous jump, low frequency infrared radiation is emitted which heats the ruby crystal so that it has to be water cooled.

Each atom then remains in an intermediate (metastable) state of excitation for about 3 milliseconds (Fig. 18–19). It is during this period that the chromium atoms can be made to produce *L*ight *A*mplification by *S*timulated *E*mission of *R*adiation (Fig. 18–20).

To increase the amplification the photons are made to fly to and fro between mirrors at either end of the ruby cylinder. The laser beam, which lasts less than a thousandth of a second, then emerges through one of the mirrors which is semi-transparent.

The spontaneous emission of one photon after excitation by the flash tube is enough to start the lasing process. Of course many photons will be emitted in all directions but only photons moving along the axis of the cylinder are amplified by successive reflection to form the laser beam. In Fig. 18–21, X represents a spontaneous emission of a photon along the axis of the ruby crystal. I reflected and at Y stimulates the emission another photon. A third stimulated emission ta place at Z and the three photons move off in ph as a coherent monochromatic beam. As this proc continues a pulse of light of immense powe rapidly developed.

The invention of the ruby laser in 1960 followed by the rapid development of many ot types of laser. Some used solids, some liquids, some gases. Some emit pulses (Fig. 18–22 shov

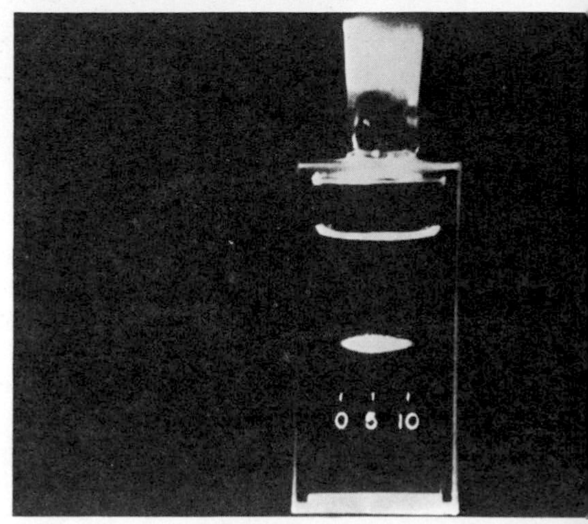

Fig. 18–22

pulse a few millimetres long travelling throug bottle of water) and others emit continuous wa Some produce power densities millions of ti greater than that on the surface of the sun, oth produce fine beams for delicate operations on retina of the eye (Fig. 18–23). In addition to el trical lasers a much more efficient chemical la is now being produced.

Fig. 18–23

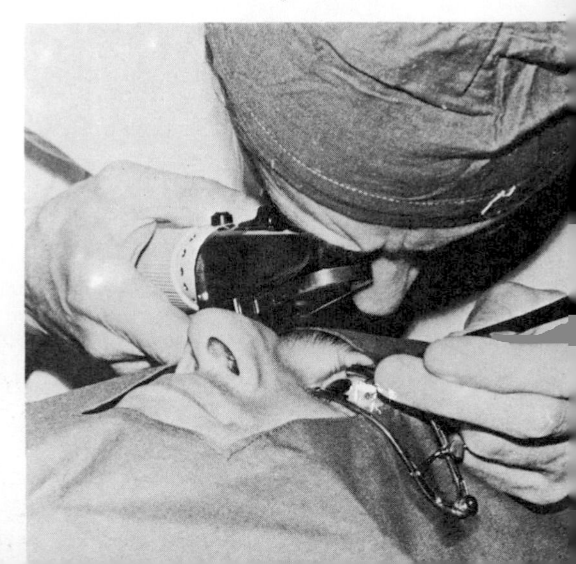

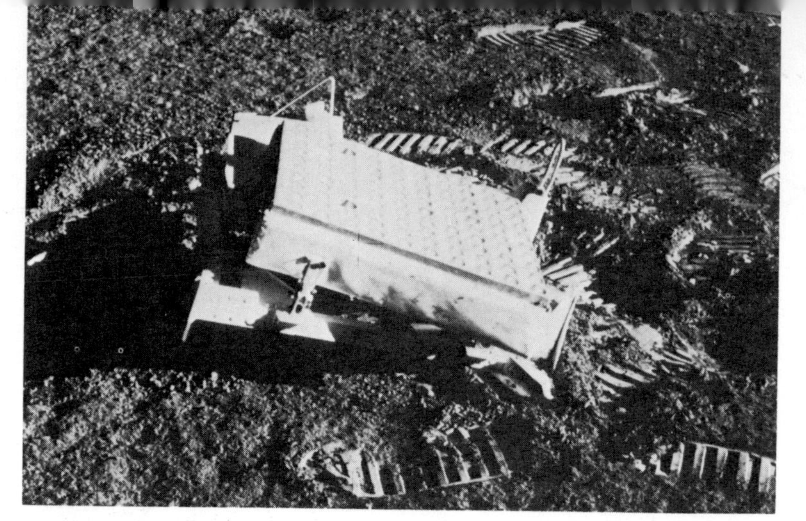

Fig. 18–24

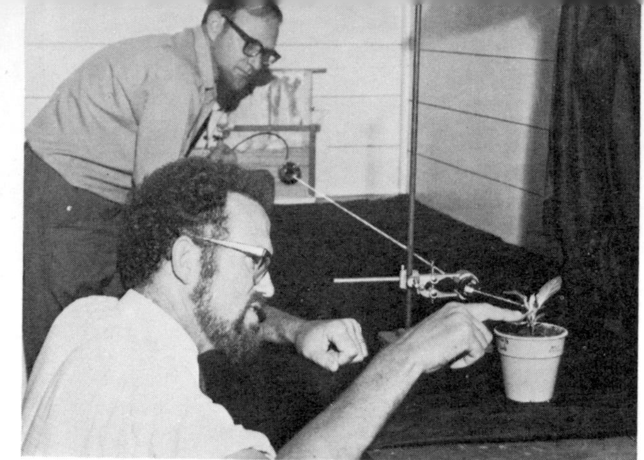

Fig. 18–26

The Laser at Work

The avalanche of photons has led to an avalanche of applications. In fact there cannot be many things around, be they molecules or moons, that have escaped the penetrating blast of the laser beam. In medicine, dentistry, and industry, in communications and research, lasers are rapidly being taken for granted.

Not all the early expectations have been realized but new ways of using lasers are being found daily. The mirror-like device in Fig. 18–24 is a laser retroreflector set up on the moon. It consists of rows of quartz cubes which act like 'cats eyes' and reflect a laser beam radiated from the Earth. By timing the round trip—about 2.5 seconds—the distance between the Earth and the Moon can be calculated to within a few centimetres. Other lasers operating on the radar principle, sometimes called *lidar*, are used to measure clouds and smoke. One form of ruby laser (Fig. 18–25) can actually indicate the concentration of solid material in smoke clouds and thus it acts as a pollution detector.

Bean seeds have been found to germinate in sixteen days when subjected to light from a ruby laser. This is ten days earlier than normal. In another experiment it was found that the floral growth of a chrysanthemum is retarded by a laser beam (Fig. 18–26).

Laser beams can rapidly drill through metals and even diamond (Fig. 18–27), the hardest mineral known to man. The solid state lasers shown in Fig. 18–28 are piercing holes in jewelled bearings for wrist watches.

Fig. 18–27

Fig. 18–25

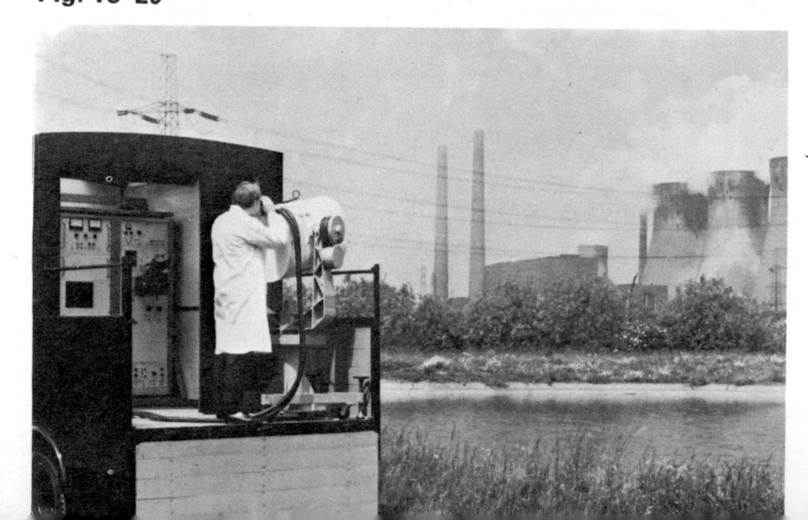

Fig. 18–28

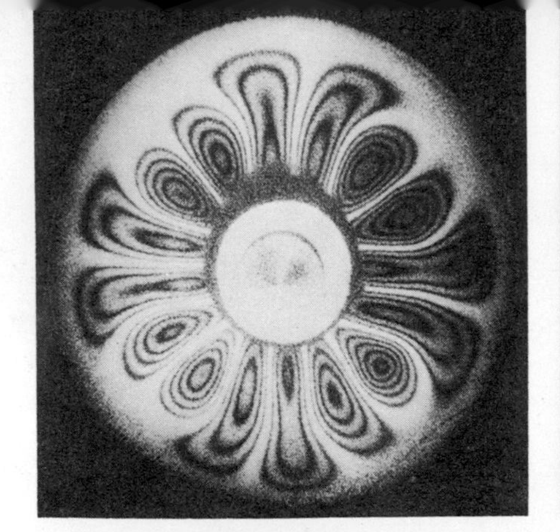

Fig. 18–29

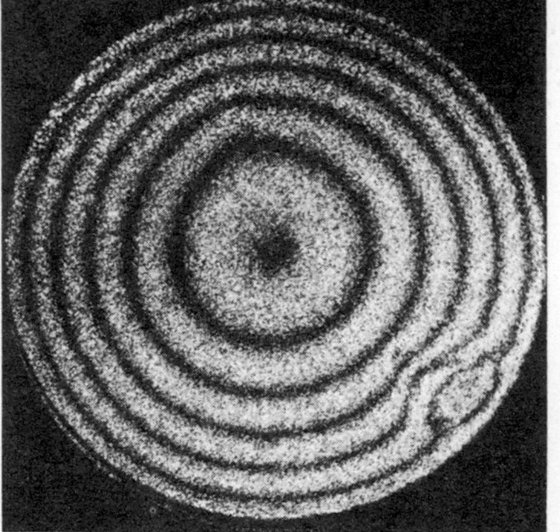

Fig. 18–30

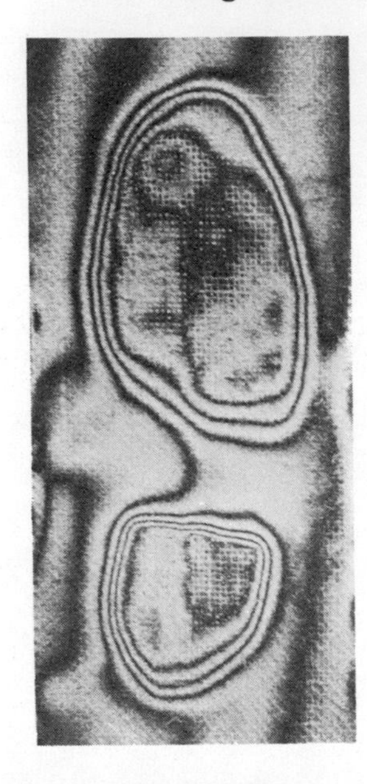

Fig. 18–31

Holography is one of the most fascinating applications of laser light. It is a form of photography without lenses, utilizing the interference of light beams. The image information, stored in a photographic plate or a crystal, can be used to produce a kind of three dimensional representation of the original object.

The interference patterns produced by holography can also be used to show vibration patterns such as that of a thin wall tube (Fig. 18–29). Flaws in a vibrating diaphragm (Fig. 18–30) or the residual movement in a carpet resulting from a recent footprint (Fig. 18–31) can be detected by the use of such photographs.

New uses for the laser in communication are continually being found. The development of tunable lasers which can be modulated is certain to bring further innovations. Lasers were used to produce a large TV picture—about 4 metres wide—at Expo 70 in Japan (Fig 18–32).

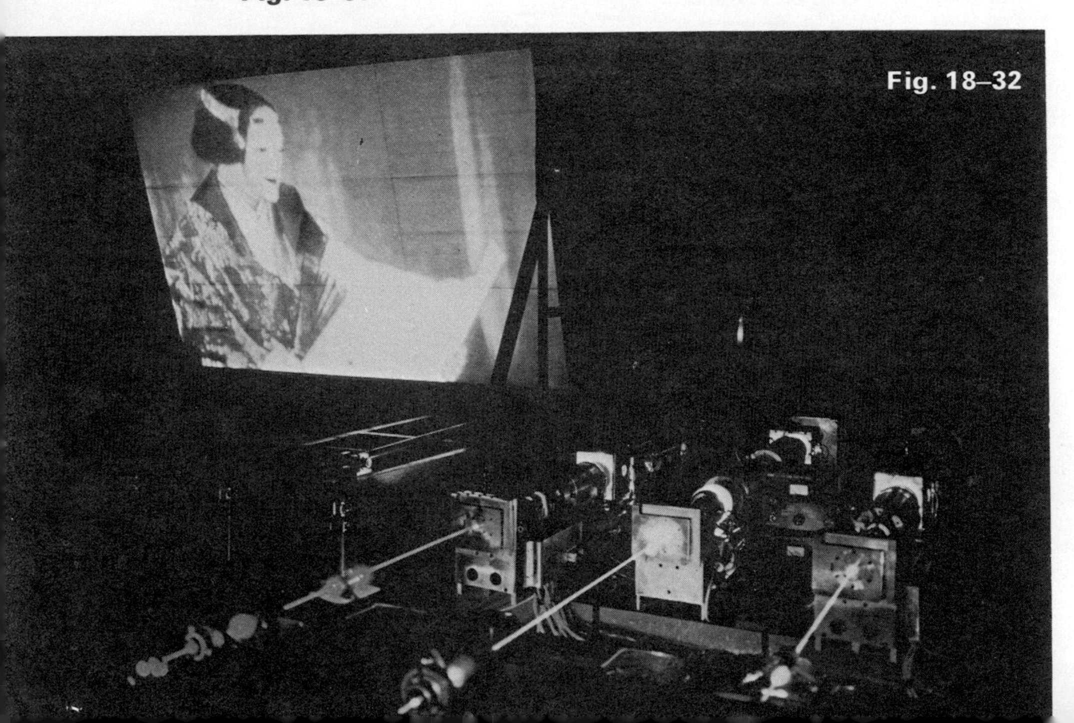

Fig. 18–32

IS NATURE SYMMETRICAL?

It is now apparent that light is schizophreni When it is being emitted or absorbed it behav like a particle; when travelling it behaves like wave. So it is understandable that sooner or lat someone was bound to ask, 'Do *all* particles beha like waves? Can cricket balls produce diffracti and interference effects? What is the waveleng and frequency of a billiard ball?'

It was the French physicist Prince Louis Broglie (pronounced de Bro-ee) who not only ask the question but went on to suggest the cons quences one might expect if all particles did in fa have wave characteristics associated with them.

As we have seen, the energy of a photon is giv by Planck's equation

$$E = hf = \frac{hc}{\lambda}$$

where c is the speed of light and λ the waveleng

From Einstein's famous equation, it is given

$$E = mc^2$$

From these two equations we have

$$\frac{hc}{\lambda} = mc^2$$

$$\Rightarrow \frac{h}{\lambda} = mc = p, \text{ the momentum of the photon}$$

$$\Rightarrow \lambda = \frac{h}{p}$$

The wavelength is therefore equal to Planc constant divided by the momentum of the partic Suppose then that $\lambda = h/p$ was applicable to pa ticles other than light photons, for example, ele trons, protons, neutrons, alpha particles, crick balls. Then a ball of mass 0.16 kg moving at 20 m would have a momentum of 3.2 kg m/s and 'wavelength' would be

$$\frac{h}{p} = \frac{6.6 \times 10^{-34}}{3.2} = 2 \times 10^{-34} \text{ metres approx.}$$

As the diameter of an atomic nucleus is of t order of 10^{-15} metres, you can see that our c culated wavelength is less than a million milli millionth of that incredibly small dimension! It hardly surprising that there is no experimen evidence to support de Broglie's theory for su 'particles'.

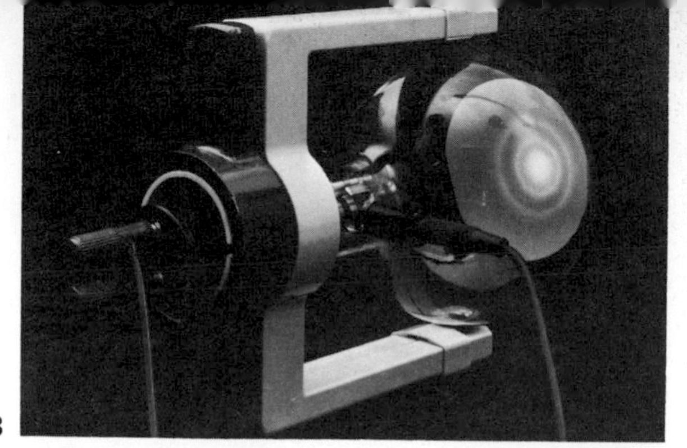

Fig. 18–33

However, in 1927 G. P. Thomson discovered, in Aberdeen, that under certain conditions *electrons* behave like waves! When a stream of electrons passes through a gold foil it produces interference rings similar to those shown in Fig. 18–33. When diffracted at, in effect, a double slit, electrons can produce a pattern (Fig. 18–34) similar to that obtained when light passes through Young's slits. The wavelength found from the diffraction patterns turned out to be the same as the de Broglie wavelength calculated from $\lambda = h/p$. Since then many experiments have been conducted with other particles including neutrons, alpha particles, and even atoms and molecules. The particles have been found to have wave-like properties and produce the diffraction and interference effects associated with a wavelength equal to h/p where p is the momentum of the particle.

Fig. 18–34

Dilemma

We have not solved the wave-particle problem but at least we have learned something by our attempts to grapple with the mysteries of light! We have learned to hold lightly to our graven images. They can too readily become false gods which conceal more than they reveal. Yet we are not all strong enough to burn those images, for without them we are left with a mathematical abstraction with meaning for only the chosen few.

We therefore retain those aspects of the wave model which are necessary to account for the transmission of light and those aspects of the quantum model which describe the emission and absorption of light. Otherwise, as far as light is concerned, we can say that any resemblance to waves on the sea or grains on the beach is entirely accidental!

SUMMARY

The smallest quantity of light energy possible is called a *quantum* of energy.

The wave model and the particle model are both needed to 'explain' experimental results.

The photon exhibits the wave characteristics of frequency f and wavelength λ and also the particle characteristics of mass hf/c^2 and momentum hf/c.

Energy quantum $= hf$ ($h =$ Planck's constant)

PROBLEMS

18.2 An electroscope A (Fig. 18–35) is charged. When ultraviolet light shines on a zinc plate as shown, the electroscope is discharged. The zinc plate is then placed on another electroscope B which is then charged. When ultraviolet light falls on the plate this electroscope is also discharged.

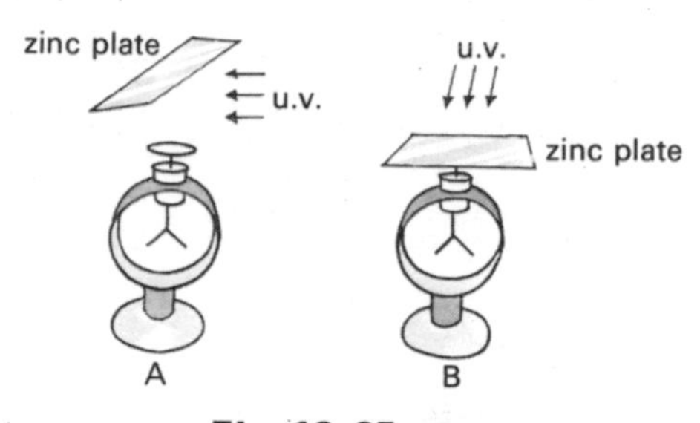

Fig. 18–35

State the original type of charge on A and on B and explain why the electroscopes are discharged.

18.3 Would you expect any difference in the behaviour if the ultraviolet source was replaced with a white light source in each case? Explain.

Why does an electron microscope require *very* high voltages to accelerate the electrons if it is to be used to study extremely small objects?

18.4 Radio 2 is broadcast on 1500 metres on a power of 400 kW. Find the frequency of the transmitter and the energy of each photon radiated. How many photons are leaving the transmitter every second?

18.5 Millikan's V_s/f graph is shown in Fig. 18–36. From it calculate Planck's constant.

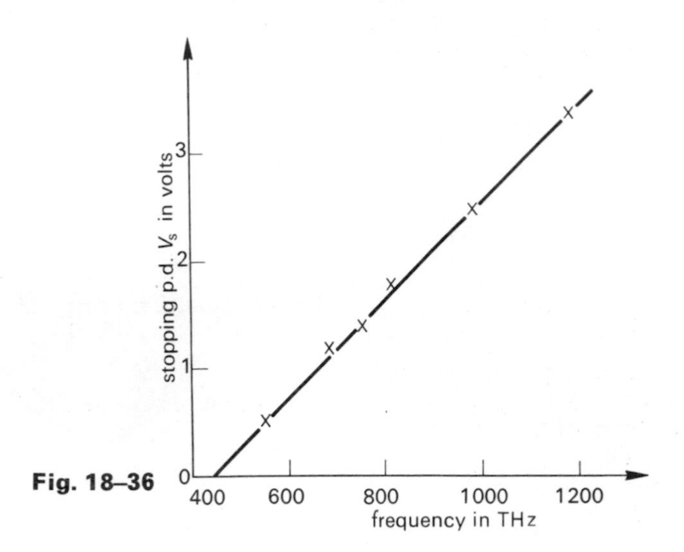

Fig. 18–36

18.6 If the work function of a metal is 5×10^{-19} J, what is the threshold wavelength? What kind of radiation is this?

What would be the maximum kinetic energy of a photo-electron emitted when radiation of wavelength 300 nm strikes this metal? If this metal were used as the cathode of a photo cell, what stopping voltage would be needed to reduce the current to zero with 300 nm radiation?

18.7 A beam of ultraviolet light falls on a suitable me plate Y, which lies on the axis of a hollow metal cylinder X and Y are connected in an electric circuit including battery and a milliámmeter as shown in Fig. 18–37.

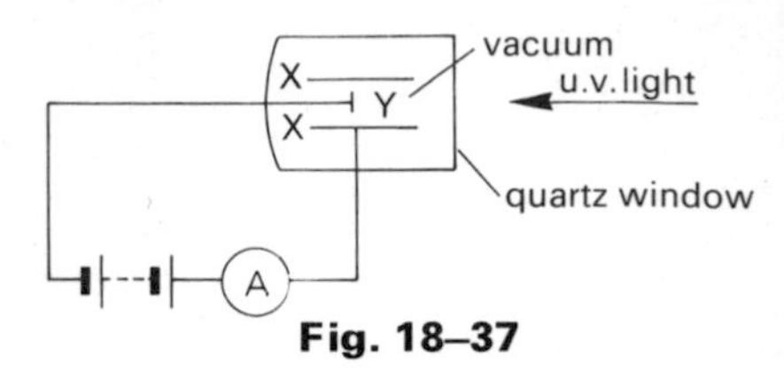

Fig. 18–37

(a) Explain why a small current is registered.
(b) What would happen to the current if the intensity the light were increased? [S.C.E. (H) 197

18.8 In Planck's Quantum Theory of Light, the energy of a quantum of light (photon) is given by the equati $E = hf$ where f is its frequency and h is Planck's consta the value of which is 6.63×10^{-34} J s. The minimum ener required to eject an electron from a certain metal 3.00×10^{-19} J. Explain whether you would expect light wavelength 5.00×10^{-7} m to eject electrons from the met What is the name given to this phenomenon? [S.C.E. (H) 197

18.9 What evidence is there that the properties of light a not fully explained by thinking of it as a wave motion? [S.C.E (H) 197

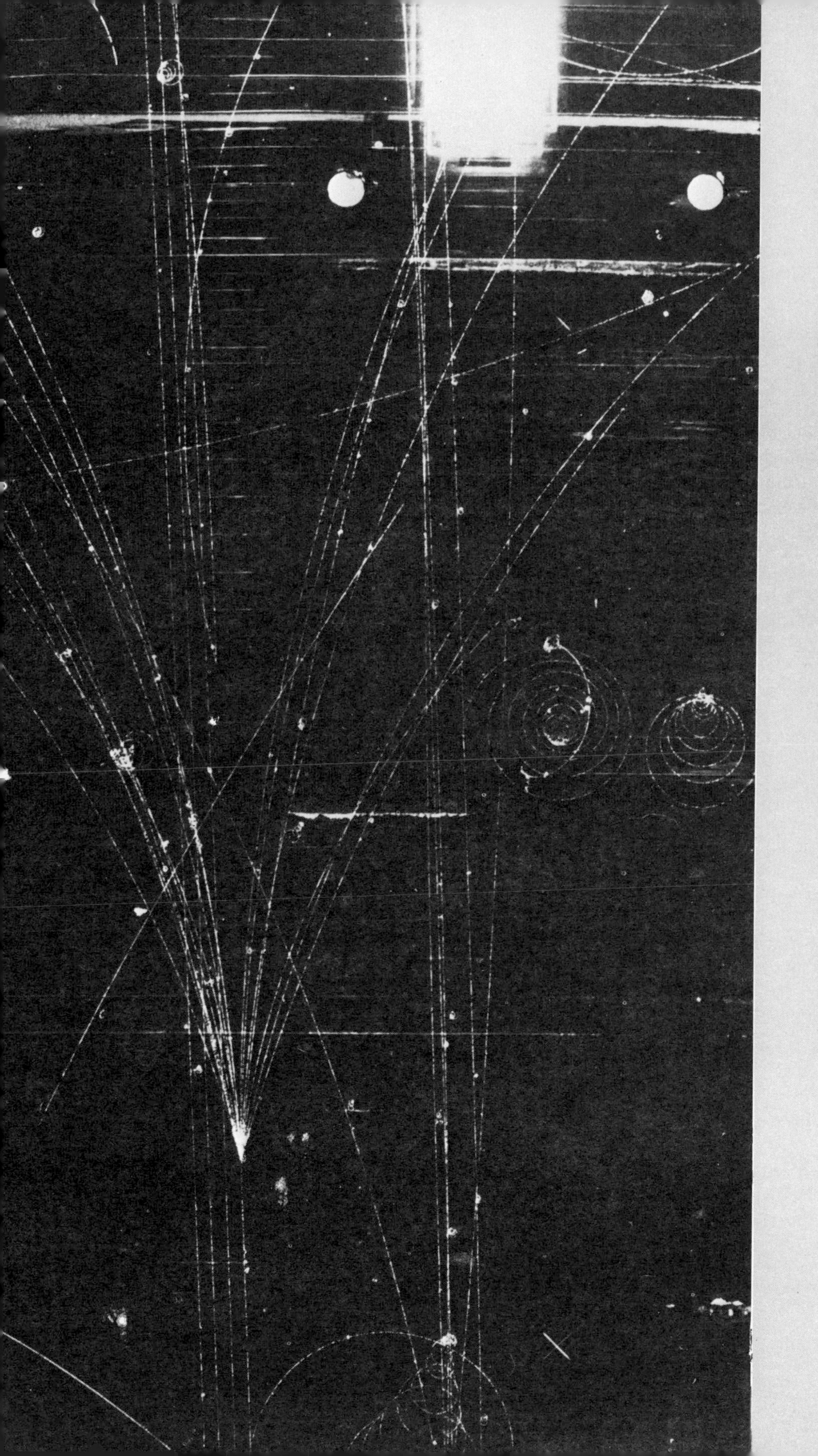

UNIT FOUR

Particles in Motion

The Atomic Model

19 The Atom

Every book should have a happy ending. We might therefore be tempted to survey the work of the course, tidy up a few ragged ends and present a complete, self-contained package. But physics is open-ended: at both ends! The summit even from this plateau that we have reached is still masked in mist. Yet the track is not uncharted. Some sign-posts remain, some milestones should not have to be passed again—even if they have to be renumbered in kilometres!

In his attempts to make sense of the natural world man has tried to find or produce some changeless (conserved) properties in a continually changing scene. He has invented concepts such as mass, force, energy, momentum, and charge which in turn have led to sophisticated ideas of waves, particles, and quanta.

Many scientists seem to have thought that the best way to understand the physical world was to discover more and more about the bits and pieces from which things are made and the current expenditure on 'atom smashers' suggests that this is still considered to be a high priority. But it is not only the *nature* of the basic particles themselves that now concerns physicists but also their *collective behaviour*. It is clear that the behaviour of a society cannot be understood in terms of the antics of a solitary cast-away on a desert island. Relationship is important also in the atomic and sub-atomic world.

We might profitably end this course with a brief historical survey of some of the more important events which led to the development of various atomic models and to nuclear science as it is known today.

THE GREEKS HAD A NAME

400 B.C. If you are tempted to think that arguments about 'science and religion' are a new phenomenon, it is worth remembering that two and a half thousand years ago two Greeks, Democritus and Plato, fought the same battle. According to Plato the material world was 'unreal' and only the spirit 'existed'. He taught his followers to desp material things and to concentrate rather the good and the beautiful. His oppone Democritus, was one of the world's fi 'materialist' philosophers. He taught th everything, including goodness and gri could be interpreted in terms of the partic of matter. He is remembered today, ho ever, more for his prophecy than his phi sophy. Without any experimental evider Democritus claimed that everything w made up of tiny hard particles which co

Fig. 19–1

not be seen although they had shape, s and weight. These particles he christer *atoma*, which means 'unsplittable' and fr which we get our word *atom*. He said th they were continually in motion and th different substances were built up from d ferent combinations of these atoms. M than 2000 years before the atomic a Democritus made this prophetic stateme 'The things which we perceive appear to reality, but the atoms and empty space the true reality'.

Aristotle, one of Plato's most brilliant pupils, also rejected the atomic theory of Democritus and accepted an idea of Empedocles that all substances are composed of four 'elements', earth, fire, air and water. Aristotle did not, however, have Plato's contempt for matter. In fact, he studied matter so carefully, particularly plant and animal life, that for the next two thousand years he was accepted as the undisputed expert on all things scientific!

Aristotle's false doctrine of matter encouraged a great many people to search for a way of mixing the 'elements' to produce gold. The alchemists (Fig. 19–2) were convinced that if they tried hard enough they would be able to change a cheap metal, such as iron, into gold. Many impostors claimed to have done this. In 1782, for example, James Price was honoured by the King and Oxford University for producing gold. When he was later forced to repeat his method of preparation before a learned audience he poisoned himself rather than admit his deceit.

Fig. 19–2

EXPERIMENTAL SCIENCE BEGINS

1589 Aristotle's two thousand year reign came to an abrupt end when a young Italian scientist climbed to the top of the famous tower in Pisa. Whether this legend of Galileo is true or not, it is certain that he attempted to destroy the erroneous theories of Aristotle. It seems incredible to us that not everyone was immediately convinced by the evidence of his eyes, but so strong was the tradition that even in Oxford University anyone making an assertion against Aristotle was fined five shillings! Galileo not only laid the foundations of experimental science: he also supported the atomic theory, and thus cleared the way for those who followed to demonstrate experimentally the existence of the atom.

1660 The search for the infinitesimally small took a new turn when a young Dutchman, Anthony van Leeuwenhoek, found a way of making and polishing a tiny spherical bead of glass which could be used as a high power microscope. He set the bead in a metal plate to which was attached his specimen (Fig. 19–3). With his microscope he was the first man to see bacteria, red blood corpuscles, and tiny crystals of snow. He watched the circulation of the blood in the capillary vessels in a tadpole's tail and calculated that there were millions of 'wee beasties' in a drop of water! His findings so astounded the Royal Society in London that

Fig. 19–3

they refused to believe him and sent a representative to Holland to see if this was a hoax. The result was his nomination as a member of the Royal Society!

Some of Leeuwenhoek's lenses had a magnifying power of 250×. With such a microscope he could study details down to about one millionth of a metre but this was still a long way from the atom!

You might like to try to make a simple bead microscope for yourself from a drop of water formed in a tiny loop of fine wire.

Chemistry Takes Shape

1661 A further attack on the alchemists came with the publication of Robert Boyle's book, *The Sceptical Chymist*. In it Boyle (Fig. 19–4) introduced the idea of chemical *elements* as substances which cannot be split into other substances by chemical means. He suggested that all elements are made up of one kind of atom and that different elements have different arrangements and movements of the atoms.

1700 By the beginning of the eighteenth century the atomic theory was becoming more popular. Newton argued that even light consisted of tiny particles or corpuscles. He demonstrated that white light could be split into 'spectral colours'. This was the beginning of modern spectroscopy, which yields important information about atomic and molecular structure.

1738 The Swiss physicist Daniel Bernouilli explained the pressure exerted by a gas by suggesting that millions of minute particles, all in rapid motion, are continually colliding with the walls of their container and with each other. The kinetic theory of gases was born.

1802 Thomas Young produced optical interference patterns and thus demonstrated the wave nature of light.

1808 The first man to put the atomic theory on a firm experimental basis became a schoolteacher at the age of twelve and a headmaster at nineteen! John Dalton claimed

Fig. 19–4

Fig. 19–5

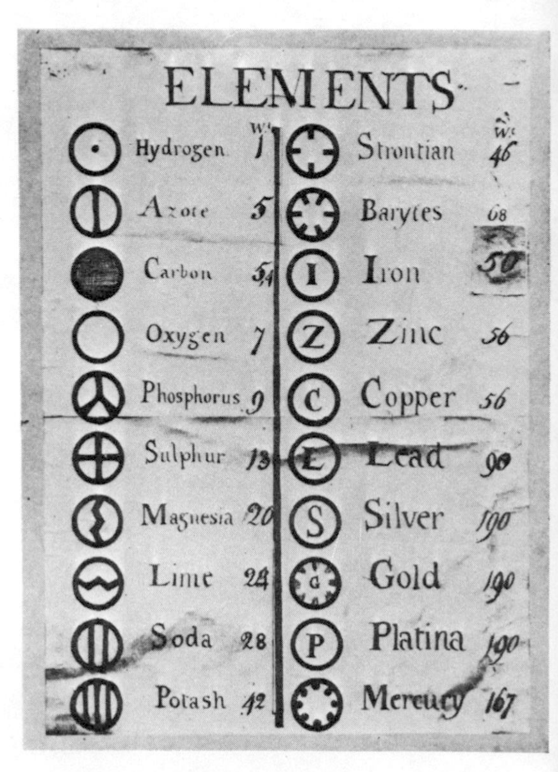

that every element had its own particular kind of atom. Such atoms were indivisible and indestructible, and all the atoms of a given element were identical. The atoms of different elements had different weights. Dalton's table of atomic weights (Fig. 19–5) produced a revolution in chemistry.

He was first to realize that in a chemical reaction elements combine in a fixed weight ratio because a certain number of atoms of one element always combine with a certain number of atoms of another. Dalton mistakenly assumed that the 'certain number' was in fact 'one' in each case. For example, he thought that one atom of hydrogen combined with one atom of oxygen to form water,

$$H + O \rightarrow HO$$

yet he knew that two volumes of hydrogen combined with one volume of oxygen to form two volumes of water vapour.

1811 The conflict was finally resolved by Gay-Lussac and Avogadro. Gay-Lussac discovered that, for a particular volume and pressure of gas, every gas expands by exactly the same amount when it is heated. Avogadro reckoned that this equal expansion must mean that the particles of all gases move apart by equal amounts when the gas is heated and that therefore the average distance between the particles must be the same for all gases at the same temperature and pressure. This, in turn, implies that equal volumes of gases at the same temperature and pressure contain the same number of particles. In stating this hypothesis Avogadro named the gas particles *molecules*. By considering the molecule of a gas rather than the atoms, he was able to describe the combination of hydrogen and oxygen in an equation which agreed with experimental results

$$2H_2 + O_2 \rightarrow 2H_2O$$

Since

$$2 \text{ volumes} + 1 \text{ volume} = 2 \text{ volumes}$$
$$2 \text{ molecules} + 1 \text{ molecule} = 2 \text{ molecules}$$

i.e. $$2H_2 + O_2 \rightarrow 2H_2O$$

Avogadro's hypothesis is also consistent with the kinetic theory of gases developed earlier. If the pressure p and the volume V of two gases are the same, then

$$pV = \tfrac{1}{3}N_1 m_1 \overline{v_1^2} = \tfrac{1}{3}N_2 m_2 \overline{v_2^2} \qquad (1)$$

and if they are also at the same temperature then

$$\tfrac{1}{2}m_1\overline{v_1^2} = \tfrac{1}{2}m_2\overline{v_2^2} \qquad (2)$$

From these two equations we see that

$$N_1 = N_2;$$

that is, the number of molecules is the same in equal volumes of gas at the same temperature and pressure.

1812 In the gradual development of the present picture of the atom, emission and absorption spectra have played an important part. Joseph von Fraunhofer, a German physicist and expert lens maker, was first to report a series of some 576 dark lines across the spectrum of the sun. He found that the same lines appeared in light from the moon and the planets, but that the spectra of starlight contained *different* lines. From this he concluded that the lines must originate in the source of light. About fifty years later Kirchhoff and Bunsen demonstrated that each element had an emission spectrum which contained a unique set of lines.

1827 The first experimental evidence of molecules in motion was produced by someone who was never to realize the implications of his discovery. The botanist Robert Brown had mixed a little pollen in a drop of water and was studying it under his microscope. To his amazement the pollen grains danced around in a completely random way without stopping or even slowing down. He was unable to explain this movement (now called Brownian motion) and said that the pollen grains behaved like tiny insects. It was not until nearly a hundred years later that the French physicist Jean Perrin made careful measurements of a suspension of gamboge and showed that the particles were constantly colliding with water molecules.

Fig. 19–6

Fig. 19–7

1834 The idea that a small but definite electric charge might be associated with each different atom was suggested by Michael Faraday (Fig. 19–6). He discovered that when a definite quantity of electricity passed through a solution, the masses of the substances liberated at the electrodes were proportional to their equivalent weights i.e. to

$$\frac{\text{atomic weight}}{\text{number of valence electrons}}$$

1869 Certain elements have such similar chemical and physical properties that they are often thought of as forming a 'family'. The halogens, inert gases, and alkali metals are striking examples. The Russian chemist Dmitri Mendeleev first showed that, if all the elements are listed in order of their atomic weights, elements with similar properties recur at regular intervals. There were one or two odd exceptions, which Moseley explained forty years later.

1885 The Swiss schoolmaster, J. J. Balmer, found that the wavelength of all the lines in the visible part of the hydrogen spectrum could be described by a single mathematical formula.

RAYS, RADIUM, RUTHERFORD

1895 Wilhelm Röntgen discovered the amazing properties of X-rays.

1896 In the following year Henri Becquerel tried to produce X-rays from fluorescent material and discovered that uranium salts emit radiations which affect photographic film and discharge electroscopes even in total darkness. Radioactivity had been discovered —the unsplittable was splitting!

1897 In the Cavendish laboratory at Cambridge, J. J. Thomson (Fig. 19–7) 'discovered' the electron; that is, he measured the charge/mass ratio for cathode rays and showed that they were composed of negatively charged particles much smaller than the lightest atom. For his contributions to atmoic physics Thomson was awarded the Nobel Prize for Physics in 1906.

Thomson later proposed a model of the atom sometimes called the 'plum pudding model', as it pictured the atom as a positively charged material with negative electrons stuck in it like plums in a pudding.

1898 The search for and the eventual discovery of radium was perhaps the most drab and yet the most dramatic in the history of science. Manya Sklodowska had won a gold medal at school in Warsaw, and wanted to go to university to study mathematics, chemistry, and physics. Unfortunately her father could not afford the fees—he was a science teacher! But this did not deter Manya. She became a governess to a rich Polish family and after four years of self-sacrifice she had saved enough money to go to Paris to study at the Sorbonne. Here she met one of

Fig. 19–8

Becquerel's colleagues, Pierre Curie. Their cooperation was not limited to the scientific field and Marie (as she was then called) soon became Madame Curie. After her graduation, Madame Curie decided to study Becquerel's mysterious rays in order to gain her doctorate. She started by investigating all the known elements to see if any others emitted these strange radiations and discovered that thorium did. She christened the phenomenon *radioactivity*.

Continuing her investigations, Madame Curie (Fig. 19–8) discovered that some minerals containing thorium or uranium were *far* more radioactive than she would have expected. Could they contain yet another substance that was more strongly radioactive than either uranium or thorium?

Pierre decided to give up his own research and join his wife in the search for this strange new substance. The first problem was to find large enough quantities of material containing the substance, at a price they could afford. Fortunately they found that, when uranium had been extracted from the salts used in the manufacture of glass, the waste was just as radioactive as the original ore. This waste material they could have for the taking and so began the long, tedious task of chemical analysis and separation.

The laboratory was an old rickety shed, in which Madame Curie spent nearly four years of unbroken toil—separating, boiling, evaporating, and filtering. She wrote, 'Often I spent whole days with some concoction or other, stirring it with an iron pole as tall as myself. By the evening I was worn out'. But later she said, 'We spent the best and happiest years of our lives entirely devoted to the task in that miserable old shed'.

After refining tonnes of pitchblende, Madame Curie produced a few crystals of two new elements. The first she named after her native Poland—*polonium*; the other was named *radium*, from the Latin word meaning 'to radiate'.

Thanks to the inexhaustible patience and tireless efforts of Madame Curie and her husband, the mountain of waste had yielded a tenth of a gramme of radium! This amazing element is two million times more radioactive than uranium. Since its discovery, its radiations have been used in thousands of ways, including the treatment of cancer. Its rays are also very dangerous, and Madame Curie's death at the age of 67 is now known to have been due to leukemia, a disease thought to be caused by gamma radiation.

1900 About the turn of the century radioactivity was found to consist of three distinct types of radiation. Ernest Rutherford (Fig. 19–9) named two of them alpha and beta rays, and Paul Villard discovered the third—gamma rays. The Curies showed that beta rays carry a negative charge.

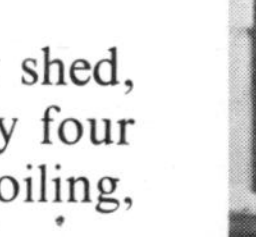

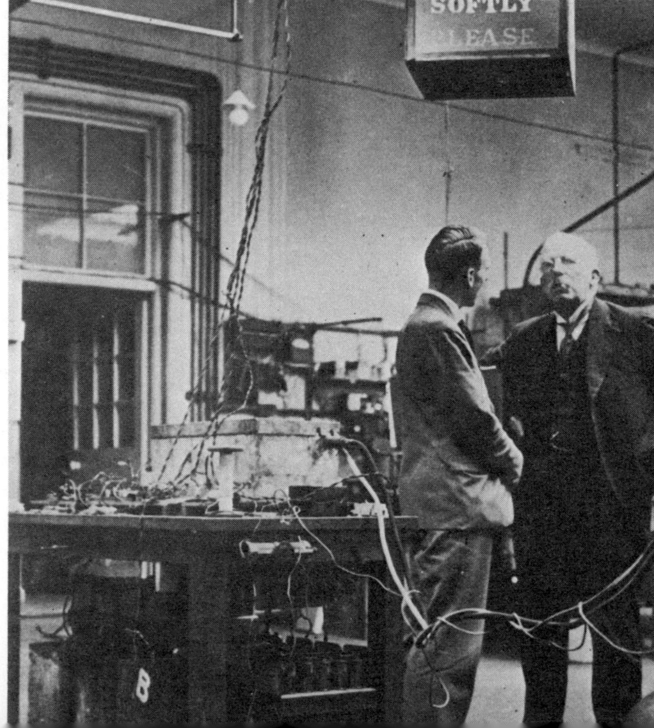

Fig. 19–9

Fig. 19–10

1903 Rutherford and Robinson then deflected alpha rays with powerful electric and magnetic fields and showed that they carried positive charges. They were then able to measure the charge/mass ratio, which showed that alpha rays consisted of positively charged particles each with about 7000 times the mass of an electron.

About the same time, the German physicist Philipp Lenard was first to show experimentally that the atom had an open structure. He made a cathode ray tube with a small aluminium window in it and found that fast-moving electrons could go straight through this 'Lenard window'. He asserted that the atom was just empty space apart from about one part in a thousand million. For his work on cathode rays he received the Nobel Prize in 1905.

1905 Einstein (Fig. 19–10) then startled the world by putting forward the theory that matter (mass) and energy are equivalent. These he related by his famous equation

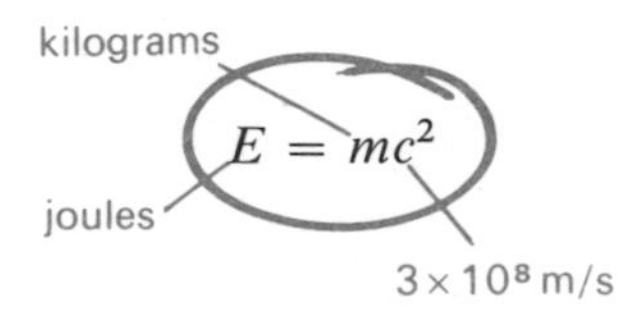

If a lump of sugar could be changed to energy, it would provide enough to keep a room warm for thousands of years!

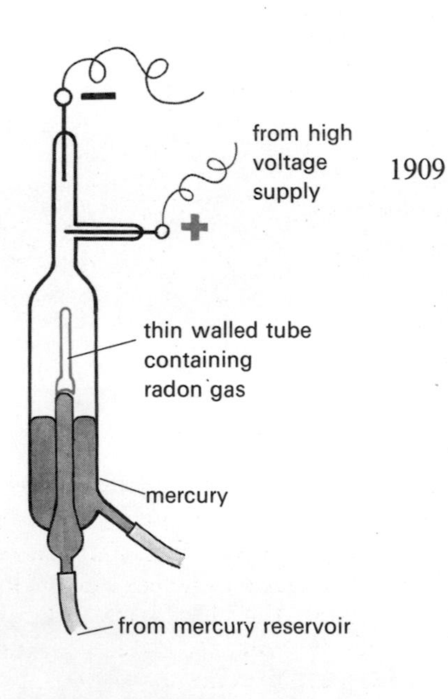

Fig. 19–11

1909 Rutherford and Royds demonstrated that an α particle was really a helium nucleus. To do this they devised a piece of equipment in which they captured some fast-moving α particles.

They used radon gas as a source of α particles, and enclosed it in a thin walled glass tube through which the particles could pass. This tube was fixed inside a much thicker glass tube through which the α particles could not escape (Fig. 19–11). The space inside the thicker tube was evacuated, and the apparatus was left for several days to allow α particles to pass through into this space.

Rutherford and Royds then pumped mercury into the tube so that its contents were compressed into a discharge tube. When a large voltage was then applie
the electrodes and the discharge exami
with the aid of a spectrometer, the hel
spectrum was observed. The α particles
each collected a couple of stray electr
and changed into helium atoms. This
accepted as conclusive evidence that α
ticles are helium nuclei.

1910 After investigating radioactive decay
atomic structure, the English chemist F
erick Soddy concluded that substances
identical chemical properties could h
different atomic weights. As these
stances must be forms of the same elem
and must therefore appear in the 's
place' in the periodic table, he propo
they be called *isotopes*. All isotopes o
element have the same atomic or pro
number but different mass or nucleon n
bers. Soddy was awarded the Nobel P
in 1921.

The development of the mass spec
graph in 1919 by F. W. Aston enabled
to separate and identify the different
topes of an element.

The Scattering Experiment

1911 Ernest Rutherford was not happy ab
Thomson's plum pudding model of
atom and suggested to his colleague H
Geiger that he might like to give one of
students, Ernest Marsden, a small resea
project. 'Why not let him see if any al
particles can be scattered through a la
angle?' Rutherford asked, although he
mitted later that he did not expect a posi
result!

Geiger and Marsden's experiment
illustrated in Figs. 19–12 and 19–13.

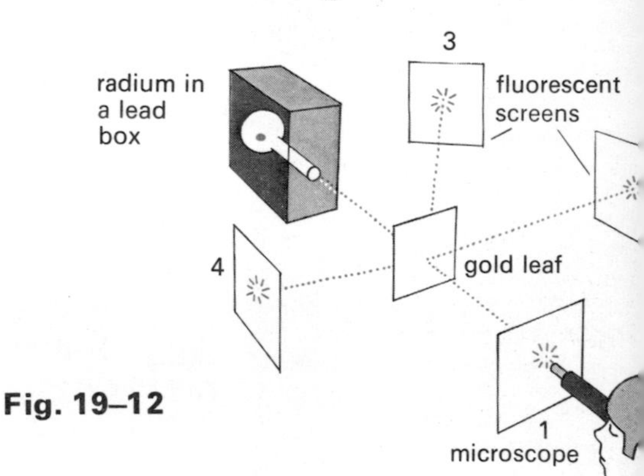

Fig. 19–12

radium source is placed inside a lead box so that it emits a narrow beam of alpha particles, which produce tiny flashes of light when they stroke a zinc sulphide screen. Without the gold leaf in position the flashes were observed through a microscope focused on screen 1 (Fig. 19–12). Even the thinnest gold foil, less than 1 μm, is still several hundred atoms thick. When such a foil was placed in the path of the alpha particles it made practically no difference. They shot through this wall of gold as if it wasn't there! If Thomson's model of the atom was correct, this was like firing a continuous stream of bullets through a mountain! But even more exciting results were to follow.

The zinc sulphide screen could be rotated to other positions (for example, positions 2, 3, and 4 in Fig. 19–12) so that any alpha particle deflected from its original course could be detected. Geiger then reported to Rutherford that they had detected some of the alpha particles 'coming backwards'. Later, Rutherford said 'It was quite the most incredible event that has ever happened to me in my life. It was almost as incredible as if you fired a 15-inch shell at a piece of tissue paper and it came back and hit you!'

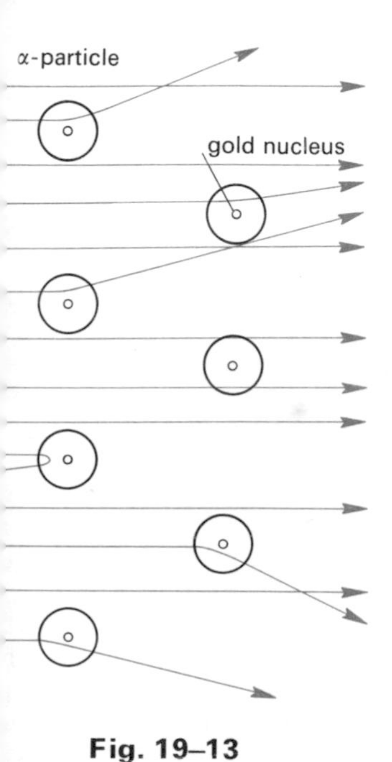

Fig. 19–13

As a result of Geiger and Marsden's experiments, Rutherford proposed a new model of the atom. He thought of it as a solid core or nucleus surrounded by distant electrons. The mass of the atom was concentrated at the core, a fact which explained the deflection of the very occasional alpha particle (Fig. 19–13). In all only about 1 in every 8000 particles was deflected more than 90°.

Fig. 19–14 illustrates a magnetic model of the scattering process. In it, ring magnets represent the nuclei of the atoms and the alpha particles are similar tiny magnets floating on carbon dioxide. The photograph was taken by a series of time exposures, each showing the path of a magnet fired from the left-hand side of the photograph.

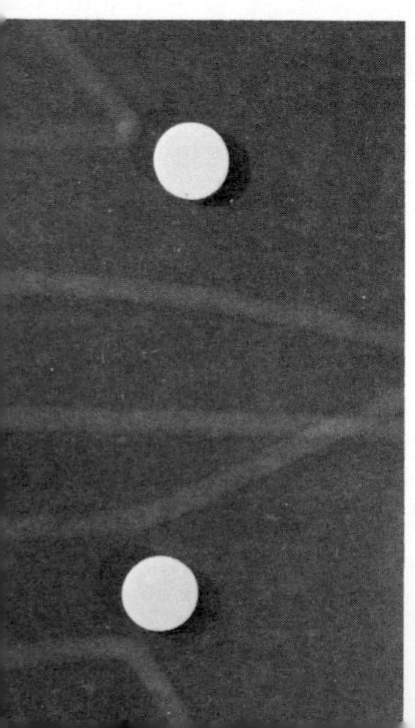

Fig. 19–14

Not only did Geiger and Marsden's experiment provide evidence for the *nuclear atom*, but it also enabled the size of the nucleus and the charge it carried to be measured. The nuclear charge was found always to be a multiple of the electronic charge, the value of which Millikan had so carefully measured, and the diameter of the nucleus was calculated as one ten-thousandth part of the diameter of the atom! As the mass of the atom is concentrated in this tiny nucleus, you can imagine what might happen if all the 'electronic space' were to disappear and a concentrated lump of nuclei were available. Fortunately such a thing is impossible on Earth but astronomers tell us that in the white dwarf stars the nuclei are so closely packed that the mass of a spoonful of matter from one of these stars would be hundreds of tonnes!

You have seen evidence for the nuclear atom in the alpha particle tracks found in a cloud chamber, which was invented by C. T. R. Wilson in 1911 (Fig. 19–15). Each track, a few centimetres long, is formed from hundreds of thousands of ions. The alpha particle must have travelled through that number of atoms without ever striking a nucleus. Only very occasionally does an alpha particle strike a nucleus and produce a forked track. Fig. 19–24 (on page 154) shows such a collision, where an alpha particle has been in collision with a nitrogen nucleus.

Fig. 19–15

ATOMIC MODELS

1913 The next Nobel Prize winner in our story is Niels Bohr, of Copenhagen. He came to England in 1911 and started work in Cambridge under J. J. Thomson. Soon he moved to Manchester to work with Rutherford on the structure of the atom.

Rutherford had described the atom as a small positively charged nucleus with electrons circling round it rather like planets in a miniature solar system. A circling electron is, however, accelerating all the time, so it ought to radiate electromagnetic waves. Yet there is no evidence to suggest that any such radiation exists or that an atom is continuously losing energy. Bohr set his mind to the problem.

He knew that atoms *did* emit electromagnetic radiation if they had been excited by collisions with other atoms, and that the frequencies of the waves emitted were different for different elements. The work of Planck (1901) and Einstein (1905) had shown that the energy changes associated with radiation could be interpreted as definite *quanta* or bursts of energy which are proportional to the frequency of the radiation.

Bohr produced a model of the atom in which electrons can move, without radiating, only in *certain* orbits or shells. If an electron jumps from one particular orbit to one nearer the nucleus, that is, from a high energy level to a lower energy level, energy is radiated as electromagnetic waves. The frequency of the radiation is then dependent on the difference between the two energy levels (Fig. 19–16).

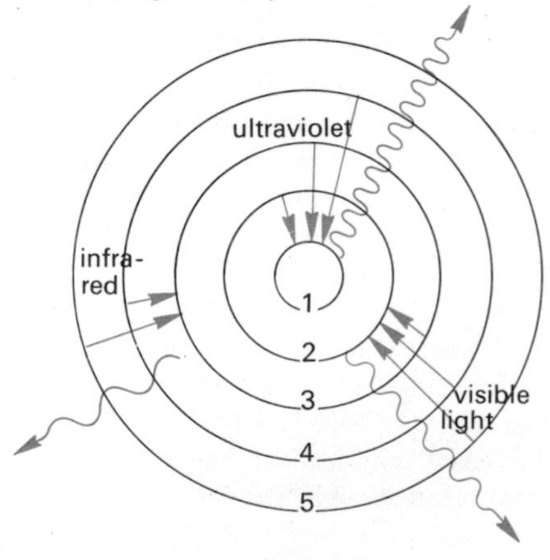

Fig. 19–16

This model was an improvement
Rutherford's model and, following the w
of Balmer and other spectroscopists, le
more satisfactory explanations of opt
and X-ray spectra.

Atomic (Proton) Number

Characteristic X-rays are known to be p
duced when electrons jump from one ene
level to another. When high speed electr
bombard a target, an inner electron in
of the target atoms may be removed (F
19–17). An electron at a higher energy l

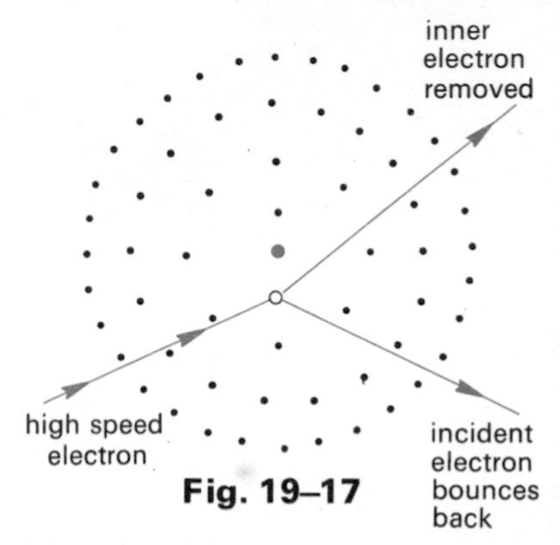

Fig. 19–17

then jumps in to the vacancy and in so d
emits X-rays. The frequencies of the X-
produced are characteristic of the ta
material used.

At first there was great speculation ab
the nature of X-rays. Since optical diff
tion gratings would not bend them, t
appeared to have a much smaller wa
length than light, if in fact they were wa
at all. Then in 1912 the German physi
Max von Laue suggested that the reg
layers of atoms in a crystal might be u
as a natural diffraction grating. If X-
are partially reflected from each of th
layers, which are about 1 μm apart, the
flected waves would be in phase (Fig. 19-

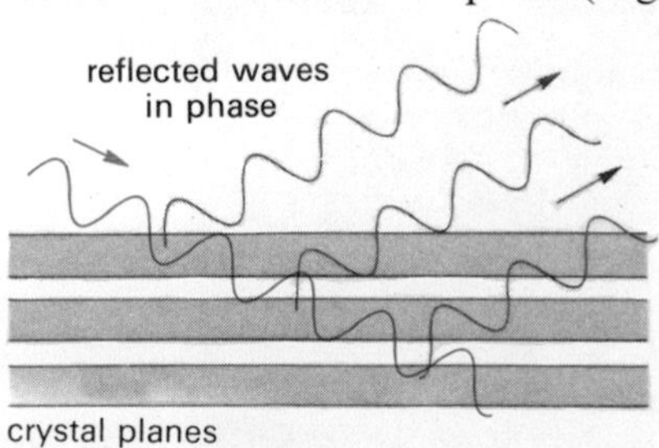

Fig. 19–18

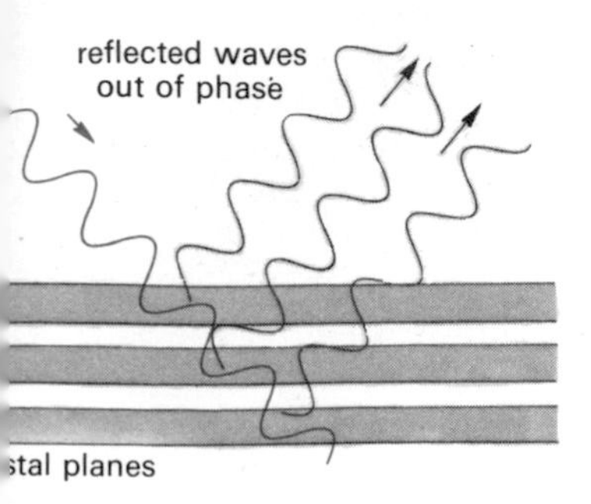

Fig. 19–19

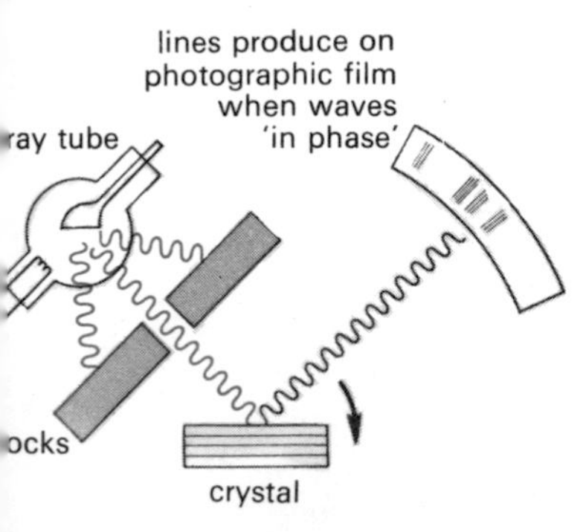

Fig. 19–20

or out of phase (Fig. 19–19), depending on the angle of incidence of the beams. Using this technique Laue's assistants were able to show that X-rays were indeed very short electromagnetic waves. If a narrow beam of X-rays is reflected from a crystal which is then turned so as to vary the angle of incidence, a line spectrum characteristic of the particular crystal can be obtained (Fig. 19–20).

One of Rutherford's associates at Manchester was a young man who was fascinated by these X-ray spectra. His name was Henry Moseley. He decided to investigate the X-ray spectra of as many elements as possible. To do this he had to build a very unusual X-ray tube. It consisted of a long wide glass tube containing a toy railway line on which a truck carried a number of different elements. Moseley used these elements in turn as targets into which he fired electrons in order to produce X-rays. In other words the elements formed the *anode* of the X-ray tube.

In addition to being a very skilful experimenter, Moseley had the same kind of devotion and enthusiasm as we have seen in Madame Curie. He could be seen at work day and night, and it was said of him that one of his specialized attainments was knowing where to get a meal in Manchester at three o'clock in the morning!

Moseley very quickly produced results which, Rutherford claimed, rank in importance with the discovery of the periodic table and spectral analysis. He found that the *frequency* of a particular line in the X-ray spectrum of an element increased regularly from element to element. The order he obtained was very similar to the order of elements in the periodic table but much better! The strange discrepancies had disappeared. Arranged in order of atomic weights potassium came before argon, and nickel before cobalt. With Moseley's arrangement these orders were reversed, and the elements then fell into the correct vertical columns with elements of similar chemical properties (see end paper). Moseley said of his results, 'We have here a proof that there is in the atom a fundamental quantity which increases by regular steps as we pass from one element to the next. This quantity can only be the charge on the central positive nucleus'. It is now called the atomic number or proton number Z. Chadwick later confirmed this conclusion by scattering alpha particles from various metals.

It is the *atomic number* then and not the *atomic weight* that determines the chemistry of an element.

When the square root of the frequency f is plotted against the proton number Z, a straight line is obtained (Fig. 19–21). Using

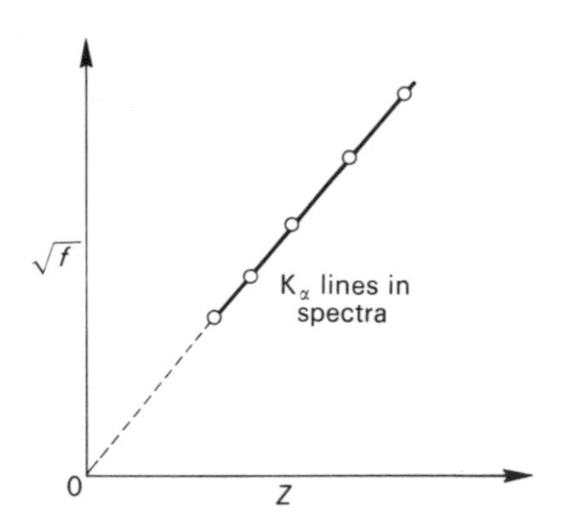

Fig. 19–21

this graph it was possible to 'call the roll' of the elements, and several 'absences' were discovered. The missing elements have now all been found.

Moseley's amazing work was done when he was only twenty-six years old. Two years later he was killed by a sniper's bullet while serving as an army officer in the Dardanelles. His unnecessary death was a major disaster for science throughout the world.

1914 Clear experimental evidence for the existence of distinct energy states within the atom was produced by the German physicists Franck and Hertz. They bombarded vaporized sodium with electrons from a heated filament. This experiment won them a Nobel prize.

ALL CHANGE

1919 One of Ernest Rutherford's ambitions was to be the world's first successful alchemist! He argued that, if an element's chemistry depends on the number of charges in its nucleus, it should be possible to change both by bombardment.

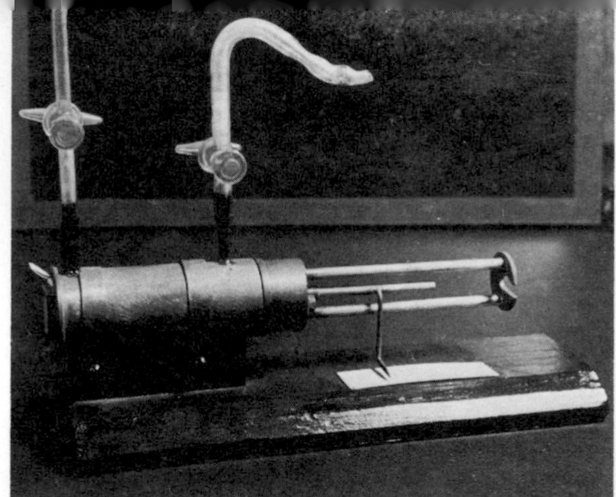

Fig. 19–22

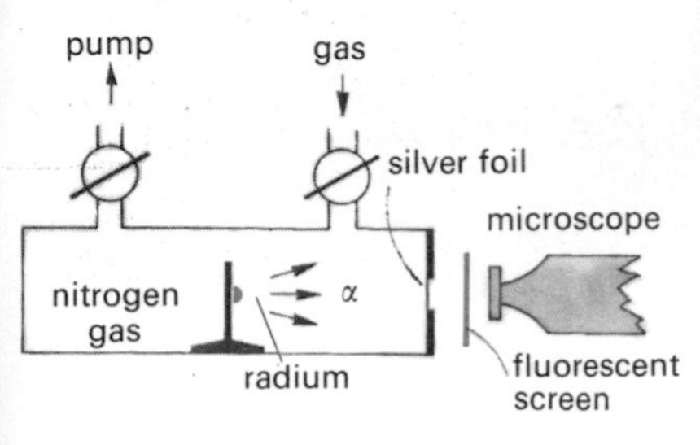

Fig. 19–23

Fig. 19–22 shows the simple apparatus which Rutherford designed to try out his theory. It is shown diagrammatically in Fig. 19–23. The small chamber was evacuated and filled with nitrogen gas. Alpha particles from the radium source were then fired at a small silver foil which was sufficiently thick to stop them. Any scintillations on the fluorescent screen beyond the foil could be observed through a microscope. When Rutherford peered through the microscope he saw nothing at first and then suddenly a tiny twinkle of light appeared . . . and then another. From measurements with applied magnetic and electric fields Rutherford identified the responsible particles as hydrogen nuclei (protons) which had passed through the silver foil.

Rutherford's student, P. M. S. Blackett, in a series of 23 000 cloud chamber photographs showing some 400 000 alpha particle tracks in nitrogen, found only eight photographs of transmutations. Fig. 19–24 is one of them. By assuming that momentum was conserved in this collision, another particle was identified. It was an oxygen nucleus. So nitrogen had been changed to hydrogen and oxygen by bombardment with an alpha particle. For the first time in man's history the atom had been split and one element changed to another.

Energy in atomic physics is sometimes measured in *electron volts*. When an electron or proton is accelerated by a difference of potential of 1 volt the *energy* gained is 1 electron volt (eV).

$$1 \text{ electron volt} = 1.6 \times 10^{-19} \text{ joules}$$

Fig. 19–24

Fig. 19–25

Fig. 19–26

The above collision may be represented by Fig. 19–25 or 19–26 or by the equation

$$^{14}_{7}N + ^{4}_{2}He \rightarrow ^{1}_{1}H + ^{17}_{8}O$$

By measuring the lengths of the two tracks it was possible to calculate the kinetic energy of each particle. It was then discovered that about 1.26 MeV (2×10^{-13} joules) of energy had been lost in the collision.

Here are the masses on both sides of the equation.

$$^{14}_{7}N = 2.325\,21 \times 10^{-26} \text{ kg}$$
$$^{4}_{2}He = 6.646\,32 \times 10^{-27} \text{ kg}$$
$$^{1}_{1}H = 1.673\,50 \times 10^{-27} \text{ kg}$$
$$^{17}_{8}O = 2.822\,71 \times 10^{-26} \text{ kg}$$

Find the gain in mass as a result of this collision. (19.1)

Calculations of this kind gave the first experimental evidence of Einstein's prediction that mass and energy are equivalent.

Use the above figures to verify the relationship $E = mc^2$. (19.2)

Fig. 19–27

Although in this particular collision energy has been changed to mass, in many other atomic collisions the total *mass is reduced* and the particles leave with much *greater energy*. A nuclear explosion is an example of mass 'changing to energy'. Nowadays particle accelerators such as the synchrotron are used to smash atoms. Fig. 19–27 shows part of the 7 GeV proton synchrotron *Nimrod* (Genesis 10.8) at the Rutherford High Energy Laboratory.

Waves and Particles

1924 In 1924 Louis de Broglie suggested that, under certain circumstances, a particle might have wave properties. Three years later the theory was confirmed experimentally when G. P. Thomson produced an electron diffraction pattern.

If a circular loop, made from a fine coil spring is fixed to a vibrator, standing wave patterns such as the one illustrated in Fig. 19–28 can be produced. By varying the vibration frequency a number of different patterns can be obtained, but a standing wave pattern can be produced *only* at these frequencies which give an integral number of complete waves on the spring.

If we consider the electron as having wave properties, including *wavelength*, we can postulate that an electron can circle the nucleus indefinitely without radiating energy, provided the circumference of its orbit is an integral number of *electron wavelengths*.

Fig. 19–28

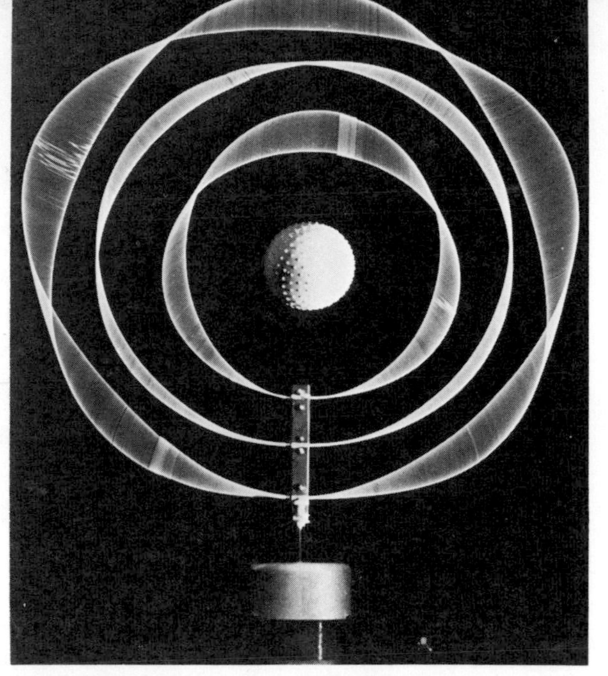

Fig. 19–29

Each of the electron orbits, Fig. 19–29, corresponds to a definite energy level.

If an electron jumps from one energy level E_4 to a lower level E_2, a quantum of energy will be radiated (Fig. 19–30). The frequency of the radiated photon will be proportional

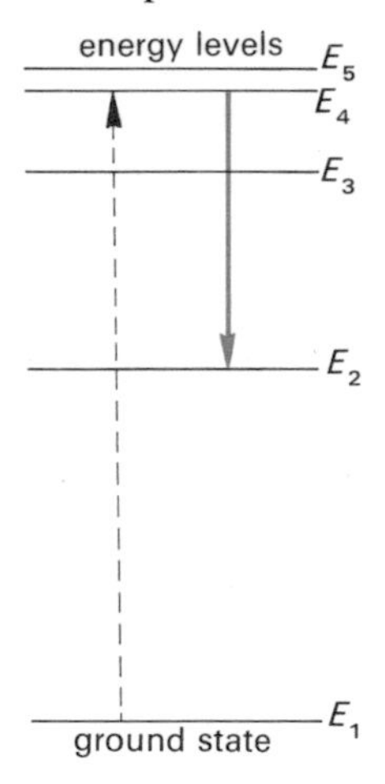

Fig. 19–30

to the difference in energy between these two levels. The constant of proportionality h is called Planck's constant.

$$E_4 - E_3 = hf$$

The most recent and comprehensive—but also the most abstract—model of the atom

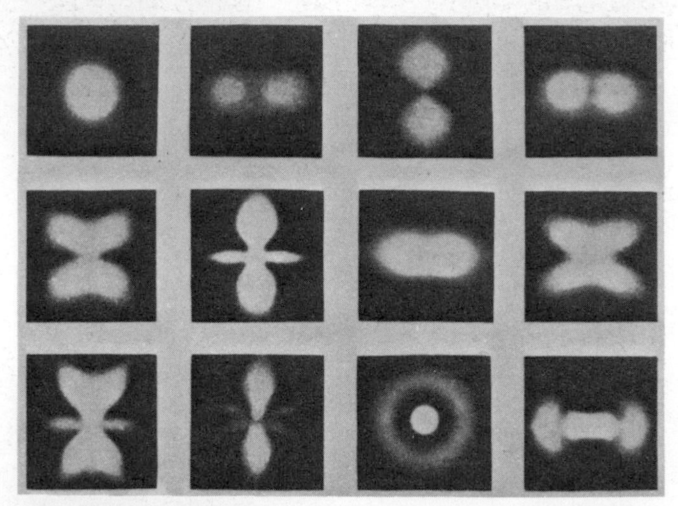

Fig. 19–31

is the *wave-mechanics model*. Fig. 19–31 shows pictorially the calculated electron-wave patterns—in order of increasing energy—for a hydrogen atom. The whiter the region, the greater the probability of finding an electron there. This model of the atom is, however, beyond the scope of this course.

1928 Geiger and Muller in Germany developed a new kind of tube for counting radioactive disintegrations. This is the type normally used now with a scaler or ratemeter.

1931 Robert Van de Graaff made his first electrostatic generator which could produce voltages of over a million volts. Fig. 19–32 shows a modern version of this generator.

Fig. 19–32

1932 James Chadwick discovered the neutron a Carl Anderson discovered the positron.

Working at Cambridge, Sir John Coc croft and E. T. S. Walton designed and bu the first 'atom smasher' or particle acceler tor. With it they hoped to produce nucle transmutations with artificially accelerat protons. A voltage multiplying device, co sisting of a step-up transformer feedi several rectifiers and capacitors, produc about half a million volts. This was appli to a long vertical tube made from gla cylinders taken from old petrol pumps. T cylinders were fastened together with met plates and plasticine! (Fig. 19–33.)

Fig. 19–33

Protons, produced in a discharge tube i the upper vessel, were accelerated down t tube by the strong electric field until th reached about 10 000 kilometres per secon They then struck a lithium plate and occ sionally a few scored direct hits on lithiu nuclei. Rutherford's well-tried zinc sulpha screen was used to detect any particl emitted. It was soon clear that alpha pa ticles were being shot off the lithium targ

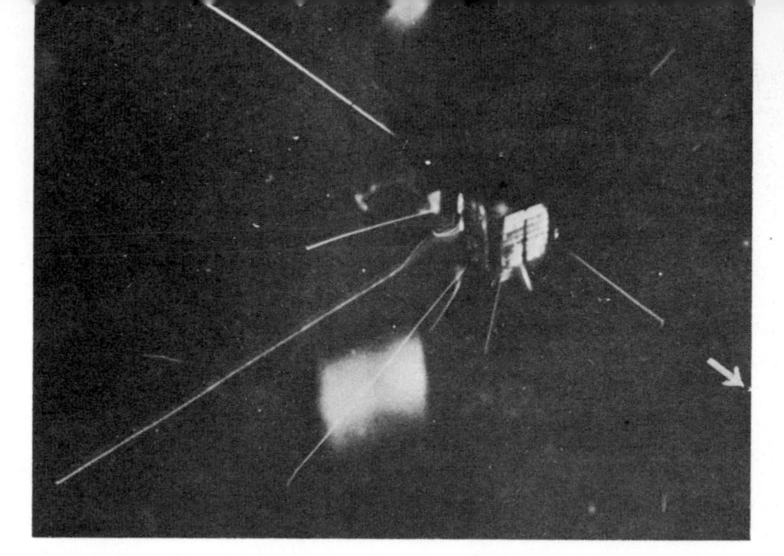

Fig. 19–34

in pairs. Later experiments using a cloud chamber (Fig. 19–34) confirmed that, when a proton struck a lithium nucleus, two alpha particles travelling in opposite directions were formed.

This, the first atomic disintegration produced by artificially accelerating particles, may be represented by Fig. 19–35 or by the following equation

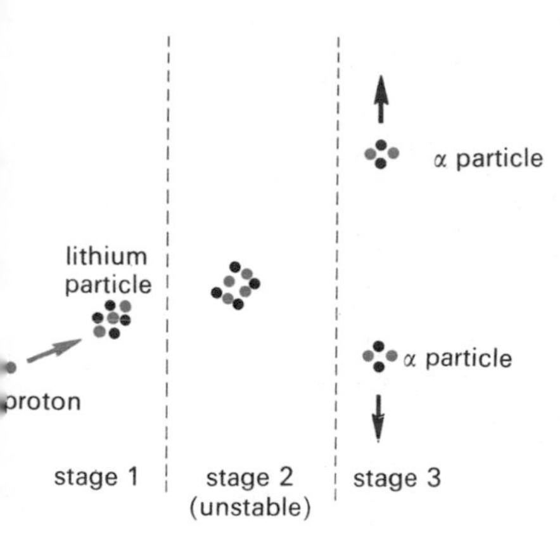

Fig. 19–35

$$^1_1H + ^7_3Li \rightarrow ^4_2He + ^4_2He$$

When the cloud chamber tracks were measured, it was clear that the energy of the alpha particles was *very* much greater than the energy of the proton which was about 0.5 MeV.

The two alpha particles were each found to have energies of about 8.6 MeV, a total of 17.2 MeV. This may seem a very small amount of energy, a few millionths of a millionth of a joule; nevertheless this energy released by splitting the lithium atom is many *millions* of times greater than the energy released per carbon atom when coal is burned.

The above nuclear reaction could therefore be more correctly written as

$$^7_3Li + ^1_1H + 0.5\,\text{MeV} \rightarrow ^4_2He + ^4_2He + 17.2\,\text{MeV}$$

Mass values for each side of the equation are given below.

$$^7_3Li = 1.165\,00 \times 10^{-26}\,\text{kg}$$

$$^1_1H = 1.673\,50 \times 10^{-27}\,\text{kg}$$

$$^4_2He = 6.646\,32 \times 10^{-27}\,\text{kg}$$

Use $E = mc^2$ to see if the sum of the mass + energy is the same on both sides of the equation. (19.3)

Fig. 19–36

Later in 1932, E. O. Lawrence in America used his Nobel-Prize-winning cyclotron to split atoms of lithium.

Fig. 19–36 shows a modern version of the Cockcroft–Walton high voltage generator in the National Acceleration Laboratory in Batavia, Illinois.

Softly Softly

1934 Irène and Frédéric Juliot–Curie discovered artificial radioactivity. They used alpha particles to bombard some light elements which then became radioactive. In the same year Lawrence in America and Enrico Fermi in Italy produced radionuclides artificially by bombarding stable elements with accelerated particles.

Fermi discovered that *slow moving* neutrons were particularly effective. During one experiment the neutron source and target were placed on a wooden table, and the rate of disintegration of the target atoms mysteriously increased! Fermi solved the puzzle and was awarded a Nobel Prize.

Some of the neutrons must have collided with the hydrogen atoms in the wood. They had then been deflected and *slowed down*.

The slower neutrons were more easily captured by the nuclei of the target atoms and so more radioactive nuclei were formed and the rate of disintegration increased.

When one particle collides elastically with another, the speed of the original particle will be altered *most* if the second particle has the same mass. In a head-on collision of this kind with a stationary particle, the original particle will be stopped. If the second particle has a much larger mass, the first will be deflected but its speed will not be greatly altered.

Ideally then, a neutron should be made to collide with another neutron, proton or hydrogen atom in order to reduce its speed. This is not always possible, as we will see later, but a number of collisions with any of the *lighter* atoms will greatly reduce the speed of the neutron.

Fig. 19–37

The chances of a *slow* neutron being captured by a nucleus are greater because it spends a larger time in the vicinity of the nucleus. The following analogies must not be taken too seriously, but they may help to illustrate the point. A magnetic dry ice puck fired at a stationary fixed puck of opposite polarity can be made to *pass by* if it is travelling quickly. If, however, it is moving slowly past, the two will be attracted to each other and coalesce. Alternatively, think of a golfer striking the ball too hard, so that it doesn't drop into the hole even though the aim is perfect! (Fig. 19–37.)

The normal random movement of molecules (heat) is sometimes called *thermal motion*. As slow neutrons are moving at speeds similar to those of molecules at room temperature, they are often called *thermal neutrons*. Fermi used such neutrons to bombard various substances, including uranium. From this he obtained new radionuclides which he could not identify. He thought he had produced elements with atomic numbers greater than uranium—the so-called *transuranic elements*.

In Germany Frau Ida Noddack suggested that, when heavy nuclei are bombarded with neutrons, they might break up into a number of large pieces which would be isotopes of other elements! This idea was, however, considered by all the physicists of the day to be completely absurd. During the next few years many other

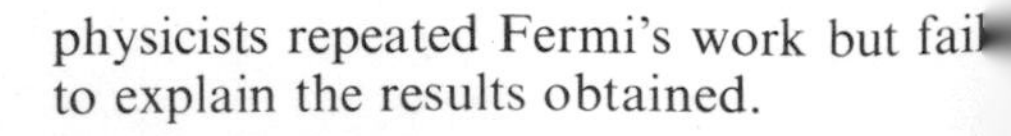

physicists repeated Fermi's work but fai to explain the results obtained.

NUCLEAR FISSION

1938 Four years after Fermi's original expe ment, the German scientists Otto Hahn a Fritz Strassmann were forced to admit th when they subjected the uranium 'radi nuclides' to careful micro-chemical analy they found *barium*. Yet they could hard believe this was possible! Uranium has protons and barium 56 in its nucleus. Nev before had a nuclear change of this mag tude been observed.

Fraulein Lise Meitner and Otto Frisc two refugees from Hitler's Reich, were Denmark at the time. They reckoned th if Hahn and Strassmann's results we correct, krypton should also be present it has 36 protons in its nucleus

$$_{92}U \rightarrow {}_{56}Ba + {}_{36}Kr$$

Further experiments confirmed this pı diction. The uranium nucleus had in fa split into two parts. The process was labell fission, a name borrowed from the biol gist's description of cell division. Its a nouncement led to a flurry of exciteme and about a hundred papers on nucle fission were published in 1939.

Niels Bohr and John Wheeler suggested model in which the nucleus is compared the oscillations of a water drop. If t amplitude of oscillations is large enough t drop will break up (Fig. 19–38). In t nuclear case the necessary energy for fissi is provided by the captured neutron.

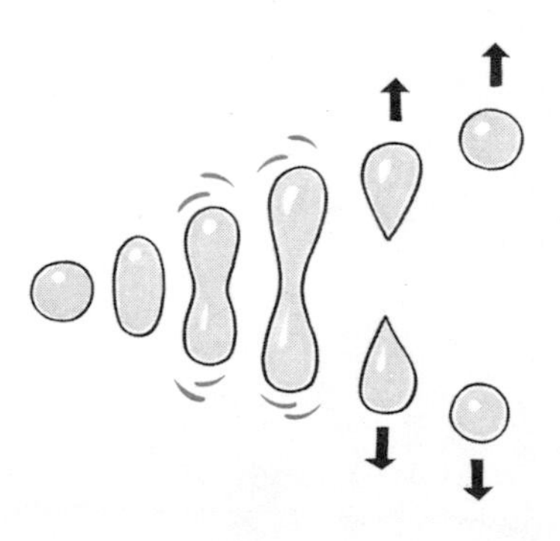

Fig. 19–38

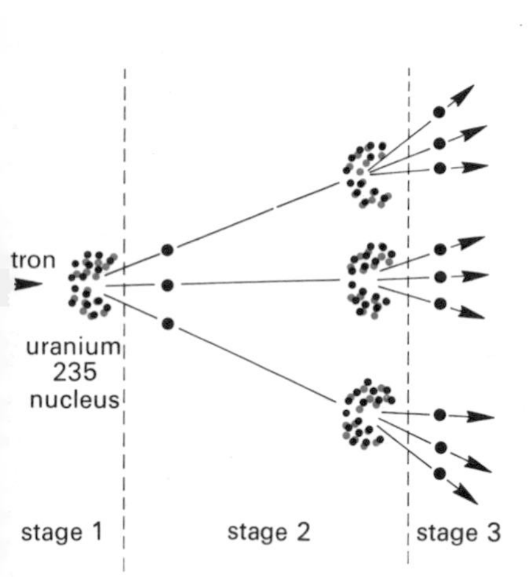

Fig. 19–39

1939 Fission of the uranium nucleus was certainly interesting, but much more exciting results were to follow. Meitner had calculated that a large amount of energy was released when the uranium nucleus split. In 1939 Fermi's associates working with a cyclotron at Columbia University were able to measure the amount of ionization produced by the fission fragments, and from this they found the energy released. It was about 200 MeV for each fission. If 1 gram of uranium-235 were used each day, it could produce energy at a continuous rate of 1 megawatt.

This was not all. Fermi suggested and later confirmed experimentally that when the uranium nucleus split, neutrons were released. These in turn might produce further fissions. A chain reaction was possible. The implications were shattering.

The most useful uranium isotope to produce nuclear fission is uranium-235 (Fig. 19–39). When it is bombarded by neutrons, the nuclear reaction may be represented by the equation

$$^{235}_{92}U + ^{1}_{0}n \rightarrow (^{236}_{92}U) \rightarrow ^{141}_{56}Ba + ^{92}_{36}Kr + ^{1}_{0}n + ^{1}_{0}n + ^{1}_{0}n$$

The unstable nuclide uranium-236 which is formed immediately splits up into barium-141, krypton-92 and three neutrons.

1942 Unfortunately natural uranium contains only 0.7% of uranium-235, while more than 99% is uranium-238 which absorbs neutrons but is not fissionable. It was found, however, that *slow* neutrons usually bounced off the uranium-238 nuclei and were captured by the uranium-235 nuclei to produce fission.

The most suitable light element to slow down neutrons was graphite. Others had been tried but they absorbed too many of the neutrons. Pure graphite slowed the neutrons down but did not absorb them.

By 1942 World War II was in its third year and most of the world's leading physicists were in America. One group led by Fermi were working night and day in a squash court under the grandstand of a football stadium in Chicago. They were building the world's first nuclear reactor or 'atomic pile' as it was then called.

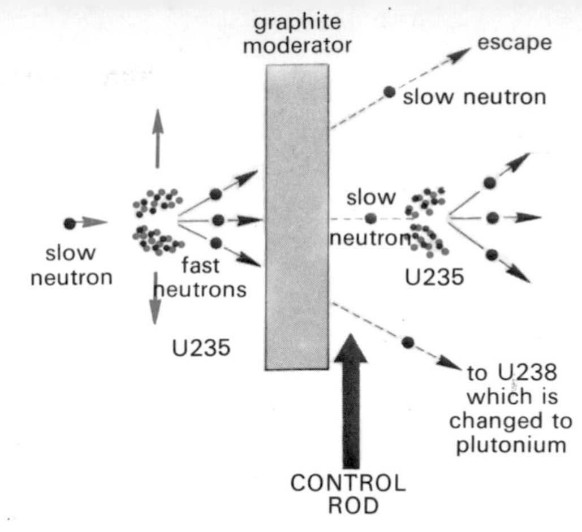

Fig. 19–40

In a nuclear reactor the object is to adjust the reaction so that, on the average, one neutron per fission will split another nucleus. The operation is outlined in Fig. 19–40. A slow neutron is first captured by a uranium-235 nucleus, which then splits to form two nuclides which fly apart at speed. Three fast neutrons are also emitted. These then pass through a graphite *moderator*, in which they collide many times with the carbon atoms and are thus slowed down. One may escape completely, another may be captured by a uranium-235 nucleus which splits and a third, which may not have been slowed down quite so much, is absorbed by a uranium-238 nucleus which changes to plutonium. If the three new neutrons behave similarly, a sustained nuclear reaction will result and heat will be produced at a steady rate.

To slow down or stop the reactor, *control rods* made of cadmium, which readily absorbs neutrons, are inserted.

Fermi's squash court reactor was made from a pile of uranium and graphite blocks. It was about 8 metres high. Fig. 19–41 shows a later form of reactor based on the same principle. The uranium fuel rods are kept in aluminium tubes and the cadmium rods are pushed into the reactor to slow it down.

Fig. 19–41

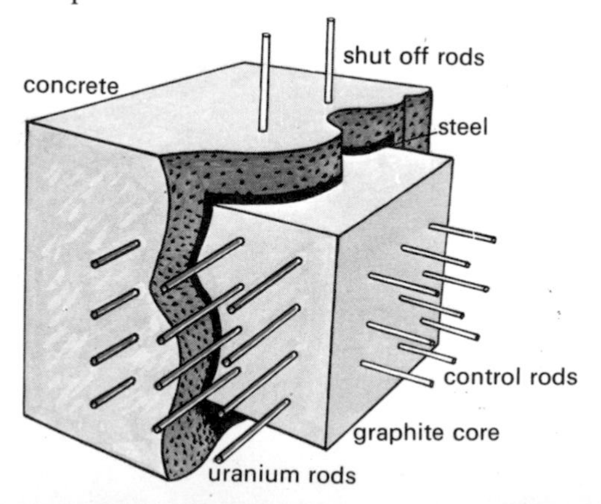

Reactors are now encased in a wall of concrete about 2 metres thick to prevent the escape of gamma radiation. This is called a biological shield.

At 3.30 p.m. on December 2, 1942, Fermi's pile produced the first nuclear chain-reaction. Later the same day the following telephone conversation took place between Chicago and Harvard University.

'The Italian sailor has landed in the New World'.

'How were the natives?'

'Very friendly . . .'.

The atomic age had begun.

$$^{238}_{92}U + ^{1}_{0}n \rightarrow ^{239}_{92}U \rightarrow ^{239}_{93}Np + ^{0}_{-1}e$$

In his reactor Fermi produced plutonium from the uranium-238. When bombarded by moderately slow neutrons a uranium-238 nucleus can capture one and become an unstable nuclide which decays ($t_{\frac{1}{2}}$ = 24 min) by the emission of a beta particle to form the first transuranic element, *neptunium.*

$$^{239}_{93}Np \rightarrow ^{239}_{94}Pu + ^{0}_{-1}e$$

Neptunium in turn decays ($t_{\frac{1}{2}}$ = 2.3 days) by beta emission to form plutonium, which is one of the most valuable radionuclides today. It has a half life of 24 300 years! Like uranium it will fission with slow neutrons and is likely to be a most important fuel for nuclear power stations in the future.

Thirteen transuranic elements have no been produced artificially. They have atom numbers ranging from 93 (neptunium) 105 (hahnium) and, like other artificial is topes produced in accelerators or reactor they are all radioactive.

Neutron beams from nuclear reactors a now being used to take 'X-ray type' phot graphs. A neutrophoto (Fig. 19–42(a)) show much more detail than an X-ray photograp (b) of the same object (c).

The Atom Bomb

Another group of atomic physicists led b Robert Oppenheimer was furiously tryi to produce the first atom bomb. For th pure uranium-235 was needed, yet natur uranium contains only 0.7% of this nuclid As U-235 and U-238 are chemically ident cal, they could not be separated chemicall Another way had to be found. A sieve fro his wife's kitchen gave Professor Fra Simon an idea! If natural uranium w vaporized and allowed to pass through very fine sieve, the lighter uranium-2 atoms could pass through faster than th heavier atoms and the percentage of U-2 to U-238 would be greater beyond the siev This diffusion process had to be repeate

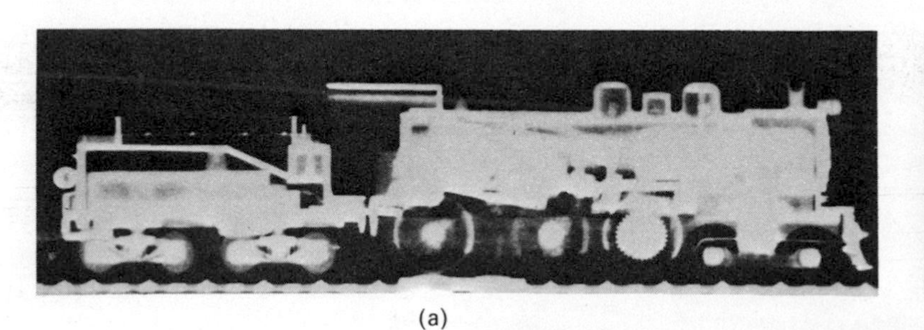

(a)

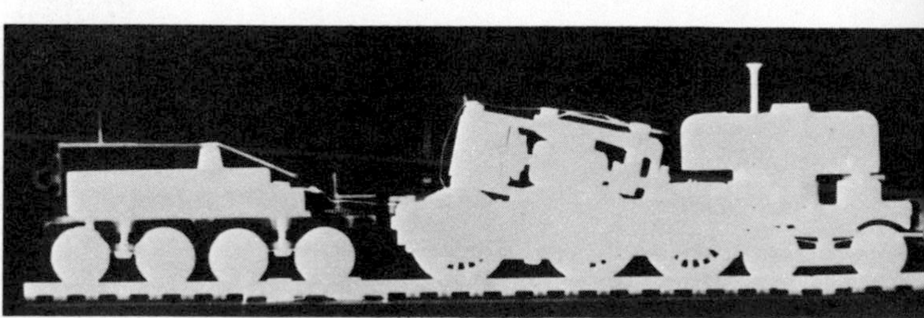

(b)

(c)

Fig. 19–42

Fig. 19–43

thousands of times in order to obtain reasonably pure uranium-235. The mammoth task—the Manhattan Project—was undertaken at Oak Ridge, Tennessee. The plant (Fig. 19–43) and 'artificial town' round it covered 70 square miles and employed about 80 000 people. Even so, the output was only a few kilograms of uranium-235 per week.

With U-235 no moderator is needed to slow down the neutrons, as the nucleus will readily accept fast neutrons. Chain reaction will not, however, take place in small chunks of U-235, as most of the neutrons produced escape through the surface.

Consider a sphere of U-235 with a radius of 2 centimetres. The surface area is $4\pi r^2$, that is, $4\pi \times 4$ and the volume $\frac{4}{3}\pi r^3 = \frac{4}{3}\pi \times 8$. The ratio of area/volume is therefore

$$\frac{4\pi 4}{\frac{4}{3}\pi 8} = \frac{3}{2} = 1.5$$

If the radius is doubled, the ratio becomes

$$\frac{4\pi 16}{\frac{4}{3}\pi 64} = \frac{3}{4} = 0.75$$

As the volume and hence the *mass* of uranium is increased, the surface area per unit volume becomes less and less, and so the chance of neutrons escaping becomes less and less. By increasing the mass, then, the chance of each new neutron producing a fission is increased, until eventually a point is reached when a chain reaction is possible. The mass of U-235 needed for this to happen is called the *critical mass*.

The principle of the atom bomb is shown in Fig. 19–44. Two hemispheres of U-235 well below the critical mass are kept apart inside a strong metal container. To explode the bomb one piece is violently forced against the other by a conventional explosive. The total mass is then greater than the critical mass.

There are always a number of neutrons flying around. Several hundreds are estimated to pass through our bodies every second, many of them produced by cosmic radiation. Thus, as soon as the mass of U-235 in the bomb exceeds the critical mass, a violent explosion takes place.

On July 16, 1945, the world's first atom bomb was exploded in a remote desert plateau in New Mexico. A small piece of uranium, smaller than a football, had changed the history of the world.

Fig. 19–44

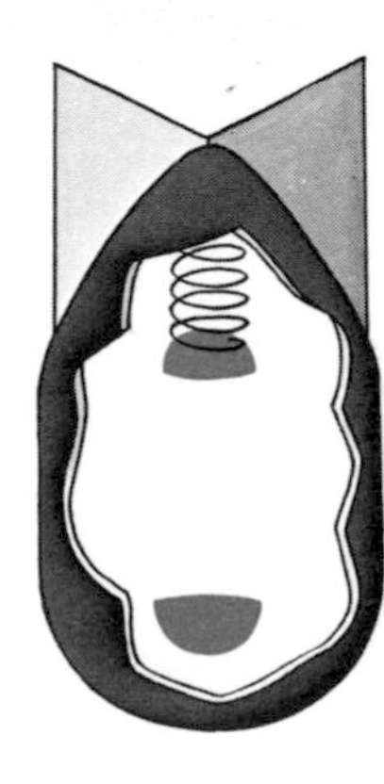

Fig. 19–45

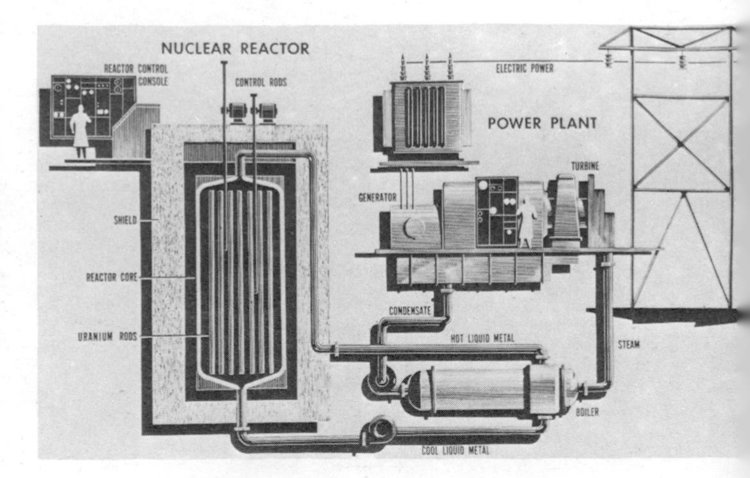

Fig. 19–46

NUCLEAR POWER

1951 After the war atomic piles were rapidly developed to produce energy for peaceful purposes. The heat produced in the reactor had to be made available to power a steam turbine, which could then drive a dynamo to produce electricity. This was first accomplished in 1951, although it was five years later before the world's first nuclear power station was opened at Calder Hall in Cumberland (Fig. 19–45).

Most modern reactors operate on the principle indicated in Fig. 19–46. Uranium fuel rods are encased in graphite blocks and cadmium rods control the reaction. The reactor is contained in a strong *pressure vessel* through which a gas or liquid metal flows in order to transfer the heat from the reactor to a steam boiler. Thereafter a nuclear power station is no different from an oil or coal burning one. Nuclear fuel simply replaces the fossil fuels used in a conventional station.

Since 1959 an experimental fast reactor has been in use in Dounreay in the North of Scotland (see back cover). It is likely that power stations of the future will be of this type, as it has many advantages. By using enriched uranium (e.g. 75% U-235 instead of 0.7%) or plutonium as fuels, no moderator is required. This means that the reactor core can be reduced from about 10 metres in diameter to 1 metre. By surrou ing the reactor with a blanket of natu uranium which catches any escaping n trons, plutonium can be produced (page 160). In fact the quantity of plutoni produced in the blanket is greater than t used as fuel! Such a system is called *breeder reactor*.

Fig. 19–47 shows the Prototype F Reactor recently opened at Dounreay. uses mixed oxides of uranium and p tonium as fuel and liquid sodium instead carbon dioxide to remove the heat from t core. The reactor produces heat at a rate 600 MW and electricity at a rate of 250 M a very high overall efficiency.

By the late 1970s it is expected that it w cost less to generate electricity from nucl fuels than from coal and oil. A much m important consideration at a time when earth's supply of fossil fuel is gradua disappearing is that fast breeder react will provide a source of almost limitl energy. New fissionable plutonium is p duced as uranium-235 is consumed.

Nuclear Powered Vessels

America launched the first nuclear power submarine, *Nautilus*, in 1954. She travell 60000 miles, using only 4 kilograms

Fig. 19–47

Fig. 19–48

uranium fuel. Three million gallons of oil would have been needed to cover the same distance using diesel motors. In 1958 the *Nautilus* made history by crossing beneath the North Polar ice cap. In 1960 her sister ship, *Triton*, travelled round the world submerged, a distance of 36 000 miles. Fig. 19–48 shows one of Britain's latest nuclear submarines, H.M.S. *Revenge*.

In 1957 Russia launched the first nuclear powered surface vessel, the ice-breaker *Lenin* (Fig. 19–49). One of America's nuclear powered carriers is shown in Fig. 19–50.

A nuclear powered airship has been proposed as one way of overcoming the difficulties associated with the massive radiation shields needed in nuclear aircraft.

Fig. 19–49

NUCLEAR FUSION

1952 On November 1, 1952, a small coral island with an area of 13 square miles disappeared. The first H bomb had been exploded. It had five times the destructive power of all the conventional bombs dropped during the whole of World War II. Ten years later an explosion equivalent to that produced by 50 million tonnes of TNT announced that an even bigger H bomb had been produced.

To understand how these bombs produce such enormous quantities of energy, study

Fig. 19–50

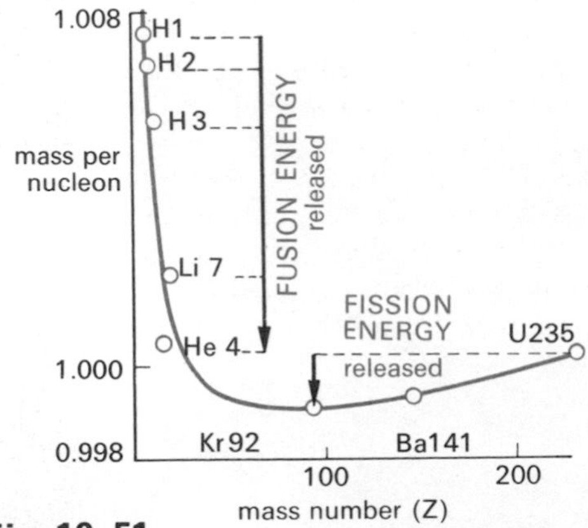

Fig. 19–51

the graph shown in Fig. 19–51. It shows how the mass per nucleon in an atom varies with the atomic number. From the graph it is clear that, if U-235 is split into Kr-92 and Ba-141, the mass per nucleon has *decreased*. We have seen that this decrease in mass shows itself as additional kinetic energy in the fission products. The heat produced in nuclear reactors comes from such a fission.

In 1932 Cockcroft and Walton fired a hydrogen nucleus (^1_1H) at a lithium nucleus (^7_3Li); the particles fused and then split into two helium nuclei ($2 \times {}^4_2He$). From Fig. 19–51 we see that the mass per nucleon has been reduced. This decrease in mass appears as kinetic energy in the two alpha particles. The process is called *nuclear fusion*.

Although the fusion of a few nuclei of light elements is possible with the help of particle accelerators, large quantities of energy can be released only if millions of nuclei are fusing at once. In the hydrogen bomb the nuclei of two isotopes of hydrogen are speeded up by raising their temperature to over a hundred million degrees Celsius. The only method known at present which is capable of producing such temperatures is the fission bomb. An H bomb therefore contains a fission bomb to produce the high pressure and temperature necessary to initiate the fusion reaction. When a collision occurs between two fast moving nuclei at such temperatures, the reaction is called *thermo-nuclear fusion*.

$$^3_1H + {}^2_1H \rightarrow {}^4_2He + {}^1_0n + \text{energy}$$

$E = mc^2$ enables us to calculate the quantity of energy released.

tritium $^3_1H = 5.008\,90 \times 10^{-27}$ k

deuterium $^2_1H = 3.344\,41 \times 10^{-27}$ k

$8.353\,31 \times 10^{-27}$ k

helium $= 6.646\,32 \times 10^{-27}$ k

neutron $= 1.674\,90 \times 10^{-27}$ k

$8.321\,22 \times 10^{-27}$ k

The mass transformed is equivalen 17.6 MeV. If 1 gram of tritium and deu ium were fused in this way, the en released would be adequate to supply average household for forty years!

In a similar reaction two deuterium nu can be fused to give about 5 MeV per fus As deuterium can be extracted very che from sea water, enough energy for at l ten thousand million years is readily a able! You can see then why so much tho and effort is at present concentrated in tr to find a way of producing a thermo-nuc fusion reaction under controlled conditi What are the difficulties?

Above a few thousand degrees Cel gas molecules disassociate into atoms temperatures around 100 000 °C even at no longer exist; they have split into p tively charged nuclei and electrons. 'fourth state of matter', consisting of e concentrations of positive and negati charged ions, is called a *plasma*.

If controlled thermo-nuclear reaction to be produced, two major problems a First, how can the required temperat which are in the region of 30000000 be achieved, and secondly how can plasma once formed be held together (tained)? These are among the most diff problems facing science and technolog the present moment.

Let us first consider the second prob that of containment. Clearly, as no stance can withstand such temperatures plasma must be kept from contact with vessel holding it. Strong magnetic f offer a solution—the 'magnetic bottle'.

Charges moving quickly in the s direction in a conductor are attracte each other. You have seen this effect two parallel wires carrying currents in same direction. If then a stream of (plasma) is moving through a tube, might expect it to 'pull itself together' a narrow beam at the centre of the tube. This is called the *pinch effect*. In practice it is about as difficult to achieve it successfully as balancing one bar magnet in mid air above another repelling magnet.

To counter this instability a coil is wrapped round the tube containing the plasma. If a current is then passed through the coil, the ions will spiral round the field lines inside the tube, thus preventing the plasma from touching the sides (Fig. 19–52). The Leybold fine-beam tube illustrated in Fig. 19–53 provides a delightful demonstration of this spiral effect with an electron beam.

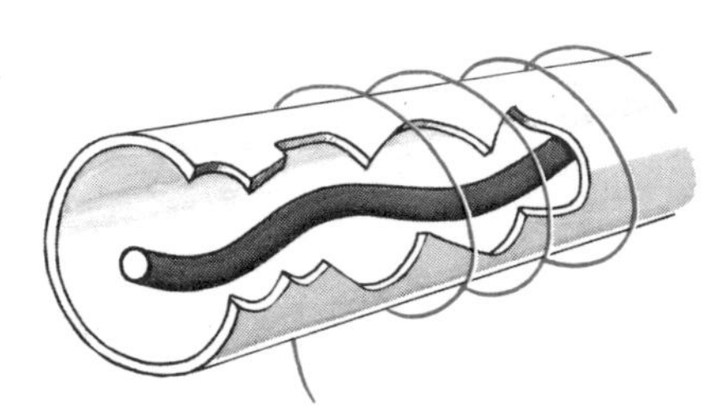

Fig. 19–52

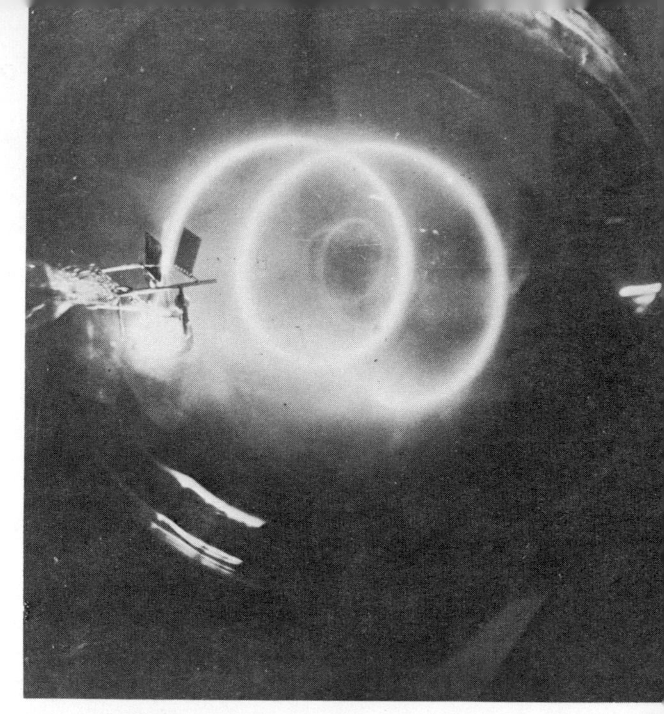

Fig. 19–53

In 1958 there were hopes that thermonuclear fusion had occurred in ZETA (Zero Energy Thermonuclear Assembly) at Harwell (Fig. 19–54). Neutrons were released,

Fig. 19–54

showing that fusion had occurred, but not, unfortunately, thermonuclear fusion. The high speed ions in the plasma had fused with stationary ions.

Zeta consists of a giant doughnut, about 3 metres in diameter. It contains deuterium gas, which acts as the single turn secondary of a double toroidal transformer (Fig. 19–55).

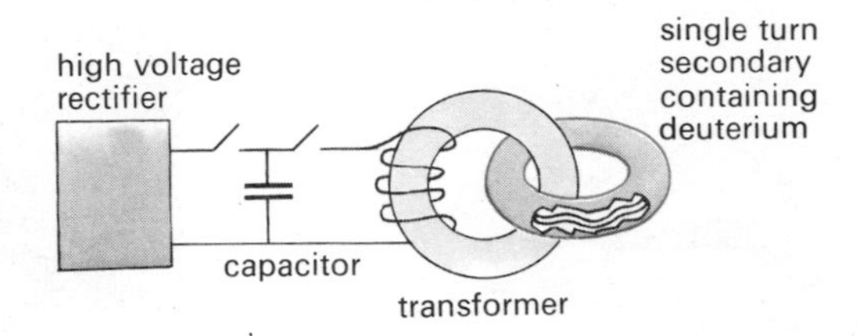

Fig. 19–55

By passing an enormous pulse of current through the primary windings, a current of the order of a million amperes flows through the deuterium. This forms a plasma at several million degrees Celsius. The giant pulse of current in the primary comes from capacitors which store several megajoules Fig. 10–27, page 81).

Other machines which have attempted to produce thermonuclear fusion at the *Stellar-atron* (Fig. 19–56) at Princeton University,

Fig. 19–56

the appropriately named *Perhapsatron* Los Alamos, and ALICE (Adiabatic Lo Energy Injection and Capture Experime at the University of California.

All the heat produced by the sun is alm certainly due to thermonuclear fusi which in effect produces helium fr hydrogen.

$$4\,{}^1_1\mathrm{H} \rightarrow {}^4_2\mathrm{He} + 2\,{}^0_1\beta + 27\ \mathrm{MeV}$$

The sun is estimated to be 'losing weig at a rate of four million tonnes per seco By doing so it hopes to 'live' for at le another ten thousand million years!

FUNDAMENTAL PARTICLES

Forty years ago it looked as though the comp structure of matter could be reduced to three sim particles—the electron, the proton, and the phot or quantum of electromagnetic radiation. In 1930s this simple picture started to change. T positron and neutron were discovered. In 1935 Japanese theoretical physicist Hideki Yuka suggested that the strong forces binding the nucl particles together might usefully be described 'particles continually being exchanged between nucleons'. Cecil Powell discovered such partic in cosmic radiation. They are called *pions*. T mass of each is about 270 times that of the electr

To account for the fact that the conservati laws of energy, momentum and angular mom tum do not appear to hold during the radioact process of beta decay, another particle was ' vented' by Pauli in 1931. It was called the *neutri* It had to have no mass and no charge and yet able to possess energy and momentum! Twen five years later the postulate was justified wh Reines and Cowan obtained experimental evider of neutrinos coming from a nuclear reactor.

Mountain tops became happy hunting groun for particle spotters. Balloons, and more recen satellites, have also been used. Up there nat provides the most powerful atom-smashers kno —the high energy protons present in cosmic rad tions. Fig. 19–57 shows the result of a prot colliding with a nucleus in a photographic pla Several pions are produced.

Work with cosmic radiation was not entir satisfactory, however. The radiation was h hazard and the energy of the cosmic particles u known. Early particle accelerators such as the V

Fig. 19–57

de Graaff generator could produce voltages up to about 7 MeV, which was not enough to produce new particles by collision. 200 MeV was needed, and this seemed out of the question.

An alternative way of providing high energy particles was, however, soon developed—the particle accelerator. These machines speed up particles to very nearly the speed of light, and produce energies of thousands of millions of electron volts. In 1971 the world's largest particle accelerator started operating in Batavia, Illinois. This synchrotron (Fig. 19–58), costing over £100 million, accelerates protons to energies of 500 GeV (5×10^{11} eV) as they whizz for nearly a million miles round the four mile circular track buried 8 metres below the ground. If electrons are accelerated in this way, there comes a point when they radiate as much energy as they gain each revolution. This difficulty, which is mainly caused by the circular motion of the electrons, can be minimized by using a linear accelerator.

In one type of linear accelerator a series of electrodes in the form of hollow sleeves are

Fig. 19–58

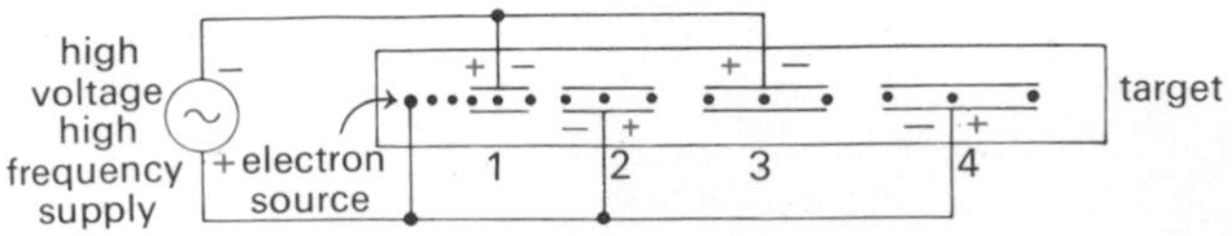

Fig. 19–59

arranged inside an evacuated tube as shown in Fig. 19–59. The sleeves are coupled to a high-frequency, high-potential source of energy. Electrons are fired from a source at one end of the tube and are accelerated towards a target at the other end. If an electron leaves the source when the sleeves are charged as shown in red, the electron will be accelerated towards the first sleeve. Once inside it the electron will coast along at a steady speed as there is no field inside the sleeve. When the electron is inside sleeve No. 1 the voltages are reversed as shown in *black* so that the electron emerges to find sleeve No. 2 attracting it and sleeve No. 1 repelling it. The electron is therefore accelerated again. As the sleeve voltages are reversed during the time the electron is inside each sleeve, the electron is accelerated between the sleeves. So that reversals take place at the right time, each sleeve must be longer than the previous one, the longest being nearest the target. A two mile long electron accelerator has been built in Stanford, U.S.A. (Fig. 19–60). With it electrons can be accelerated to 20 GeV.

The 500 GeV synchrotron in Batavia (Fig. 19–5
is fed by an 8 MeV accelerator—the small circul
structure at the bottom right of the picture. Th
in turn is fed from a linear accelerator rated
0.2 GeV.

When accelerated particles reach their maximu
energy, they are fired into a target to produce ne
particles by collision. These particles are th
allowed to pass into a bubble chamber.

The chance of a nuclear collision occurring in
gas is very remote, as the molecules are so far apa
The Wilson cloud chamber is not therefore partic
larly well suited to the study of collisions. A liqui
on the other hand, has about one thousand tim
as many molecules per cubic centimetre. If, the
particles are passed into a transparent liquid, su
as liquid hydrogen, a stream of bubbles may
formed under certain circumstances.

The liquid is kept under pressure at a temperatu
above its normal boiling point. If a charged partic
is then passed through the liquid and if the pressu
is reduced suddenly, a stream of tiny vapo
bubbles will form on the ions produced by t
particle. Fig. 19–61 shows a photograph of t
bubble tracks of a number of different sub-nucle
particles. If the bubble chamber, invented in 19
by D. A. Glaser, is situated in a strong magne
field, the curvature of the track gives a clue to t
charge on the particle. The length of the track oft
indicates its energy or the duration of the particl

Fig. 19–60

Fig. 19–61

life before it decays into another particle! The expectation of life for many of the new particles is of the order of 10^{-10} seconds or less.

The study of these incredible fragments of matter and anti-matter and their creation out of energy is perhaps the most exciting and the most expensive form of research at the present moment. Fig. 19–62 shows the bubble track of an electron and its 'anti-particle', the positron. These have been formed from a photon.

How many fundamental particles are there? The number depends to some extent on how you define the particles, but it is of the order of 200. This makes nonsense of the idea of 'fundamental particles', and many attempts are being made to find a pattern into which they will fit. The more important particles in order of mass are

(i) photons.	Quanta of electromagnetic radiation.
(ii) leptons.	The electron, muon and neutrino.
(iii) mesons.	The pion and kaon.
(iv) baryons.	These are heavy particles such as the proton, neutron and a few heavier particles.

Research in fundamental particles is one of the spearheads of physics today. One interesting theory suggests that there may be particles which are more elementary than the elementary particles! These have been called *quarks*, and it has been suggested that three such particles may, when tightly bound together, produce all the fragments which are at present called 'fundamental particles'. The first man to find experimental evidence for such particles will make history!

EPILOGUE

During this course you have caught a glimpse of some of the exciting and challenging developments that are taking place in modern physics. These developments confront you with a double challenge: first to achieve a deeper understanding of basic physical principles and secondly to ensure that this knowledge is used for the good of mankind in agriculture, education, industry, medicine and research. The vision of Teilhard de Chardin provides a fitting goal.

> 'The true physics is that which will one day achieve the inclusion of man in his wholeness in a coherent picture of the world'.

Fig. 19–62

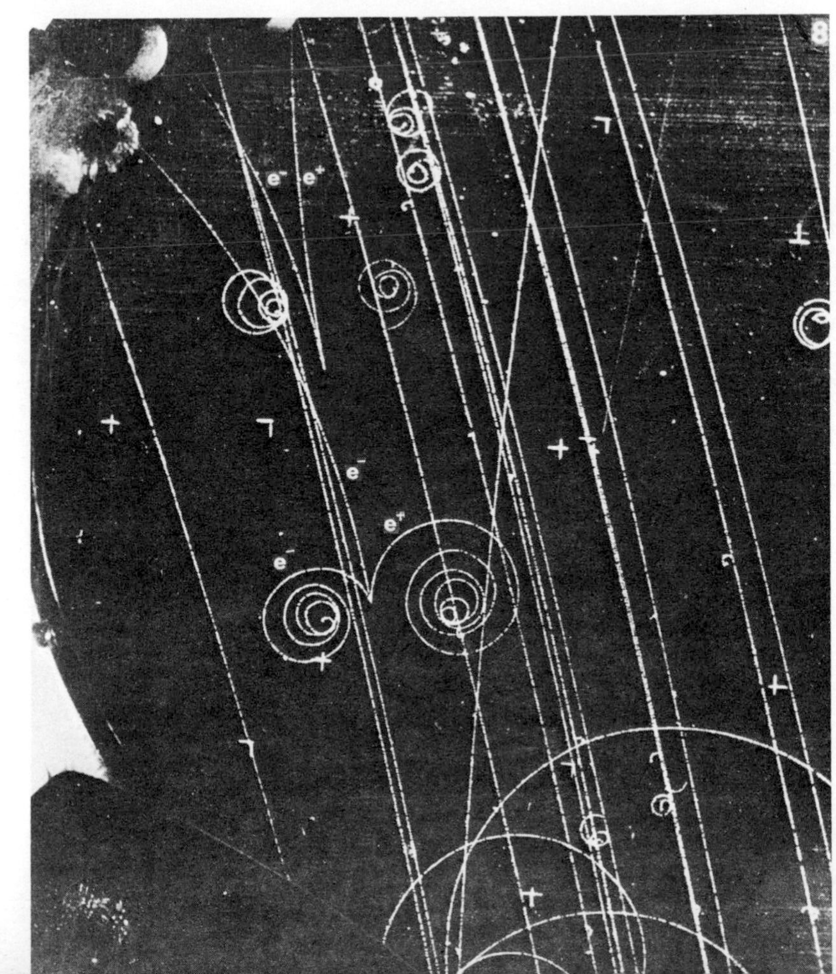

SUMMARY

1895 Röntgen discovered X-rays.
1896 Becquerel discovered radioactivity.
1897 J. J. Thomson measured charge/mass for electron
1905 Einstein suggested equivalence of mass and energy $E = mc^2$
1909 Rutherford and Royds demonstrated that alpha particles are helium nuclei.
1911 Geiger and Marsden's scattering of alpha particles showed that the atom is largely empty space with its mass concentrated in a minute fraction of its volume. This led to the Rutherford–Bohr model of the atom.
1919 Rutherford produced artificial transmutation.
1927 G. P. Thomson demonstrated wave properties of electrons.
1942 Fermi and co-workers produced first nuclear reactor.
1956 First nuclear power station opened at Calder Hall.
1971 Prototype fast reactor opened at Dounreay.

PROBLEMS

19.4 (a) A radioactive source gives a count rate of 120 per second and has a half life of 90 seconds. How long will it take for the count rate to reduce to 15 per second?
(b) The half value thickness of an absorber is that thickness which reduces the radiation to one half its original value. A gamma absorption experiment gave the following data.

Background count rate = 1 per second

Count rate (counts per second)	*Thickness of lead Absorber (units)*
46	0
38	4
32	7
26	11
16	22
8	36

Using a suitably large scale, draw a graph of count rate due to the source against thickness of lead absorber, and from the graph determine the half value thickness of lead. What thickness of lead will be necessary to reduce the count rate to 3 per second?
(c) Describe an α particle experiment which supported the idea of the nuclear atom. What development of this idea was necessary to explain the existence of isotopes?
[S.C.E. (H) 1968]

19.5 (a) Describe briefly any type of detector which will respond to alpha particles but not to beta particles or gamma rays.
(b) Another type of detector can be obtained in two forms, one which will detect *both* beta and gamma radiation, and the other which will respond to gamma rays only. What is likely to be the difference between these two forms of detector?
(c) What changes of charge and mass occur in the nucleus as the result of the emission of (i) an alpha particle, (ii) a beta particle, and (iii) a gamma ray?
(d) The nucleus $^{238}_{92}U$ decays by the emission of eight alpha particles and six beta particles. Give the atomic number and mass number of the new nucleus.
(e) A gold leaf electroscope, with a scale calibrated in volts, has negligible leakage. Arrangements are made so that it can be charged quickly and radioactivity measured by the rate at which the leaf falls thereafter. When some thoron gas is introduced, the leaf is found to fall at the rate of $200\ V\ s^{-1}$ immediately afterwards. Assuming that background radiation is negligible, and that the half-life of thoron is 50 seconds, what rates of fall (in $V\ s^{-1}$) would be obtained if they were measured at times (i) 50 s, (ii) 100 s, (iii) 150 s, and (iv) 200 s after the introduction of the gas?
(f) By considering these results, derive a formula giving the rate of fall at any time t after the introduction of the gas. Use it to find the expected rate of fall 25 seconds after the start.
(g) Would it matter if the electroscope was charged positively or negatively for this experiment? Explain your answer.
[S.C.E. (H) 1969]

19.6 It has been calculated that it takes 4.5×10^9 years for half of the uranium in a rock-sample to disintegrate radioactively and turn into lead. The metals uranium and lead are contained in a particular rock-sample. 30% of the atoms present in the metallic content are uranium and 70% of the atoms are lead. By drawing a graph of mass of uranium present against time, or otherwise, estimate the age of the rock, assuming that 50% of the atoms in the original metallic content were uranium. [S.C.E. (H) 1970]

19.7 When a certain radioactive source is tested with a scaler connected to a G-M tube with a thin window, a high count-rate is obtained which drops appreciably when a sheet of paper is placed between the source and the window. A sheet of lead, a few millimetres thick, in place of the paper

causes no further appreciable drop in the count-rate, but a sheet of lead, several centimetres thick, does cause a further drop in the count-rate.
(a) State what this experiment tells you about the nature of the radiation and give reasons for your conclusions.
(b) Why is there still a very small count-rate to be observed even when the thick lead screen is in position?
[S.C.E. (H) 1970]

19.8 (a) Fig. 19–63 is a copy of a cloud chamber photograph of α-particle tracks.

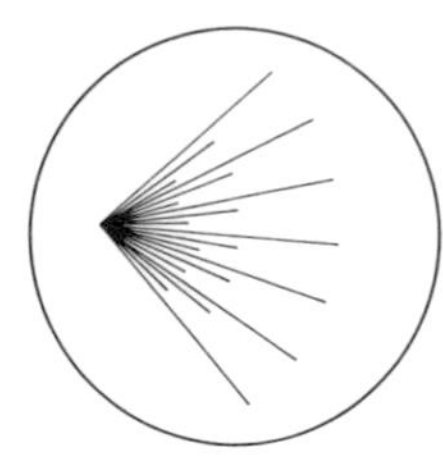

Fig. 19–63

(i) Account briefly for the formation of the tracks.
(ii) State with reasons what the photograph suggests about the energy of the α-particles emitted by the source in the chamber.

(b) When a β-particle and an α-particle of the same initial kinetic energy are released in a cloud chamber, one produces a short, thick track and the other a long narrow track.
(i) Identify the tracks and explain why they differ in appearance.
(ii) What simple additional piece of equipment could be used to assist in the identification of the tracks? How would it be used?

(c) The essential parts of Rutherford's famous α-particle scattering experiment are represented in Fig. 19–64.
(i) What kind of 'target' did Rutherford use?
(ii) What type of detector was used?

The detector could be rotated in a circle with the mid-point of the target as centre.
(iii) In which position (A, B, C or D) were *most* events detected? Account for this.

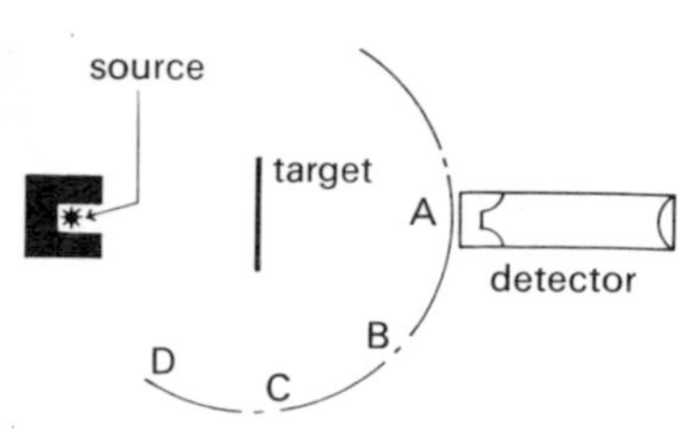

Fig. 19–65

(iv) In which position were the *most significant* events detected?
(v) How did Rutherford explain the occurrence of these latter events?
(vi) Why was it necessary to do the experiment in a vacuum?
[S.C.E. (H) 1971]

19.9 A sample of a radioactive element gives the following results.

Time in seconds	*Count Rate per second*
0	56
30	41
60	29
90	23
120	17
150	13
180	10
210	7
240	6

Draw a graph of the count rate against time and from the graph determine:
(a) the background count rate;
(b) the half life of the sample.
(c) Explain why a background count rate of about 20 counts per second would have made it difficult to determine accurately the half life of this sample. [S.C.E (H) 1972]

Answers

CHAPTER ONE

1.1 A dry-ice puck, attached by a thread to a fixed central point, is moving in a circular path. When the thread is burned through the puck moves off in a straight line, at a tangent to the circle.

1.2 (i) 300 m, (ii) 100 m south, (iii) 0.83 m s^{-1} south, (iv) 2.5 m s^{-1}.

1.3 (a) 343 km hr^{-1}, (b) 0.

1.4 100 minutes.

1.5 Constant velocity. Bench. 0.3 s, 45 cm, 1.8 m s^{-1}. Same; the vertical acceleration does not affect speed in horizontal direction.

1.6 5000 m s^{-1}, 1.05×10^6 m.

1.7 33 km.

1.8 See graphs below.

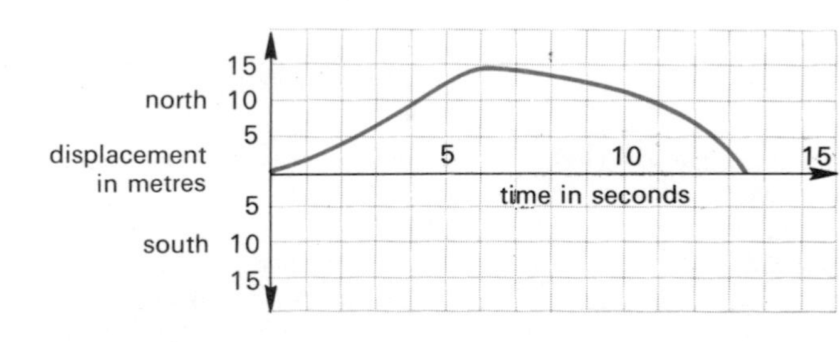

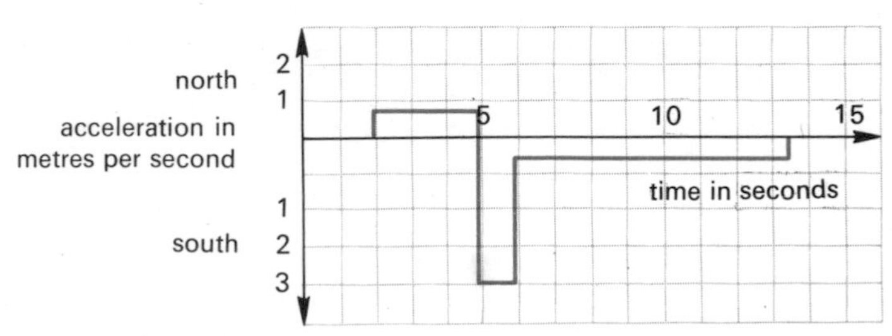

1.9 A represents the motion of a body moving eastwards and slowing down uniformly until it stops instantaneously and then accelerates uniformly westwards. Throughout the motion the body has a constant negative acceleration east, i.e. a positive acceleration west (cf. throwing a ball in the air).

The first part of the motion in B is similar to that of A but the acceleration westwards is much greater. This is followed by an eastwards constant acceleration which continues until the body attains the same velocity as it had originally.

In C the body accelerates uniformly eastwards and then decelerates eastwards (i.e. accelerates westwards) uniformly until it stops. The westwards acceleration is three times as great as the eastwards acceleration.

The distance travelled is the same in A, B and C. In A and B the total displacement is zero as in each case the body goes equal distances east and west. In C the displacement is eastwards and it is *twice* the eastwards displacement in A or B.

1.10 See graph of movements below. No.

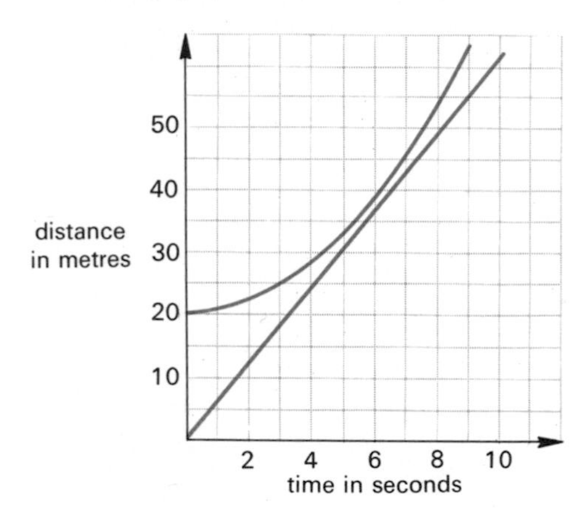

1.11 20 m s^{-1}, 30 m s^{-1}, 40 m s^{-1}.

1.12 5 m.

1.13 Motion right to left. K.E. less after each bounce. Interval between bounces gets less. Horizontal component of velocity practically constant throughout.

1.14 See vector diagram below.

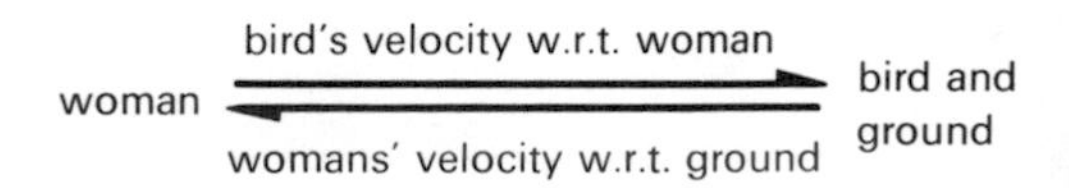

1.15 15.0 m s^{-1}, 86° N. of W.

1.16 44.7 km h^{-1}, 26.5° E. of N.

1.17 15.8 km h^{-1}, 18.5° S. of E.

1.18 240 m.

1.19 3 s.

1.20 (a) 157 m, (b) 100 m E., (c) 15.7 m s^{-1}, (d) 10 m s^{-1} E., (e) 3.14 m s^{-2} S.

1.21 (a) 4.24 m s^{-1} N.W., 10 m s^{-2} N.W.; (b) 424 m s^{-1} N.E.

CHAPTER TWO

2.1 Unbalanced force of 1 N acting on 1 kg produces acceleration of 1 m s^{-2}. Unbalanced force of F acting on 1 kg produces acceleration of F m s^{-2}. Unbalanced force of F acting on m produces acceleration of $\frac{F}{m}$ m s^{-2}.

$$\text{acceleration } a = \frac{F}{m}$$

$$\Leftrightarrow \quad F = ma$$

2.2 Acceleration = 2 cm per fifth second per fifth second = $0.02 \times 5 \times 5$ m s^{-2} = 0.5 m s^{-2}.
Force = $m \times a$ = 1.5 N.

2.3 Spring balances are calibrated for use in upright position. Allowance necessary if used horizontally. Best done in conjunction with friction-compensated runway so that constant speed is achieved when balance reads zero. Very sensitive spring balance is needed, but then it is difficult to keep reading constant.

2.4 Sand, as it gives more.

2.5 Bend knees. As in 2.4 this increases Δt and reduces F.

2.6 Anything which *increases* the *duration* of the drivers deceleration will reduce the average force. Springy bumpers, engine and front end collapsing *under* the driver, and of course seat belts, can all reduce the average deceleration of the driver and so reduce the average force exerted on him.

2.7 Units in $F = ma$ are N = kg × m s^{-2}, hence $\frac{\text{N}}{\text{kg}}$ = m s^{-2}.

2.8 X = 10 N, Y = 5 N, Y.

2.9 On Earth air resistance will cause paper to float down more slowly. On the Moon, with no atmosphere, both will fall together.

2.10 350 N, 31.5 m.

2.11 3338 kg.

2.12 20 m s^{-2}, 10 m s^{-2}.

2.13 If D = drag, force as it rises = $mg + D$, force as it falls = $mg - D$. There is therefore a greater force and therefore a greater (negative) acceleration as it rises. It will therefore take longer to fall.

2.14 Where g is less, e.g. on a high mountain, the range would be greater.

2.15 If you are standing upright in a bus when it suddenly accelerates, you 'fall backwards', as a force is applied to your feet. Your centre of mass (centre of gravity) tends to stay put (inertia) but your feet are forced forward and you lose balance.
If you had been leaning forward at the time, your weight, mg, acting through your centre of mass would have tended to make you topple forwards. This could have prevented your toppling backwards!
To enable the sprinter to exert a large force on the ground—which in turn will cause the ground to exert a large accelerating forward force on his feet—he must also lean forward. The forward directed force (exerted by the ground) then acts through the sprinter's centre of mass.

2.16 2 m s^{-2} up, 0.67 s.

2.17 a = 3 m s^{-2}, F = 150 N.

2.18 100 N.

2.19 $x = vt$, $y = \frac{1}{2}gt^2 = \frac{1}{2}g\left(\frac{x^2}{v^2}\right) = \left(\frac{g}{2v^2}\right)x^2 = kx^2$.

2.20 20 s, 6 km.

2.21 3.33 m s^{-2}, 5 m s^{-2}, 10 m s^{-2}.

2.22 In 1-metre arm: 98 N. In 4-metre arm: 13 N.

2.23 10 sin θ = 4, Frictional force = 4 N.

2.24 600 cos 20° = 564, Work = 56 400 J.

2.25 800 N. 150 m, 2000 N.

2.26 3.3 mg.

2.27 Very nearly. (a) Total momentum is conserved. (b) *Change* of momentum for each vehicle is the same size but opposite in direction. (c) Average force acting on each vehicle is same size but the forces act in opposite directions.

2.28 5 m s^{-1}, 0.25 N.

2.29 8 kg m s^{-1}, 160 N.

2.30 120 N.

2.31 See vector diagram below. 1.2 kg.

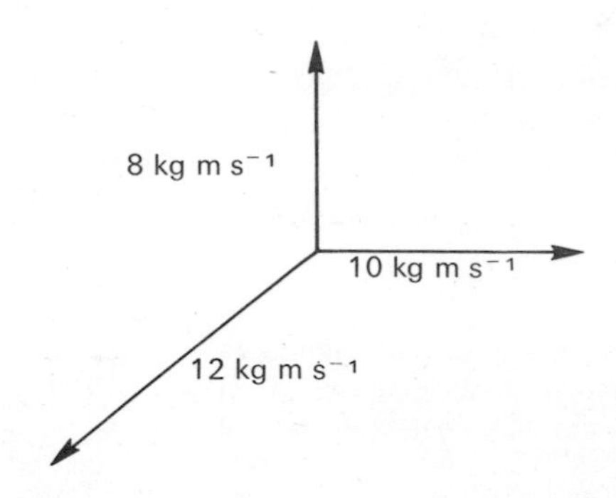

2.32 All materials give, to some extent, when a force is applied. Even the small force exerted on the front of a train when a fly strikes it will cause it to yield slightly. To appreciate what happens, imagine a coil spring attached to the front of the train. The fly approaches (A) and is slowed down when it comes in contact with the spring. It then stops briefly with respect to the Earth (B) but the train goes on (C) and the spring is compressed. The spring then exerts a force on the fly and pushes it in the opposite direction (D). In practice both the fly and the front of the train are deformed —not necessarily both elastically!

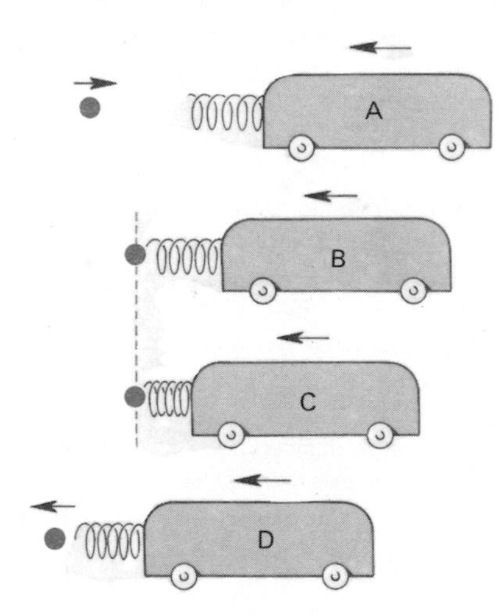

2.33 $4\,m\,s^{-2}$, 200 N, $40\,m\,s^{-2}$, 2000 N, $400\,m\,s^{-2}$, 20 000 N, $F \times \Delta t$ = constant (impulse).

2.34 375 N, 7.5 N s (kg $m\,s^{-1}$).

2.35 5 s.

2.36 2000 N, 40 000 N, 20 000 N s.

2.37 $2.22\,m\,s^{-2}$, 4 N, 0.51 kg.

2.38 (a) $0.5\,m\,s^{-2}$, 5 N, 15 N, 20 N. (b) 20 N. (c) 10 m newtons. (d) (10 m–20) N. (e) 2.1 kg.

2.39 $6.67\,m\,s^{-2}$.

2.40 (a) 0.5 s, (b) $3.0\,m\,s^{-1}$, (c) 9×10^{-2} kg m s (d) 9×10^{-2} N s, (e) $6 \times 10^{4}\,m\,s^{-2}$, (f) 1.8×10^{3} (g) $5.83\,m\,s^{-1}$, 31° to vertical.

2.41 1.5 kg.

2.42 (a) $5\,m\,s^{-2}$, (b) $10\,m\,s^{-1}$, (c) 40 N s, (d) 4C $m\,s^{-1}$, N s (impulse) = kg $m\,s^{-1}$ (change of momentu

2.43 They arrive together. The boy pulls and increases tension in the rope; the increased tension acts on bc They have same mass and are acted on by same unbalan force; therefore both move from rest with same upw acceleration.

2.44 $1\,m\,s^{-1}$, (a) increased, (b) not affected.

2.45 (a) 100 N, (b) $3.75\,m\,s^{-1}$, (c) $2.7\,m\,s^{-1}$.

2.46 (a) 6 kg $m\,s^{-1}$, (b) 11.4 kg $m\,s^{-1}$ W., (c) 11.4 N s E., (d) 228 N E., (e) 228 N W.

2.47 (a) 10 N s E., (b) 10 kg $m\,s^{-1}$ E., (c) 500 N (d) If the ball were kicked horizontally along a frictionl surface, the displacement would be 60 m east. If kicked fr the edge of a cliff the displacement would be $(60^2+45^2)^{\frac{1}{2}}$ = 75 m E, at an angle of 37° below the horizontal. If kick at an angle to a level surface so that it *landed* after 3 s, horizontal displacement would be 39.5 m east. The effe of air resistance and wind have been ignored; they co alter the displacement considerably.

2.48 (a) 10^6 N, (b) $1.75\,m\,s^{-2}$.

2.49 (a) 360 N, (b) as (a) but opposite, (c) 720 N s.

2.50 $4.67\,m\,s^{-2}$, (b) 2.17 m, (c) 34.7 m, (d) 217 m.

CHAPTER 3

3.1 The alpha particle and helium nucleus have the sa mass.

3.2 1449 J.

3.3 (a) Speed is constant; therefore K.E. is the sar (b) Momentum is different. Velocity changes direction a momentum is a vector quantity. Fixed puck (and Ear experience corresponding change in momentum. (c) Duri collision some energy is stored in the magnetic field (P.E This is released as K.E. as pucks separate. Total ener (P.E.+K.E.) in the system is conserved.

3.4 17.5 J.

3.5 $1\,m\,s^{-1}$.

3.6 (a) 400 MW, (b) Increased kinetic and internal energy of water. Air warmed. Noise.

3.7 (a) 10 N, (b) 25 N.

3.8 Yes. Very nearly 90°. Scalar.

3.9 (a) 1.5 units, (b) 70.3%, (c) heat, noise, deformation.

3.10 3000 N.

3.11 (a) 0.5 s, (b) $4\,m\,s^{-1}$, (c) $v = (25+16)^{\frac{1}{2}} = 6.4\,m\,s^{-1}$. Angle with vertical $= \tan^{-1} 0.8 = 39°$, (d) $0.16\,kg\,m\,s^{-1}$ horizontal, (e) $0.16\,N\,s^{-1}$ horizontal, (f) $400\,m\,s^{-2}$, (g) 16 N, (h) 0.01 s, (i) 0.32 J, (j) 0.5 J, (k) 0.82 J, (l) 0.

3.12 4.8 J, 240 N, $Ft = m\,\Delta v$ or $F = ma$ where a is found from $v^2 = 2as$.

3.13 (a) $8\frac{\sqrt{1.6}}{0.02} = 504\,m\,s^{-1}$, (b) 2560 J, 6.4 J, (c) the bullet transfers most of its energy as it penetrates the sand. The sand and bullet become warmer, (d) 0.25%.

3.14 1.02×10^{-8} J.

3.15 9×10^{13} J $= 2.5 \times 10^{7}$ kWh^{-1}, £250 000.

3.16 $F = 0.7$ N, $T = 1.2$ N. Work done depends on length l of the pendulum thread. Bob raised $(l - 0.82l) = 0.18l$ metres, hence work $= 0.18l$ joules.

3.17 (a) $0.6\,N\,s^{-1}$, (b) 10 m s, (c) 3.05 J.

3.18 (a) $500\,N\,s^{-1}$, (b) $125\,m\,s^{-2}$, (c) 2500 J, (d) 12 500 N. Ignored losses due to heating at impact, and noise produced.

3.19 (a) $1\,m\,s^{-1}$ approach, (b) $1\,m\,s^{-1}$ separation, (c) $0.1\,kg\,m\,s^{-1}$ before and after, (d) elastic.

CHAPTER 4

4.1 Archimedes Principle.

4.2 Upthrust = difference between spring balance readings in Figs. 4–13 and 4–14. Use a displacement can and catch the liquid displaced when the brick is lowered into it. Weigh this liquid.

4.3 The brick would float.

4.4 They must displace their own weight of water.

4.5 Weight and pressure.

4.6 (a) not affected, (b) the height of the column would be greater to produce the same pressure.

4.7 5 GPa.

4.8 3.6 kPa.

4.9 2.7×10^{5} N, 2×10^{5} Pa.

4.10 762 mm taking g as $9.8\,N\,kg^{-1}$. Reduced pressure $= 0.12 \times 13\,600 \times g = h \times 1.2 \times g$. Hence $h = 1360$ m.

4.11 3.5 m.

4.12 The Dead Sea contains about 25% dissolved salt and is therefore very dense. To obtain the same upthrust much less of the body has to be immersed in it than in fresh water.

4.13 See graphs below.

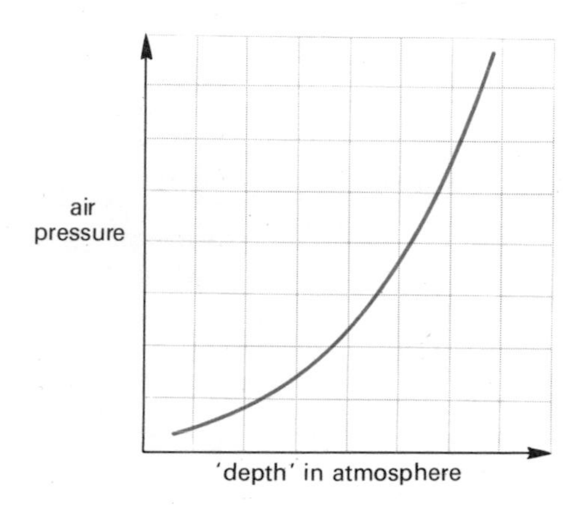

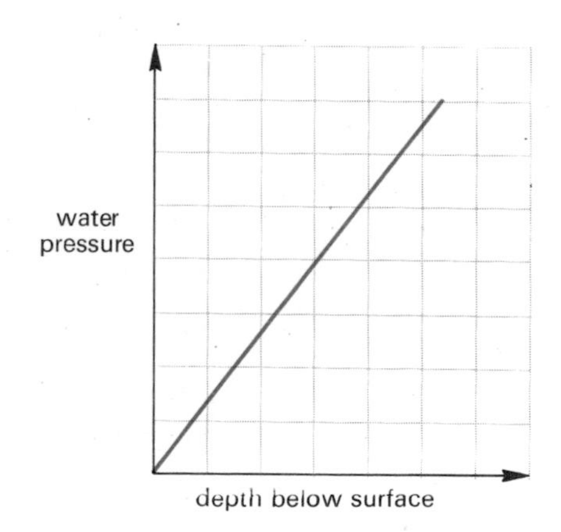

4.14 Just over 2 atmospheres. The same.

4.15 Either (a) a bubble of air will run up to the vacuum and the mercury will then run down the tube into the cup at the bottom, or (b) the mercury thread will break and part of it will be forced up to the top of the tube and the rest will run down.

4.16 and **4.17** See graphs below.

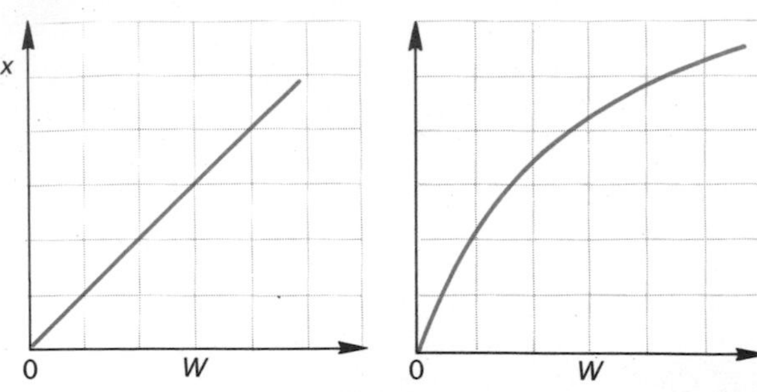

4.18 Yes, for pressures in excess of atmospheric pressure. The spring exerts a force which is proportional to its extension (Hooke's Law) and the area of the washer is constant. The extension will therefore be proportional to the excess pressure.

4.19 Jill is correct. There will be the same **percentage** change in the volume of air in each case but this will produce a much greater movement of the mercury thread in (b). Its main disadvantage is that it will be affected by change in temperature as well as pressure.

4.20 The upthrust exerted on the air filled containers assist the diver.

CHAPTER 5

5.1 Taking relative mass as m and r.m.s. speed as $(\overline{v^2})^{\frac{1}{2}}$ we find that $\frac{1}{2}mv^2$ is nearly constant throughout.

5.2 As the speed increases the frequency will rise.

5.3 A high note would be produced.

5.4 As the speed of sound in helium is much greater than in air, the resonant frequencies of the mouth and throat cavities will be higher and higher notes will be produced.

5.5 2.7×10^{25}, 4.7×10^{-26} kg.

5.6 500 m s^{-1}, 500 m s^{-1}.

5.7 (a) 460 m s^{-1}, (b) 5.6×10^{-21} J, (c) 11.2×10^{-21} J, (d) 10^6 N m^{-2}.

5.8 Hydrogen. Same kinetic energy—smaller mass.

5.9 Greater. Tyre will be warmer and temperature is proportional to average kinetic energy of molecules. Faster molecules produce greater pressure.

5.10 (a) Solid molecules are held together by strong attractive forces between the molecules (atoms). At any temperature above absolute zero the molecules are vibrating but cannot free themselves from the attractions of their neighbours. The molecules are thus kept in position and solid keeps its shape.

In a liquid the molecules have greater K.E. The fa vibrations have *partly* broken down the forces which h each molecule in place so that they are now free to m about, colliding with each other: but they are still abou close together as in the solid state.

In a gas the molecules have so much K.E. that the bo between them are almost completely broken and they m much more freely. They collide elastically with other mo cules and do so with such vigour that they are kept m further apart. The attractions between them are now m too weak to make them 'stick' together when they coll
(b) Pressure is force on unit area and force is the rate change of momentum. When a gas molecule colli elastically with a container wall it rebounds with the sa speed in the opposite direction, so that the change momentum is $mv-(-mv) = 2mv$. If there are n molecu hitting 1 cm^2 of wall every second the pressure will be $2r$ since this is the rate of change of momentum on unit a
(c) Raising the temperature raises the average K.E. and the average speed of the molecules. If they are to col with the container walls with the same effect as be (exerting the same pressure as before) the walls will hav be farther apart. Thus the gas must expand as the tempe ture rises—assuming it is to be kept at the same press
(d) There is no change in temperature when a substa changes state, so the average K.E. of the molecules is same. But work has to be done against the intermolec forces which bind the molecules together as a solid liquid, i.e. the molecule has to be pulled away against attraction of its neighbours. The energy needed for this the latent heat.

5.11 The Gas Board. For the same volume (same co the lower the pressure, the less the density of the gas. T means fewer molecules and therefore less heat—since ea molecule gives up the same quantity of energy in burni

5.12 The same number of molecules occupy 750 times much space in the gaseous state as in the liquid state. T means that their spacing is about $\sqrt[3]{750}$ times as grea the gas, i.e. between 9 and 10 times. (Think of eight b arranged in a cube. If the spacing between them is doub the volume is increased by a factor of 8, i.e. 2^3.)

5.13 6×10^{23}.

5.14 1.[illegible]3 kg m^{-3}.

5.15 2.7×10^7.

5.16 1 nm Maximum. The oil film is one molecule thi and the molecules are standing upright on the water surfa (One end of each molecule is strongly attracted to the wat

5.17 $p = \frac{1}{3}\rho v^2$ but the temperature and therefore $\overline{v^2}$ constant. $760/630 = \rho_{S.L.}/\rho_{1.5\,km}$ or $\rho_{1.5\,km} = 0.83\,\rho_{S.L.}$

5.18 Area of piston $= A$. $F\Delta x = \frac{F}{A}\Delta x \times A = p\,\Delta V$. Work done (energy transfer) raises the temperature of the gas. As the piston moves inwards, gas molecules which collide with it are speeded up. The average kinetic energy of the molecules—and therefore the temperature—will increase.

5.19 In a gas containing n molecules, we assume that on average $\frac{1}{3}n$ molecules move to and fro parallel to each of three mutually perpendicular axes.

5.20 As temperature is proportional to the average kinetic energy of the molecules, those with twice the mass (B) will have twice the kinetic energy and therefore the absolute temperature will also be double. Pressure depends on momentum changes when the molecules strike the walls. Those of double mass (B) will have double momentum —as speed is the same—and therefore the pressure they produce will be double. This is consistent with $p \propto T$.

CHAPTER 6

6.1 They carry energy.

6.2 The masses fall into 5 groups: approximately 26, 39, 52, 65 and 78 g, i.e. in regular steps, indicating that there is a definite number of articles in each bag, e.g. 2, 3, 4, 5 and 6. Marbles are more likely than marmalade.

6.3 1.6×10^{-19} C.

6.4 9.1×10^{-31} kg, J. J. Thomson was correct.

6.5 $\frac{1}{2}mv^2 = eV$, $v = 1.3 \times 10^7$ m s^{-1}.

6.6 6.25×10^{18}, 6.25×10^{18} electrons per second.

6.7 6.25×10^{11} electrons.

6.8 (i) 6×10^{-8} N, weight $= 3 \times 10^{-8}$ N, (ii) 1 m s^{-1}, (iii) $\frac{1}{2}mv^2 + mgh = 3 \times 10^{-9}$ J $= QV$.

6.9 Weight of drop $= 1.6 \times 10^{-14}$ N, Charge $= \frac{1.6 \times 10^{-14}}{2 \times 10^4} = 8 \times 10^{-19}$ C, 5 electronic charges.

6.10 2.8 eV. About ninety times the average kinetic energy of an air molecule at room temperature.

6.11 Force on electron $= 6.4 \times 10^{-17}$ N. Acceleration $= 7 \times 10^{13}$ m s^{-2}. In gravitational field, acceleration $=$ 10 m s^{-2}. approx.

6.12 1.6×10^{-15} kg.

CHAPTER 8

8.1 Will read only p.d.s of the order of hundreds or thousands of volts. Difficult to calibrate and read accurately. Gold leaf electroscopes are easily damaged, particularly in transit.

8.2 Expensive. Many have to be calibrated before use. Not direct-reading—calculations have to be performed. Not easily transported.

8.3 Method 1 (Fig. 8–17) 101 Ω. Method 2 (Fig. 8–18) 98 Ω.

8.4 'Adjust zero ohms' or similar label.

8.5 The meter will normally have a linear current scale and the cell a constant e.m.f. The ohms reading will therefore be *inversely* proportional to the *total* resistance in the circuit.

8.6 $\frac{I_A R_1}{I_A R_2} = \frac{I_B R_3}{I_B R_4}$, hence $R_1/R_2 = R_3/R_4$.

8.7 66.7 cm.

8.8 $\frac{X}{3} = \frac{25}{75}$ or $\frac{75}{25}$ giving $X = 1$ or 9 Ω. The ratio must be taken as $\frac{X}{K} = \frac{I_1}{I_2}$ in Fig. 8–24 page 67.

8.9 2 Ω

8.10 If when key is pressed, there is *no change* in the galvanometer reading then the bridge is balanced.

8.11 e.m.f. $= 1.5$ V, t.p.d. $= 1.2$ V, $I = 0.12$ A, internal resistance $= 2.5$ Ω, low.

8.12 $X = 3.33$ Ω. Resistance in series with galvanometer to protect it from overload. It would not.

8.13 $1.5 - 0.2 = 1.3$ V. As R is increased the current will decrease. The t.p.d. will then increase.

8.14 Their internal resistance is relatively high. Also, any attempt to take a high current from them would lead to rapid polarization and consequent increase in internal resistance. Bulk and expense prohibitive.

8.15 (a) 2.5 V, (b) 3 V, 1000 Ω voltmeter is better.

8.16 (a) 0.3 A, (b) 0.43 A.

8.17 0.9 Ω, 0.2 V.

8.18 If the other sources of e.m.f. cause a current of 1.5 A to flow through the cell in the *forward* direction, the t.p.d. across the cell will be zero.

8.19 3.4 Ω.

8.20 Series: 1.4 A. Parallel 3.75 A. Small internal resistance.

8.21 *Dry cell* $r=\frac{1.5}{3}=0.5\Omega.$

Accumulator $r=\frac{2}{200}=0.01\Omega.$

If an accumulator is short-circuited, the extremely high current flowing will lead to the generation of much heat in the accumulator plates as well as in the external circuit. The plates may be buckled badly or even caused to disintegrate.

8.22 By connecting batteries in series. 6V. There would be a large current.

8.23 2Ω.

8.24 Moving iron meter requires greater circuit for operation than the moving coil meter and this produces a larger voltage drop across the internal resistance of the battery.

8.25 (a) 2000Ω (2 kΩ), (b) 100 V, (c) 115 V, (d) 118 V.
N.B. If an instrument has a basic movement of 1 mA and is designed to have an f.s.d. as a voltmeter of n volts the current will be 1 mA and the voltmeter resistance (i.e. meter+series multiplier) will be $\frac{n}{0.001}$ ohms, i.e. $1000\,n$ ohms. Such a voltmeter is said to have a resistance of 1000 ohms per volt.

8.26 Each pair of terminals is connected directly to the poles of the battery so that any appliances used are in parallel with each other.

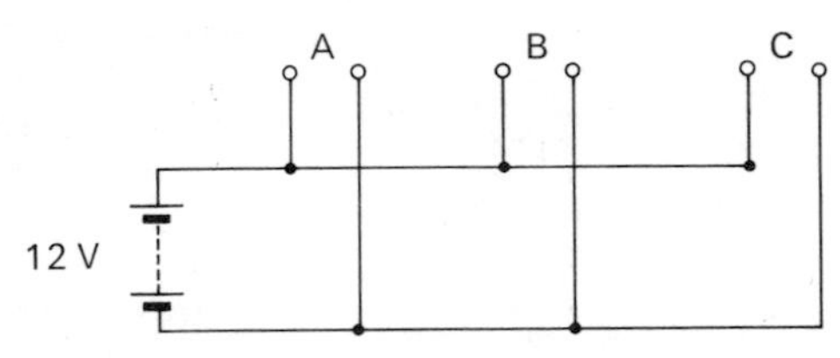

12 V is available at A, B and C.

Total current through the battery=10 A.
Therefore p.d. inside the battery (lost volts)=10×0.2=2 V.
Therefore t.p.d. of battery, i.e. p.d. at the third pair of terminals=12−2=10 V.

8.27 The lamp could be dimmed but the arrangement is not satisfactory.

The resistance of a quartz-iodine headlamp bulb is about 1.5 ohms. If the resistance of the variable resistor is adjusted to be much more than that, then the effective resistance will be that of the bulb. Then nearly all the current will pass through the bulb, making it light brightly.

If the variable resistor is adjusted to have a value about the same as that of the lamp, the effective resistance, ignoring the resistance of the battery, will be half that of the lamp. Then twice as much current will be taken from the battery. The lamp will take half of it and will be just as bright as before. The battery may be damaged by a sustained current of about 17 A. The variable resistor will have to dissipate about 100 W.

If the variable resistor were reduced further, the lamp would receive less current and would become dimmer, but much more corrent would be taken from the battery and much more power would have to be dissipated by the variable resistor.

The variable resistor should be in series with the lamp.

CHAPTER 9

9.1 Square waves.

9.2 The deflection depends on the *direction* of the current.

9.3 During one half-cycle, charges flow through two of the rectifiers and through the meter. During the other half-cycle, charges flow through the *other* two rectifiers and in the *same direction* through the meter.

9.4 7.07 V.

9.5 5 ms cm^{-1}.

CHAPTER 10

10.1 Twice.

10.2 Half.

10.3 5×10^{-3} C, 2.5 A.

10.4 1.2×10^{-7} C.

10.5 When the battery is momentarily connected to the electroscope plate, charge flows so that the electrophorus plate and electroscope case on the one hand and the electroscope plate and leaf on the other, have equal and opposite charges. The p.d. will not be enough to produce a deflection of the leaf.

The two plates and the conductors attached to them, with the polythene between them form a capacitor. When the upper plate is lifted, the capacity is decreased and so the p.d. is increased; the leaf rises.

Work has been done in pulling the charged plates further apart. More energy has been stored in the electric field between the plates. This energy has come from the work done in raising the upper plate.

10.6 The plate-to-earth capacitance of the electroscope is increased. The charge remains the same and so the p.d. is reduced.

10.7 Less.

10.8 Greater.

10.9 (a) If the sheet is of negligible thickness, the capacitance is unchanged. Two capacitors have been formed, each half the thickness and so twice the capacitance of the original. Being in series their combined capacitance equals that of the original capacitor.

(b) As the thickness of each capacitor is now less than half that of the original, the capacitance of each is more than twice that of the original. The total capacitance is increased.

(c) The capacitance is increased.

10.10 Deflection increases. As vanes open the capacitance decreases and the p.d. increases as there is no change in charge.

10.11 (a) P.D. will decrease. Yes, a pulse of electrons will flow in the direction indicates by the arrow. (b) Electrons will again flow from the supply charging the electroscope and larger capacitance to the original p.d. The deflection of the leaf will be the same as it was when originally charged, and a further pulse of electrons will be indicated in the galvanometer in the direction shown by the arrow. (c) When the vanes are opened the p.d. across the capacitor will be greater than the p.d. across the E.H.T supply terminals, and electrons will flow *from* the capacitor *to* the supply until the p.d. and deflection of leaf are again the same as originally.

10.12 4000 μF. See graphs below.

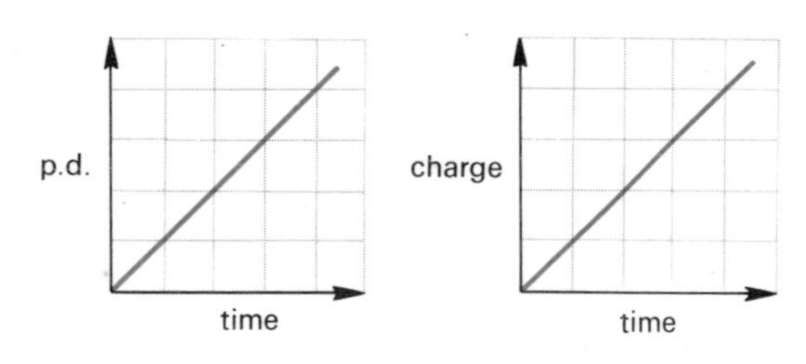

10.13 2.4 μF, both have 24 μC, 6 V, 4 V.

10.14 (a) 4×10^{-4} C, (b) 2 mA.

10.15 (a) 10^{-2} C. (b) 2×10^{-4} C. (c) The capacitor takes 0.01 s to charge and the same time to discharge. In the first 0.01 s no charge is flowing through the meter. In the second 0.01 s *all* the charge is flowing through it. Thus the charge flowing *to* the capacitor and that flowing *from* it both equal the charge through the meter in 0.02 s, i.e. 2×10^{-4} C. (d) 10 V. (e) 20 μF.

10.16 Same.

CHAPTER 11

11.1 An a.c. voltmeter.

11.2 The current through the coil on the core would be small, as its reactance would be greater than that of the other coil and the bulb would probably not light up.

11.3 The sum of the two inductances.

11.4 Less

11.5 (a) A current pulse would charge the capacitor. No current would then flow; (b) an alternating current would flow; (c) the current would increase.

11.6 (a) steady d.c., (b) a.c., (c) a.c. would decrease, (d) much larger a.c.

11.7 114 Ω, 456 W (min) resistor in series—extravagant as much energy lost as hot air.

0.36 H inductor (coil on core) in series—better, expensive as coil would have to be large to carry 2 A.

28 μF capacitor capable of carrying 2 A a.c. (350 V peak)—suitable capacitor expensive as electrolytic not suitable. All above have the disadvantage that mains voltage is on the bulb—shock danger. A 20:1 step down transformer would be safest and best solution.

CHAPTER 12

12.1 No. 5.

12.2 The supply current to a parallel resonant circuit is minimum at resonance. The volts drop across R will then be minimum.

12.3 (a) The ticker-timer method, i.e. place the blade in a solenoid, fed from a low voltage a.c. supply, so that one end is clamped and the other end is free to move between the poles of a permanent magnet; use the electric bell principle. (b) A variable speed motor driving a crank which is connected to the oscillating mass by a rubber band in such a way as to pull down on the mass on each downward swing. See also Fig. 12.12. (c) The same sort of arrangement but with the crank revolving in a horizontal plane and driving an arm clamped to the torsion bar (or wire). See also Fig. 12.13. (d) Some arrangement for tilting the curved track bodily each time the ball reaches the extremity of the track; some more sophisticated methods may employ photo-cells to switch on electromagnets to attract a (steel) ball at the appropriate part of the track. (e) See Fig. 12-6.

12.4 (a) Rotational K.E. to strain P.E. stored in the twisted wire or rod. Damping is due to 'internal friction' and air resistance and energy is used to overcome these. This energy is finally transformed into heat. (b) K.E. of the moving string to P.E. of the stretched string. Damping is due to the same factors as discussed in (a).

12.5 The mains driving frequency is 50 Hz. If the tuner is tuned to this frequency we get maximum amplitude of vibration. The natural frequency could be reduced by loading the strip with, say, a piece of metal fixed on with Sellotape. See graph.

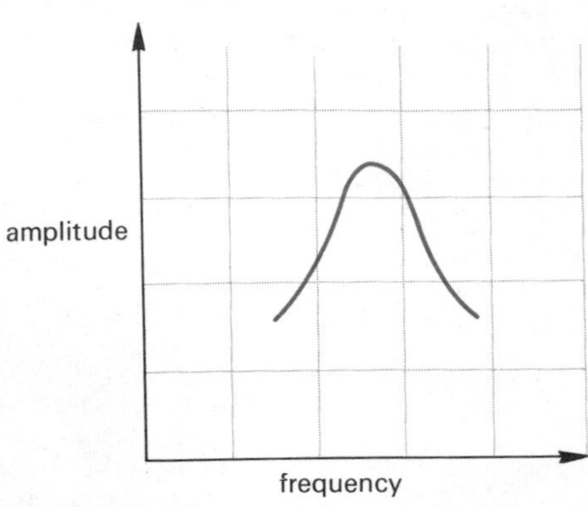

12.6 He could stand further back, nearer the point of support.

12.7 At the same frequency as that at which it swings on its own, i.e. its *natural frequency*. Yes, one-half, one-third etc. of the natural frequency of the swing.

12.8 The greater his mass, the lower will be the frequency.

12.9 (a) Not at all. (b) Frequency decreases. (c) Very little difference. (i) At lowest point of swing (centre). (ii) As (i). (iii) At end of its swing at either side, i.e. highest point.

12.10 (a) minimum, (b) maximum, (c) more, (d) less.

12.11 (a) maximum, (b) maximum, (c) minimum, (d) less, (e) more.

12.12 See graph below.

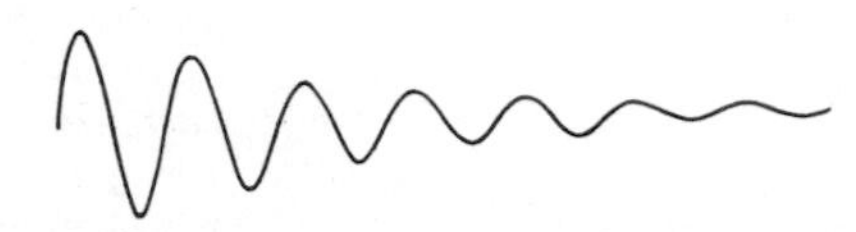

12.13 A capacitor. With the switch at A the total impedance of the two capacitors in series would be greater than that of the capacitor in the 'black box' alone. With the switch at C the current is greater because the capacitor in the black box together with the coil form a series circuit resonant at a frequency near to that of the supply.

12.14 The figure could represent either the change in current through a series circuit as the frequency is taken through resonance OR the change of impedance of a parallel circuit.

12.15 R_1 should be small. Energy is lost from the circuit as heat is produced in R_1.

R_2 should be large. To prevent any energy loss from the circuit no current should be shunted through R_2; to minimize the loss the current through R_2 should be kept as small as possible.

12.16 The circuits shown in Figs. 12–45 and 12–46 are both parallel circuits and therefore both will show a minimum current at resonance.

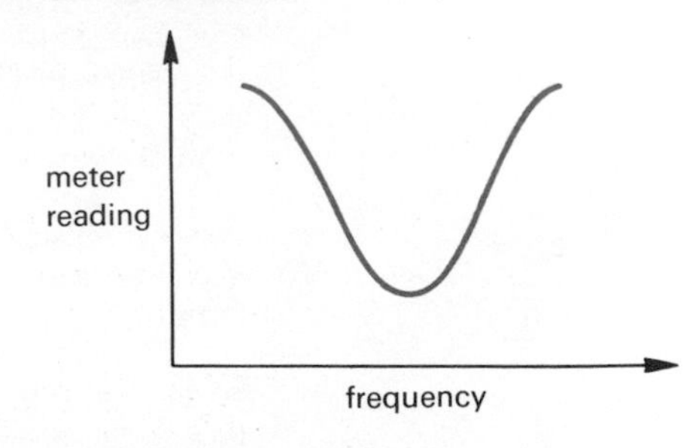

12.17

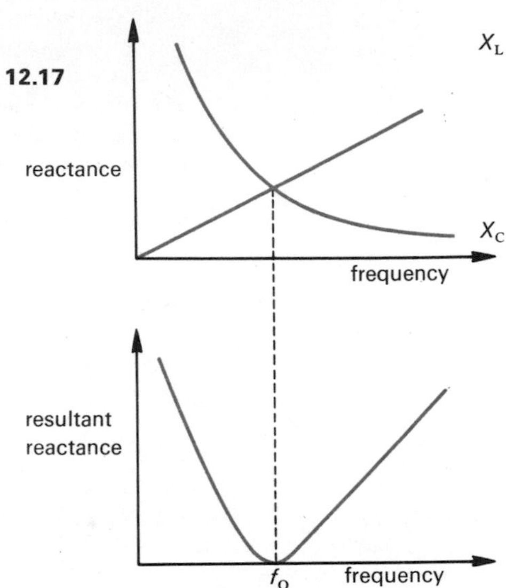

The graph represents the resultant of a series circuit. It is a minimum at the resonant frequency. Since the voltages across the capacitor and the inductor are in antiphase at any frequency the resultant reactance is the difference between their separate reactances. The resonant frequency is indicated by the point at which the two reactances are equal (and opposite), i.e. at the point where the two curves cross.

12.18 A would increase, B would decrease, C would remain constant and D would get brighter until the input frequency was equal to the resonant frequency of the series circuit. As the input frequency was increased further the bulb would dim.

CHAPTER 13

13.1 2.14 m.

CHAPTER 14

14.1 *Empedocles theory*: What then have light sources to do with vision? We should be able to see in the dark. How

about photography? Does the film 'radiate a beam' as well? *Pythagoras' theory*: Why don't they 'shoot out' particles in the dark? If they require light to do this the theory is not very different from current theories. The particles are 'reflected' from the object and therefore appear to be 'shot out' from it.
In support of the theories, it could be held that they suggest ways in which information might be transmitted from an object to the eye.

14.2 The beam of starlight is bent *towards* the Earth.

14.3 (a) $n_3 > n_2$, (b) $\theta_4 = \theta_1$, (c) no change.

14.4 As we are accustomed to seeing things by light rays which travel in straight lines, we tend to interpret all visual images produced on the retina as if they had been formed by such rays. The rays entering the eye from A in Fig. 14–22 then appear to have come from a point in the water, i.e. the water seems to be shallower.

$n_w = \sin 8°/\sin 6° = 1.33 = v_a/v_w$

$v_w = \dfrac{3\times10^8}{1.33} = 2.25\times10^8\,\text{m s}^{-1}$ $\sin\theta_c = 1/1.33 \Rightarrow \theta_c = 49°$

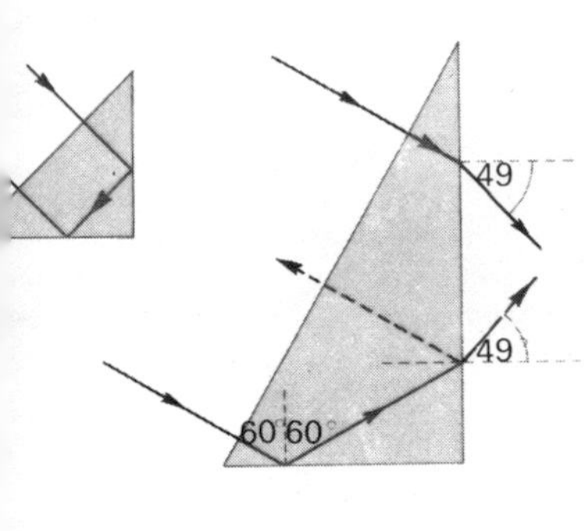

14.5 See diagram opposite.

14.6 Pyrex rods in mixture would be seen to disappear!

14.7 Centre ray $\sin\theta_a/\sin\theta_g = \sin 39°/\sin 24° = 1.5$.

14.8 $n_g = 1/\sin 45° = 1.414$.
$n_g > 1.414$ for total internal reflection.

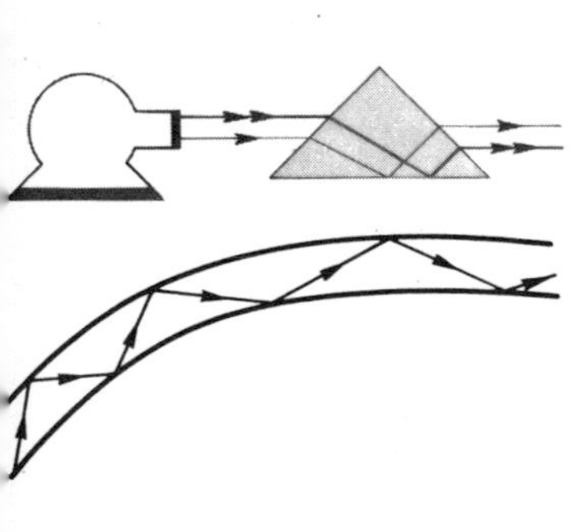

14.9 See diagram opposite.

14.10 See diagram opposite.

14.11 As refractive index of glass > 1.414, critical angle > 45° (c.f. question 14.8). For a ray to pass into end of cylinder θ_A must be less than 45°. θ_B must therefore be always greater than 45°. Total internal reflection will always occur.

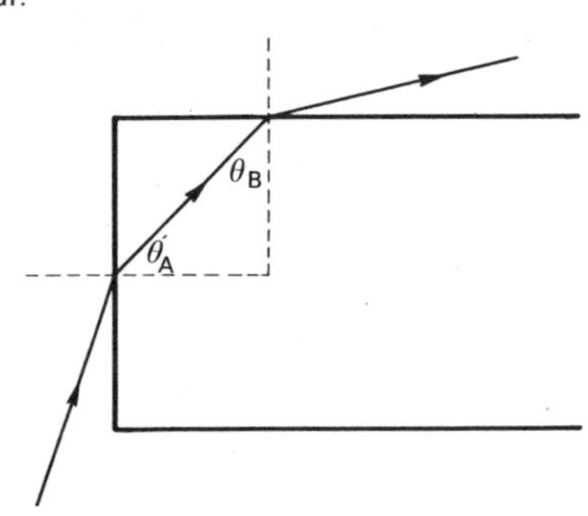

14.12 $n_d = 2.2$. Diamond bends and disperses light more than most other transparent materials. It produces vivid 'rainbows' of colour. As the critical angle is small, total internal reflection takes place over a wide range of incident angles within the diamond. There is therefore less transmission and more reflection of light in general.

14.13 Colour depends on the *frequency* of light. When a ray of monochromatic light enters a glass block, for example, it is bent, but the frequency is not altered in any way. When white light enters a triangular prism, the different constituent colours are bent by different amounts. The deviation of each light ray depends on its frequency.

CHAPTER 15

15.1 In Fig. 15–3 the source is further from the lens than the principal focus ($u > f$). In Fig. 15–5 the source is between the principal focus and the lens ($u < f$).

15.2 14.3 cm, 10 cm, 5.9 cm, −10 cm, −5.9 cm.

15.3 Upside down. Retinal image is inverted, but we *interpret* this to mean 'right way up'.

15.4 To produce a real image, the distance between the two telescope lenses should be *increased*.

15.5 No!

15.6 Inverted, diminished, real. Formed at common principal foci of both lenses. This image is the 'object' for the eyepiece lens. Final image is at infinity. It is inverted, magnified, virtual.

15.7 7.5 cm from the lens on the object side of it. Upright, diminished, virtual. 1 cm tall.

15.8 *Note:* focal length and object-image scales are not the same. Object is 1.6 cm tall 12 cm from the lens.

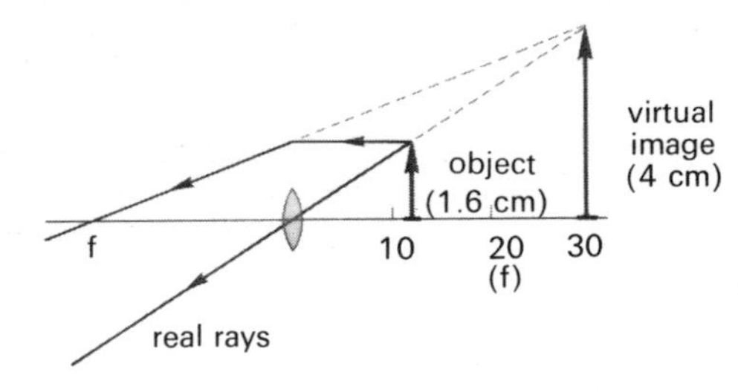

15.9 (a) Move projector further from screen, (b) move lens towards the slide.

15.10 2.02 D, 60 cm.

15.11 It has a small aperture (e.g. f16) and therefore great depth of field.

15.12 Converging (convex) lenses.

15.13 Light radiated from the lamp over a large solid angle is bent so that it passes through the film. A bright image of the film is then formed on the screen by the other lens.

15.14 1.4 m, 0.1 m.

15.15 Diameter of virtual image is 5 cm. It is 24 cm from the lens. *Note*: scales not same for focal length and object-image sizes. By calculation $v = -24$ cm, $m = 5$.

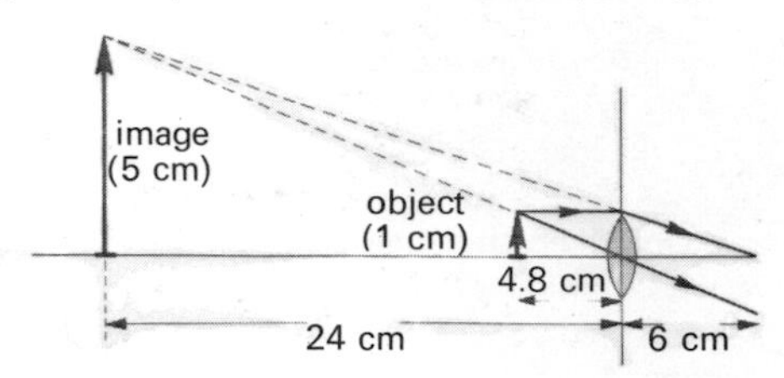

15.16 Binoculars, telescope, reading glass, spectacles, movie and slide projectors, camera, torch, slide viewer, microscope, kaleidoscope, watch with date, spy lens in door, bathroom scales.

CHAPTER 16

16.1 109 nm.

16.2 The wavelength is so short that the small amount of bending due to diffraction is not normally observed.

16.3 No. The two 'sources' must produce waves which are 'in step'. The two filaments are quite independent and the light waves they produce would not be in phase. The two filaments would also be too large and would radiate a wide range of light frequencies from these large areas. Monochromatic light is needed to produce reasonably sharp bands.

16.4 590 nm. Sodium flame or lamp. No, they would be closer together.

CHAPTER 17

17.1 The one on the left.

17.2

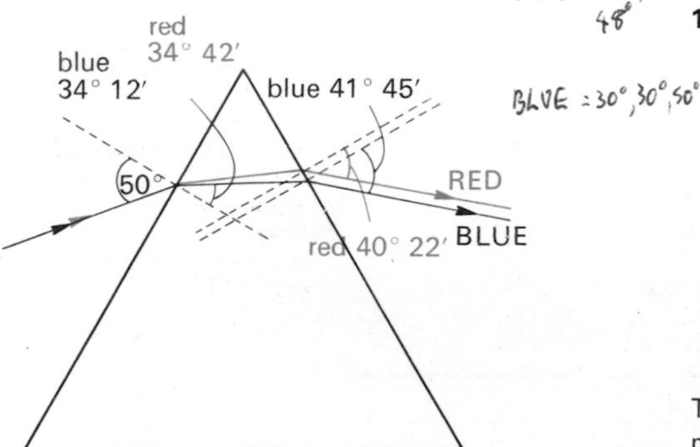

17.3 Sodium flame, various vapour lamps, electrodes of spark gap.

17.4 589 nm, 45°.

CHAPTER 18

18.1 If the air were ionized, we would expect positively or negatively charged electroscopes with brass or zinc plates to be discharged. If the ultraviolet radiation were a stream of positive charges, we would expect a negatively charged electroscope with a brass plate to be discharged. The suggestion that the zinc plate loses electrons when ultraviolet radiation falls on it seems most probable.

18.2 A was positively charged and then discharged by electrons emitted from the zinc plate. B was negatively charged and discharged as electrons left the zinc plate. Neither A nor B would be discharged as no electrons are emitted from the zinc when visible light shines on it.

18.3 $\lambda = h/p$, where p is the momentum of the electron. To produce the necessary *short* wavelengths the momentum of the electron must be very great, i.e. the *speed* must be great. High voltages are needed to produce the necessary accelerating force.

18.4 200 kHz, 1.32×10^{-28} J, $3 \times 10^{33}\ s^{-1}$.

18.5 $h = 6.5 \times 10^{-34}$ J s^{-1}.

18.6 400 nm ultraviolet, 1.6×10^{-19} J, 1 volt.

CHAPTER 19

19.1 2.18×10^{-30} kg.

19.2 Energy lost $= 2.18 \times 10^{-30} \times 9 \times 10^{16}$
$= 1.96 \times 10^{-13}$ J.

19.3 Before the reaction.

Total mass $= 13.3235 \times 10^{-27}$ kg
$= 119.9115 \times 10^{-11}$ J
Energy $= 0.5$ MeV $= 0.008 \times 10^{-11}$ J
Total mass + energy $= 119.9195 \times 10^{-11}$ J

After the reaction
Total mass $= 13.29264 \times 10^{-27}$ kg
$= 119.63376 \times 10^{-11}$ J
Energy $= 17.2$ MeV $= 0.2752 \times 10^{-11}$ J
Total mass + energy $= 119.90896 \times 10^{-11}$ J

Taking a reasonable number of significant figures, the mass + energy is the same on both sides of the equation.

Index

Periodic Table of Element

1 **H** Hydrogen 1								
7 **Li** Lithium 3	9 **Be** Beryllium 4							
23 **Na** Sodium 11	24 **Mg** Magnesium 12							
39 **K** Potassium 19	40 **Ca** Calcium 20	45 **Sc** Scandium 21	48 **Ti** Titanium 22	51 **V** Vanadium 23	52 **Cr** Chromium 24	55 **Mn** Manganese 25	56 **Fe** Iron 26	27
85·5 **Rb** Rubidium 37	88 **Sr** Strontium 38	89 **Y** Yttrium 39	91 **Zr** Zirconium 40	93 **Nb** Niobium 41	96 **Mo** Molybdenum 42	98 **Tc** Technetium 43	101 **Ru** Ruthenium 44	R 45
133 **Cs** Cæsium 55	137 **Ba** Barium 56	139 **La** Lanthanum 57	178·5 **Hf** Hafnium 72	181 **Ta** Tantalum 73	184 **W** Tungsten 74	186 **Re** Rhenium 75	190 **Os** Osmium 76	77
223 **Fr** Francium 87	226 **Ra** Radium 88	227 **Ac** Actinium 89						

139 **La** Lanthanum 57	140 **Ce** Cerium 58	141 **Pr** Praseodymium 59	144 **Nd** Neodymium 60	147 **Pm** Promethium 61	**S** Sa 62
227 **Ac** Actinium 89	232 **Th** Thorium 90	231 **Pa** Protactinium 91	238 **U** Uranium 92	237 **Np** Neptunium 93	P 94